Introduction to Ordinary Differential Equations

Introduction to Ordinary Differential Equations

Roger C. McCann

MISSISSIPPI STATE UNIVERSITY

HBJ HARCOURT BRACE JOVANOVICH, INC.

NEW YORK   SAN DIEGO   CHICAGO   SAN FRANCISCO   ATLANTA
LONDON   SYDNEY   TORONTO

ISBN: 0-15-543485-3

Library of Congress Catalog Card Number: 81-85997

Printed in the United States of America

Technical art on the cover and in the text by Mel Erikson Art Service

PREFACE

Differential equations play an enormously important role in engineering, physics, chemistry, the life sciences, and many areas of applied mathematics. Virtually any phenomenon that varies in a continuous or nearly continuous fashion can be modeled using differential equations. It is the goal of this text not only to describe how to solve elementary differential equations, but also to illustrate how differential equations are actually used to model real-world processes.

Traditionally, most of the examples illustrating the use of differential equations involved simple mechanical systems or electrical circuits. I have drawn examples and exercises from numerous books and journals illustrating some of the diverse areas in which differential equations are used. To the best of my knowledge, many of these examples are appearing in an elementary text for the first time. Among the more interesting and unusual of these examples are the rate at which ocean water circulates (Exercise 3, Section 1.3), the concentration of silica in the sediment of the floor of the North Sea (Section 2.4), the stabilization of production in a closed economy (Section 3.5), the relationship between rainfall and runoff in a watershed (Example 3, Section 4.4, and Example 4, Section 4.5), the Lancaster war model (Example 1, Section 5.7), the Ross model for the way malaria affects a community (Example 3, Section 5.10), the water level in a canal that empties into the open sea (Example 2, Section 6.5), and a nonparametric description of a cycloid (Exercise 21, Section 7.1). One-sixth of the 156 examples and one-eighth of the 774 exercises involve applications.

Throughout the text I have tried to keep the explanations and derivations as simple and straightforward as possible. Virtually every concept is illustrated by an example before it is used. Discussions begin with elementary situations and progress to the more complex. For example, the treatment of linear differential equations begins by considering only second-order equations (Chapter 2). This greatly simplifies the concept of linear independence, since only two functions need to be considered. In the following chapter higher-order equations are considered. It is here that the general definition of linear independence is given. This delay in presenting the general definition enables the student to have experience with this essential concept before working in a general setting.

I believe that numerical methods do not constitute a separate area of study, but are merely another technique for finding solutions. For this reason numerical methods appear as sections in Chapters 1 and 5, and not as a separate chapter. I also believe that the numerical methods should be discussed along with other techniques so that the reader realizes that equations that do not have a "nice" form can still be "solved." Of course there is no detailed discussion of numerical methods, but only an indication of the types of methods that are commonly used.

With the exception of Chapter 7 (Boundary-Value Problems) and Sections 1.6, 1.7, and 1.10 (Exact Differential Equations, Integrating Factors, and Existence of Solutions), only a knowledge of calculus of a single variable is assumed. At various points in the text certain elementary properties of determinants and complex numbers are needed; hence, a discussion of these properties is contained in two appendixes.

Chapters 1 and 2 constitute the basis for any introductory course in elementary differential equations. Once these chapters have been covered, the remaining chapters may be studied in any order, with the exception that Chapter 4 should precede Section 5.9. The independence of Chapters 3 through 7 makes the book flexible enough to meet the needs of almost any instructor. My personal view is that Chapter 5 is exceptionally important and should be at least partially covered.

Before writing this text, I did not realize how much of a team effort is required by such a project. I was blessed with exceptionally knowledgeable reviewers who made significant contributions to the final form of the text. I would like to thank them for their assistance: Donald Blevins, Trinity College, Washington, D.C.; John Brothers, Indiana University; Murray Cantor, University of Texas at Austin; J. R. Dorroh, Louisiana State University; Richard Koch, University of Oregon; and Hugh Maynard, University of Texas at San Antonio.

Without a coauthor to share the blame, I alone am responsible for any misprints, weaknesses, or errors that appear in the text. I hope that readers will bring to my attention not only errors and shortcomings of the text, but also its features that they particularly like.

I am grateful to Marilyn Davis, Editor at Harcourt Brace Jovanovich, Publishers, for her guidance and cooperation. I would also like to thank Andrea Haight and Christopher Lang for their excellent editorial work, and Anna Kopczynski for the book's handsome design. Various portions of the manuscript were expertly typed by Linda Brent and Pam Bost. Robert Finley read the manuscript and assisted in the preparation of the answer section. Last, and certainly not least, I would like to thank my wife, Susan, without whose help and understanding this book could never have been written.

Roger C. McCann

CONTENTS

3 Higher-Order Linear Differential Equations   152

4 Laplace Transforms   191

5 Systems of Linear Differential Equations   226

1 First-Order Differential Equations

1.1 Introduction

Differential equations play an enormously important role in engineering, physics, chemistry, and various areas of applied mathematics. Virtually any phenomenon that varies in a continuous or nearly continuous fashion can be modeled using differential equations. Systems as varied as electrical circuits, economic growth, rainfall-runoff in a watershed, and the metabolism of glucose have been modeled using differential equations. It is the goal of this book not only to describe how to solve elementary differential equations, but also to illustrate how differential equations are used to model various processes. In this section we will present some of the basic definitions from the theory of differential equations, while the later sections will be devoted to determining solutions of elementary differential equations and investigating their applications.

We will consider equations such as

$$\frac{d^3y}{dx^3}(x) - 2\frac{d^2y}{dx^2}(x) + 4\frac{dy}{dx}(x) - 3y(x) = 0 \tag{1.1}$$

and

$$\frac{d^2z}{dt^2}(t) + 4z(t) = 0 \tag{1.2}$$

Our goal is to find functions that satisfy the given equation. For example, the function $y(x)=e^x$ satisfies equation (1.1) since

$$\frac{d^3}{dx^3}(e^x)-2\frac{d^2}{dx^2}(e^x)+4\frac{d}{dx}(e^x)-3e^x=e^x-2e^x+4e^x-3e^x=0$$

Likewise, the function $z(t)=\cos 2t$ satisfies equation (1.2) since

$$\frac{d^2}{dt^2}(\cos 2t)+4\cos 2t=-4\cos 2t+4\cos 2t=0$$

Equations in which the unknown is a function of a real variable and which contain not only the functions themselves, but also certain of their derivatives, are called **ordinary differential equations**. For example, in equation (1.1), x is the real variable and y is the unknown function. In equation (1.2), t is the real variable and z is the unknown function. Other examples of ordinary differential equations are

$$\frac{dy}{dx}(x)+x^3y(x)=x^{2-}$$

$$\left(\frac{d^3w}{dz^3}(z)\right)^2+z\frac{dw}{dz}(z)+4w(z)=\sin z \qquad \textbf{(1.3)}$$

$$\frac{d^2\theta}{dt^2}(t)+\sin\theta(t)=0$$

Relations that do not involve derivatives, such as

$$x^2+y^2=x+y$$

or relations that contain partial derivatives, such as

$$\frac{\partial^2 y}{\partial x^2}(x,t)=\frac{\partial^2 y}{\partial t^2}(x,t)$$

do not define ordinary differential equations.

For brevity, an ordinary differential equation will be called a differential equation. In order to simplify the notation,

$$y, \qquad \frac{dy}{dx}, \qquad \frac{d^2y}{dx^2},\ldots, \qquad \frac{d^ny}{dx^n}$$

will frequently be used in place of

$$y(x), \qquad \frac{dy}{dx}(x), \qquad \frac{d^2y}{dx^2}(x),\ldots, \qquad \frac{d^ny}{dx^n}(x)$$

respectively, when writing differential equations. With this understanding, the differential equations in (1.1), (1.2), and (1.3) can be rewritten as

$$\frac{d^3y}{dx^3} - 2\frac{d^2y}{dx^2} + 4\frac{dy}{dx} - 3y = 0$$

$$\frac{d^2z}{dt^2} + 4z = 0$$

$$\frac{dy}{dx} + x^3y = x^2$$

$$\left(\frac{d^3w}{dz^3}\right)^2 + z\frac{dw}{dz} + 4w = \sin z$$

$$\frac{d^2\theta}{dt^2} + \sin\theta = 0$$

The **order** of a differential equation is the order of the highest derivative appearing in the equation. Thus the differential equations in (1.1), (1.2), and (1.3) are of third, second, first, third, and second orders, respectively. Throughout this book we will always assume that an nth order differential equation can be written in the form

$$\frac{d^ny}{dx^n} = f\left(x, y, \frac{dy}{dx}, \ldots, \frac{d^{n-1}y}{dx^{n-1}}\right) \tag{1.4}$$

Equations (1.1) and (1.2) are of this type since they can be rewritten as

$$\frac{d^3y}{dx^3} = 2\frac{d^2y}{dx^2} - 4\frac{dy}{dx} + 3y$$

and

$$\frac{d^2z}{dt^2} = -4z$$

The second equation in (1.3) is a third-order differential equation that can not be written in the form in (1.4).

A **solution** on an open interval I of the differential equation in (1.4) is a function u defined on I such that

$$\frac{d^nu}{dx^n}(x) = f\left(x, u(x), \frac{du}{dx}(x), \ldots, \frac{d^{n-1}}{dx^{n-1}}u(x)\right)$$

for every x in the interval I.

▶ **Example 1**   The function $u(x)=e^x$ is a solution on $(-\infty,\infty)$ of dy/dx $=y$ since $du/dx=e^x=u(x)$ for every x in $(-\infty,\infty)$.   ◀

▶ **Example 2**   The function $u(x)=x^{-1}$ is a solution on $(0,\infty)$ of $dy/dx=$ $-y^2$ since $du/dx=-x^{-2}=-[u(x)]^2$ for every x in $(0,\infty)$.   ◀

A relation $g(x,u)=0$ is called an **implicit solution** on an open interval I of equation (1.4) if it determines at least one real function on I and this function is a solution of the differential equation.

▶ **Example 3**   The relation $x^2+u^2-1=0$ is an implicit solution of dy/dx $=-x/y$ on $(-1,1)$ since it determines the function $u_1(x)=\sqrt{1-x^2}$ on $(-1,1)$ and

$$\frac{du_1}{dx}=\frac{-x}{\sqrt{1-x^2}}=-\frac{x}{u_1(x)}$$

for every x in $(-1,1)$. The relation $x^2+u^2-1=0$ also determines the function $u_2(x)=-\sqrt{1-x^2}$ on $(-1,1)$. A calculation similar to that done for u_1 shows that the function u_2 is also a solution of the differential equation $dy/dx=-x/y$ on $(-1,1)$.   ◀

▶ **Example 4**   One of the most important problems in mechanics is that of determining the motion of an object constrained to move along a straight line and acted upon by a force F. If the mass of the object is m and the acceleration and position of the mass at time t are $a(t)$ and $y(t)$, respectively, then by Newton's second law of motion we have

$$F=ma$$

or, since $a(t)=d^2y/dt^2$,

$$\frac{d^2y}{dt^2}=\frac{F}{m} \tag{1.5}$$

The force acting on the object may vary with time, with the position of the object, or with the velocity of the object. For example, the force due to gravity is a function of the distance from the center of the earth, and the force acting on the moving object due to air resistance is a function of the velocity of the object. Thus the equation of motion in (1.5) has the form of equation (1.4) for the case $n=2$.

The simplest form of equation (1.5) occurs when the force F is constant. In such a case the equation can be solved by integrating twice. Doing this gives

$$y(t) = \frac{F}{2m}t^2 + c_1 t + c_2 \qquad \text{(1.6)}$$

where c_1 and c_2 are constants from the two integrations. If we wish to determine the position $y(t)$ at time t of the object, we must have some additional information in order to evaluate the constants c_1 and c_2. Such information may be given in terms of the initial position, $y(0)$, and the initial velocity, $dy(0)/dt$. If we require that $y(0)=y_0$ and $dy(0)/dt=y_1$, then using equation (1.6) we find that

$$y_0 = y(0) = c_2$$

$$y_1 = \frac{dy(0)}{dt} = c_1$$

Thus with these initial values we are able to uniquely determine the position $y(t)$ at time t of the object:

$$y(t) = \frac{F}{2m}t^2 + y_1 t + y_0$$

Notice that it took two pieces of initial data to determine a unique solution of equation (1.5).　　　　　　　　　　　　　　　　　◀

This example illustrates two fundamental properties of differential equations. First, a differential equation usually has infinitely many solutions. Second, if values of the solution and certain of its derivatives are preassigned, then exactly one solution may be determined. One set of such values which usually determines a unique solution of the differential equation in (1.4) is

$$y(a) = y_0, \qquad \frac{dy}{dx}(a) = y_1, \qquad \ldots, \qquad \frac{d^{n-1}y}{dx^{n-1}}(a) = y_{n-1} \qquad \text{(1.7)}$$

where $y_0, y_1, \ldots, y_{n-1}$ are constants. Such conditions are often called **initial conditions** because x frequently measures time and a is often taken as the starting time of the process involved. Notice that in Example 4 the initial conditions determined uniquely the position of the moving object. The problem of finding a solution of equation (1.4) that satisfies the initial conditions in (1.7) is called the **initial-value problem** for equation (1.4).

Common problems involving differential equations are: (1) to find a solution that satisfies preassigned conditions, such as initial conditions; (2) to find all of the solutions; (3) to determine properties of solutions without actually computing the solutions; and (4) to approximate a solution numerically. All of these problems will be discussed in varying detail throughout the remainder of the text.

In general, only differential equations that have special forms can be solved explicitly. We will now define one special form of a differential equation that arises frequently in applications and can often be solved explicitly.

Let $a_{n-1}, \ldots, a_1, a_0$, and r be functions defined on an open interval I. If equation (1.4) can be rewritten as

$$\frac{d^n y}{dx^n} + a_{n-1}(x)\frac{d^{n-1} y}{dx^{n-1}} + \cdots + a_1(x)\frac{dy}{dx} + a_0(x)y = r(x) \qquad \text{(1.8)}$$

then it is called a **linear differential equation**. If the differential equation can not be written in this form, it is called a **nonlinear differential equation**. Equations (1.1), (1.2), and the first equation in (1.3) are linear differential equations. The remaining two equations in (1.3) are nonlinear differential equations. The differential equation in Example 1 is linear, while those in Examples 2 and 3 are nonlinear.

If the coefficient functions $a_0, a_1, \ldots, a_{n-1}$ are constant functions, then, as we will see later in the text, there are techniques that will determine all the solutions of the linear differential equation in (1.8).

Exercises   SECTION 1.1

In Exercises 1–8 determine the order of the differential equation and whether it is linear.

1. $\left(\dfrac{dy}{dx}\right)^2 + y = x^3$

2. $\dfrac{d^2 y}{dx^2} + \left(\dfrac{dy}{dx}\right)^4 + x^5 y = 0$

3. $\dfrac{d^5 y}{dx^5} + \dfrac{d^3 y}{dx^3} + y + x^7 = 0$

4. $x^2\dfrac{d^2 y}{dx^2} + x\left(\dfrac{dy}{dx}\right)^3 + y = e^x$

5. $\sin\left(\dfrac{d^3 y}{dx^3}\right) + y = 0$

6. $\dfrac{d^3y}{dx^3}+x\dfrac{d^2y}{dx^2}+e^y=x^3$

7. $\dfrac{dy}{dx}+y^2+y=x+e^x$

8. $\dfrac{d^2y}{dx^2}+(\cos x)\dfrac{dy}{dx}+y=3x$

In Exercises 9–16 verify that the given function is a solution of the differential equation.

9. $\dfrac{d^2y}{dx^2}+y=0,\quad u(x)=\cos x$ on $(-\infty,\infty)$

10. $\dfrac{d^2y}{dx^2}+4y=0,\quad u(x)=\sin 2x$ on $(-\infty,\infty)$

11. $\dfrac{dy}{dx}+2y^{3/2}=0,\quad u(x)=x^{-2}$ on $(0,\infty)$

12. $\dfrac{d^2y}{dx^2}+3\dfrac{dy}{dx}+2y=0,\quad u(x)=e^{-x}+e^{-2x}$ on $(-\infty,\infty)$

13. $\dfrac{d^2y}{dx^2}-2\dfrac{dy}{dx}+y=0,\quad u(x)=2e^x+3xe^x$ on $(-\infty,\infty)$

14. $\dfrac{d^3y}{dx^3}-2\dfrac{d^2y}{dx^2}-\dfrac{dy}{dx}+2y=0,\quad u(x)=3e^x+2e^{-x}+e^{2x}$ on $(-\infty,\infty)$

15. $\dfrac{d^3y}{dx^3}+3\dfrac{d^2y}{dx^2}+3\dfrac{dy}{dx}+y=0,\quad u(x)=e^{-x}(3+5x+4x^2)$ on $(-\infty,\infty)$

16. $\dfrac{d^4y}{dx^4}+4\dfrac{d^3y}{dx^3}+6\dfrac{d^2y}{dx^2}+4\dfrac{dy}{dx}+y=0,\quad u(x)=x^3e^{-x}$ on $(-\infty,\infty)$

In Exercises 17–20 find numbers r such that $u(x)=e^{rx}$ is a solution of the given equation.

17. $\dfrac{d^2y}{dx^2}+5\dfrac{dy}{dx}+4y=0$

18. $\dfrac{d^2y}{dx^2}+3\dfrac{dy}{dx}-4y=0$

19. $\dfrac{d^3y}{dx^3}-2\dfrac{d^2y}{dx^2}-\dfrac{dy}{dx}+2y=0$

20. $\dfrac{d^4y}{dx^4}-5\dfrac{d^2y}{dx^2}+4y=0$

In Exercises 21–24 find numbers r such that $u(x)=x^r$ is a solution of the given differential equation.

21. $x^2\dfrac{d^2y}{dx^2}+2x\dfrac{dy}{dx}-6y=0$

22. $2x^2\dfrac{d^2y}{dx^2}+5x\dfrac{dy}{dx}+y=0$

23. $x^2\dfrac{d^2y}{dx^2}-x\dfrac{dy}{dx}-2y=0$

24. $x^3\dfrac{d^3y}{dx^3}+x\dfrac{dy}{dx}-y=0$

In Exercises 25–28 determine the constants c_1 and c_2 such that $u(x)=c_1e^x+c_2e^{-x}$ is a solution of $d^2y/dx^2-y=0$ that satisfies the initial conditions.

25. $y(0)=0,\quad \dfrac{dy}{dx}(0)=1$

26. $y(0)=1,\quad \dfrac{dy}{dx}(0)=0$

27. $y(0)=0,\quad \dfrac{dy}{dx}(0)=0$

28. $y(0)=2,\quad \dfrac{dy}{dx}(0)=4$

In Exercises 29–32 find the solution of each initial-value problem.

29. $\dfrac{dy}{dx}=x^2,\quad y(0)=2$

30. $\dfrac{dy}{dx}=x^3+1,\quad y(0)=-1$

31. $\dfrac{dy}{dx}=\sin x,\quad y(\pi)=0$

32. $\dfrac{d^2y}{dx^2}=\cos x,\quad y(0)=1,\quad \dfrac{dy(0)}{dx}=1$

1.2 First-Order Linear Differential Equations

In the following section we will see that various physical, chemical, biological, and economic processes can be modeled by the first-order linear differential equation (where $'$ denotes d/dx)

$$y'+p(x)y=q(x) \tag{1.9}$$

In the case that p and q are continuous functions on an open interval I, it is possible to find all solutions of this equation by means of a trick: we multiply each side of the equation by $e^{P(x)}$, where $P(x)$ is a function such that $P'(x)=p(x)$. (The function $e^{P(x)}$ is called an **integrating factor** for equation (1.9).) Doing this, we have

$$y'e^{P(x)}+yp(x)e^{P(x)}=q(x)e^{P(x)}$$

We now notice that

$$\left(ye^{P(x)}\right)'=y'e^{P(x)}+yP'(x)e^{P(x)}$$

$$=y'e^{P(x)}+yp(x)e^{P(x)}$$

$$=q(x)e^{P(x)} \qquad \text{(1.10)}$$

If we now integrate each side of equation (1.10), we obtain

$$ye^{P(x)}=\int q(x)e^{P(x)}\,dx+c$$

where c is an arbitrary constant and $\int q(x)e^{P(x)}\,dx$ is an indefinite integral of $q(x)e^{P(x)}$ on I. Thus

$$y(x)=e^{-P(x)}\left[\int q(x)e^{P(x)}\,dx+c\right] \qquad \text{(1.11)}$$

Thus we are able to obtain a formula for the solutions of equation (1.9) on any open interval on which p and q are continuous. The expression given in equation (1.11) is called the **general solution** of equation (1.9). In the following examples we will carry out the calculations leading to equation (1.11) with specific choices for the functions p and q. For simplicity the interval I will often not be given. In such a situation it will be understood that I is any open interval on which p and q are continuous.

▶ **Example 1** Consider the differential equation

$$y'+\frac{2x}{x^2+1}y=x$$

If we multiply each side of this equation by the integrating factor

$$e^{\int 2x/(x^2+1)\,dx}=e^{\ln(x^2+1)}=x^2+1$$

we obtain

$$(x^2+1)y'+2xy=(x^2+1)x$$

or, equivalently,

$$((x^2+1)y)'=x^3+x$$

Thus

$$(x^2+1)y=\tfrac{1}{4}x^4+\tfrac{1}{2}x^2+c$$

so that

$$y(x)=(x^2+1)^{-1}\left(\tfrac{1}{4}x^4+\tfrac{1}{2}x^2+c\right) \qquad \blacktriangleleft$$

► **Example 2**   Consider the differential equation

$$y'+2y=e^{-3x}$$

For this differential equation an integrating factor is

$$e^{\int 2\,dx}=e^{2x}$$

Multiplying each side of the equation by this integrating factor yields

$$e^{2x}y'+2e^{2x}y=e^{-x}$$

or, equivalently,

$$\left(e^{2x}y\right)'=e^{-x}$$

Thus

$$e^{2x}y=-e^{-x}+c$$

so that

$$y(x)=-e^{-3x}+ce^{-2x} \qquad \blacktriangleleft$$

If an initial condition $y(a)=y_0$, a in I, is given along with equation (1.9), then the definite integral, instead of the indefinite integral, may be used when integrating each side of equation (1.10). When doing this, it is convenient to choose the antiderivative $P(x)$ of $p(x)$ so that $P(a)=0$; i.e., choose $P(x)$ to be $\int_a^x p(s)\,ds$. Then, integrating each side of equation (1.10) from a to x, we have

$$y(x)e^{P(x)} - y(a) = \int_a^x \frac{d}{ds}\left(y(s)e^{P(s)}\right) ds$$

$$= \int_a^x q(s)e^{P(s)}\,ds$$

so that

$$y(x) = y_0 e^{-P(x)} + e^{-P(x)}\int_a^x q(s)e^{P(s)}\,ds \qquad \textbf{(1.12)}$$

whenever x is in I and $P(x)$ is an antiderivative of $p(x)$ on I such that $P(a)=0$. Thus we have found the solution on I of the initial-value problem

$$y' + p(x)y = q(x), \quad y(a) = y_0$$

This is an important result and we will state it explicitly as a theorem.

THEOREM 1.1 Let p and q be continuous functions on an open interval I containing the number a. Then for each number y_0, the initial-value problem

$$y' + p(x)y = q(x), \quad y(a) = y_0$$

has the unique solution on I

$$y(x) = y_0 e^{-P(x)} + e^{-P(x)}\int_a^x q(s)e^{P(s)}\,ds$$

where $P(x)$ is the antiderivative of $p(x)$ on I such that $P(a)=0$.

▶ **Example 3** In many elementary applications the special form

$$y' + p(x)y = 0$$

of equation (1.9) arises. For example, in the following section and its accompanying exercises, this equation will be used to date archaeological finds, investigate the circulation of the oceans, and study the nature of chemical reactions. Usually this equation is accompanied by an initial condition, $y(0)=y_0$. The solution of this initial-value problem may be obtained directly from equation (1.12) by noting that $q(x)\equiv 0$ on the interval I:

$$y(x)=y_0 e^{-P(x)}$$

where $P(x)$ is a function such that $P'(x)=p(x)$ on I and $P(0)=0$. ◀

In practice, we may encounter differential equations that are not first-order linear equations, but that can be made into such by an appropriate change of variables. One of the simplest cases that illustrates this technique is an equation of the form

$$y''+p(x)y'=Q(x) \tag{1.13}$$

The change of variable $w=y'$ transforms equation (1.13) into

$$w'+p(x)w=Q(x)$$

which is a first-order linear differential equation for the variable w.

▶ **Example 4**   Consider the differential equation

$$y''+2y'=x+1$$

Setting $w=y'$, we obtain

$$w'+2w=x+1$$

$e^{2x}w' + 2e^{2x}w = (x+1)e^{2x}$

$(e^{2x}w)' = e^{2x}(x+1)$

Using equation (1.11), we find that the general solution of this equation is

$$w(x)=\frac{x}{2}+\frac{1}{4}+ce^{-2x}$$

Since $y'=w$, we may obtain y by integrating w. Doing this, we have

$$y(x)=\frac{x^2}{4}+\frac{1}{4}x+c_1 e^{-2x}+c_2$$

where $c_1=-\frac{1}{2}c$. ◀

Another type of equation that may be transformed into a first-order linear equation, but not in such an obvious fashion, is **Bernoulli's equation**

$$y'+p(x)y=q(x)y^n \qquad (n\neq 1) \tag{1.14}$$

which can be rewritten in the form

$$\frac{1}{y^n}y'+p(x)\frac{1}{y^{n-1}}=q(x)$$

This equation can be transformed into a first-order linear equation by changing variables according to $v=1/y^{n-1}$. Doing this, we have

$$v'=(1-n)y^{-n}y'$$

Elementary algebra now enables us to write equation (1.14) in the form

$$v'+(1-n)p(x)v=(1-n)q(x) \qquad\qquad \textbf{(1.15)}$$

which is a first-order linear differential equation for the variable v. We may now solve equation (1.15) for v and then find y since $y=v^{1/(1-n)}$.

▶ **Example 5**   Consider the equation

$$y'+y=y^4$$

which can be rewritten as

$$\frac{1}{y^4}y'+\frac{1}{y^3}=1$$

The change of variables $v=1/y^3$ enables us to obtain the following equation for v:

$$v'-3v=-3$$

Solving this equation using equation (1.11) yields

$$y^{-3}=v=1+ce^{3x}$$

or, equivalently,

$$y(x)=\frac{1}{(1+ce^{3x})^{1/3}}$$

provided $1+ce^{3x}\neq0$.   ◀

Exercises   SECTION 1.2

In Exercises 1–10 find the general solution of each of the following equations on $(0, \infty)$.

1. $y' + y = 1$

2. $y' + 2y = e^{-2x}$

3. $y' + \dfrac{1}{x}y = \dfrac{e^x}{x}$

4. $y' + \dfrac{2}{x}y = x + 1$

5. $y' + \dfrac{2x}{x^2+1}y = x^2 + \dfrac{1}{x}$

6. $xy' + (x+1)y = 1$

7. $x^2 y' + y = -1$

8. $y' - xy = x$

9. $y' + 3x^2 y = xe^{-x^3}$

10. $x^2 y' - xy = 1$

In Exercises 11–20 find the solution of the given initial-value problem.

11. $y' + 3y = 4$, $y(0) = 5$

12. $y' - 5y = e^x$, $y(1) = 2$

13. $xy' + y = \ln x$, $y(1) = 2$

14. $xy' - 2y = -x^3 \cos x$, $y(\pi) = 0$

15. $y' - y = \dfrac{e^x}{x}$, $y(e) = 0$

16. $xy' + y = e^x$, $y(2) = e$

17. $y' + (\cot x)y = -1$, $y(\pi/4) = \pi^2$

18. $(x+1)y' + (x+2)y = 5$, $y(0) = 2$

19. $y' + (\tan x)y = \cos x$, $y(0) = 0$

20. $y' + (\tan x)y = \dfrac{1}{\cos x}$, $y(0) = 1$

In Exercises 21–38 find a general solution of the given equation.

21. $y'' + x^{-1}y' = x^2 + 1$

22. $y'' + xy' = x$

23. $y'' + y' = x^2$

24. $y'' - 2xy' = e^{x^2}$

25. $y'' + (\sin x)y' = \sin x$

26. $y'' + x^{-2}y' = x^{-2}$

27. $y''' - x^{-1}y'' = xe^x$

28. $y'' + \dfrac{1}{x+1}y' = \dfrac{e^x}{x+1}$

29. $y''' + y'' = 1$

30. $y^{(4)} + y^{(3)} = e^x$

31. $y' + y = y^2$

32. $y' + 2y = y^{1/2}$

33. $y' + x^{-1}y = xy^4$

34. $y' - x^{-1}y = y^{-3}$

35. $xy' + 3y = 3e^x y^{2/3}$

36. $y' + y = e^{-x}y^2$

37. $y' + (\tan x)y = -y^2$

38. $y' + 2x^{-1}y = -\tfrac{1}{2}xy^3$

39. An equation of the form

$$\frac{dy}{dx} = P(x)y^2 + Q(x)y + R(x)$$

is called **Riccati's equation**. Show that if f is any particular solution of this equation, then the transformation $y = f + 1/v$ transforms Riccati's equation into

$$\frac{dv}{dx} + [2P(x)f(x) + Q(x)]v = -P(x)$$

which is a first-order linear differential equation for v. Thus knowing one particular solution enables us to find a general solution.

In Exercises 40 and 41 use the previous exercise to find the solution of the given equation.

40. $\dfrac{dy}{dx} = -y^2 + 2xy - x^2 + 1, \quad y(0) = 1$

  Hint: $f(x) = x$ is a solution.

41. $\dfrac{dy}{dx} = y^2 - xy + 1, \quad y(0) = 3$

  Hint: $f(x) = x$ is a solution.

1.3 Applications of First-Order Linear Differential Equations

First-order linear differential equations have been used to model a wide range of phenomena. Several of these models are based on the assumption that the rate of change of a quantity with respect to time, dN/dt, is proportional to the amount of the quantity, N, present at time t. That is,

$$\frac{dN}{dt} = kN$$

or, equivalently,

$$\frac{dN}{dt} - kN = 0$$

From Example 3 of Section 1.2 we have

$$N(t) = N_0 e^{kt}$$

where N_0 is the amount present at time $t = 0$.

For example, let $N(t)$ denote the number of atoms of a radioactive substance present at time t. The physicist Rutherford showed that the number of atoms dN/dt that disintegrate per unit time is proportional to N; that is,

$$\frac{dN}{dt} = kN \tag{1.16}$$

Thus

$$N(t) = N_0 e^{kt} \tag{1.17}$$

where N_0 is the number of atoms present at time $t = 0$. The most common measure as to how fast a radioactive substance decomposes is its half-life, which is the amount of time required for half of a given number of atoms of the radioactive substance to decompose. The constant of proportionality k in equation (1.17) may be determined in terms of the half-life T of the radioactive substance:

$$\tfrac{1}{2}N_0 = N(T) = N_0 e^{kT}$$

Hence

$$k = -T^{-1}\ln 2 \tag{1.18}$$

and we may rewrite $N(t)$ as

$$N(t) = N_0 e^{-(T^{-1}\ln 2)t} \qquad\qquad \text{(1.19)}$$

One of the most interesting applications of radioactive decomposition is that of radiocarbon dating. Carbon consists of three isotopes that are chemically indistinguishable but have different atomic weights. The most common isotope has an atomic weight of 12, while the remaining two have atomic weights of 13 and 14. For our purposes, the most important fact about these three isotopes is that carbon-14 is radioactive. W. F. Libby hypothesized: (1) that the proportion of carbon-14 in the carbon of the atmosphere is constant; and (2) that every living thing continually renews its carbon-14 by absorption of atmospheric carbon-14. From these hypotheses he concluded that every living thing has the same ratio of carbon-14 to carbon-12. When an organism dies it ceases to obtain carbon-14 from the air, and the decomposition of carbon-14 in the organism is no longer counterbalanced by new carbon-14 from the air. Hence the ratio of carbon-14 to carbon-12 will decrease at a fixed rate. There are intricate procedures for measuring how many atoms have decomposed in a given period of time. In order to obtain accurate measurements, the item under consideration must undergo a number of processes to eliminate any possible contamination. (Among these processes are washing with hydrochloric acid and controlled combustion of the item. Thus items with intrinsic or historical value are not dated in this manner.) The number of atomic decompositions in a given period of time is then measured by a highly sensitive counter, thus approximating dN/dt. The half-life of carbon-14 is approximately 5730 years, so that (after processing) a sample of wood from a tree that died 5730 years ago would produce only half as many counts per minute as a tree cut down today.

For simplicity of notation we will denote dN/dt by $N'(t)$. We will show that if we know $N'(0)$ and $N'(s)$, where s is unknown, then it is possible to determine s.

From equations (1.16) and (1.17) we have

$$N'(s) = kN(s)$$

$$N'(0) = kN(0) = kN_0$$

$$N(s) = N_0 e^{ks}$$

Hence

$$N'(s) = kN(s)$$

$$= kN_0 e^{ks}$$

$$= N'(0) e^{ks}$$

If we now solve this equation for s, we find that

$$s = \frac{1}{k} \ln \frac{N'(s)}{N'(0)}$$

Recalling that k can be written in terms of the half-life of carbon-14, equation (1.18), we have

$$s = -\frac{5730}{\ln 2} \ln \frac{N'(s)}{N'(0)} \tag{1.20}$$

▶ **Example 1**   A piece of charcoal excavated in Nippur in 1950 registered an average of 4.029 counts per minute, while a piece of living wood registered an average of 6.68 counts per minute [W. F. Libby, *Radiocarbon Dating*, 2nd ed., University of Chicago Press, Chicago, 1955, page 81]. By equation (1.20) we are able to compute the number of years s since the tree (from which the charcoal came) was cut down:

$$s = -\frac{5730}{\ln 2} \ln \frac{4.029}{6.68} \simeq 4180 \text{ years}$$

Hence the tree was alive in approximately 2200 B.C.   ◀

The above description of radiocarbon dating is greatly simplified, but the general idea is correct.

We will now present two models for population growth. Since a population changes by integer amounts, it might seem absurd to try to find a continuous function that represents a population at time t. However, if the population is relatively large, then the change due to the addition or deletion of one individual is small when compared to the entire population. These small changes enable us, in certain cases, to approximate the population by a continuous function.

One of the simplest models for population growth is one in which the rate of change of the population at time t is proportional to the entire population $P(t)$ at time t. Hence

$$\frac{dP}{dt} = kP \tag{1.21}$$

so that

$$P(t) = P_0 e^{kt} \tag{1.22}$$

where P_0 is the population at time $t = 0$.

We will now use equation (1.22) to model the population of the United States. In 1810 the population of the United States was 7.24 million, while in 1820 the population was 9.64 million. We will take the year 1810 to be the time $t=0$ and measure t in years. From equation (1.22) we have

$$9.64 = 7.24e^{10k}$$

so that

$$k = \frac{1}{10}\ln\frac{9.64}{7.24} \approx .0286$$

Equation (1.22) may now be rewritten as

$$P(t) = 7.24e^{(.0286)t} \tag{1.23}$$

Table 1.1 compares the population of the United States at ten-year intervals with the predictions of equation (1.23).

Table 1.1

Year	Population* (in millions)	Predicted Population (in millions)
1810	7.2	7.2
1820	9.6	9.6
1830	12.9	12.8
1840	17.1	17.1
1850	23.2	22.7
1860	31.4	30.2
1870	39.8	40.3
1880	50.2	53.6
1890	63.0	71.4
1900	76.2	95.0
1910	92.2	126.4
1920	106.0	168.3
1930	123.2	224.0
1940	132.2	298.1
1950	151.3	396.9
1960	179.3	528.3
1970	203.2	703.2

*Encyclopedia Americana, Vol. 27, Americana Corporation, New York, 1976, page 531.

For the years between 1810 and 1870, equation (1.23) gives estimates of the population that are accurate to within four percent. After 1870 the difference between the predicted population and the actual population grows until by 1970 the predicted population is nearly four and one-half times as large as the actual population.

In 1837 the Dutch mathematical biologist Verhulst proposed a model for population growth that can be expressed in our notation by the differential equation

$$\frac{dP}{dt} = aP - bP^2 \qquad\qquad (1.24)$$

where a and b are positive constants. This equation is called the **logistic equation** for population growth. In general, the constant b will be very small relative to the constant a. Hence if P is relatively small, the term $-bP^2$ will be negligible when compared to aP and the population will grow nearly as fast as if we used

$$\frac{dP}{dt} = aP$$

as our model. When P is large, the term $-bP^2$ is no longer negligible and considerably decreases the rate at which the population grows.

Equation (1.24) is a Bernoulli equation, which may be rewritten as

$$P^{-2}\frac{dP}{dt} - aP^{-1} = -b \qquad\qquad (1.25)$$

Proceeding as in Section 1.2, we introduce the change of variable $v = P^{-1}$. Then

$$\frac{dv}{dt} = -P^{-2}\frac{dP}{dt}$$

and equation (1.25) may be rewritten as

$$\frac{dv}{dt} + av = b \qquad\qquad (1.26)$$

A short calculation shows that solutions of equation (1.26) have the form

$$v(t) = \frac{b}{a} + ce^{-at}$$

where c is an arbitrary constant. Hence solutions of equation (1.24) have the form

$$P(t) = \left(\frac{b}{a} + ce^{-at} \right)^{-1}$$

If P_0, $P_0 > 0$, is the population when $t = 0$, then

$$P_0 = \left(\frac{b}{a} + c \right)^{-1}.$$

so that

$$c = \frac{1}{P_0} - \frac{b}{a}.$$

Thus

$$P(t) = \frac{1}{\dfrac{b}{a} + \left(\dfrac{1}{P_0} - \dfrac{b}{a} \right) e^{-at}} = \frac{aP_0}{bP_0 + (a - bP_0)e^{-at}} \qquad (1.27)$$

Since $a > 0$, we observe that

$$\lim_{t \to \infty} P(t) = \frac{a}{b}$$

regardless of the value of P_0. Thus the logistic equation gives a model for the population in which the estimates of the population remain bounded and in fact approach a limiting population of a/b as $t \to \infty$.

A straightforward calculation shows that

$$\frac{dP}{dt} = \frac{a^2 P_0 (a - bP_0) e^{-at}}{\left(bP_0 + (a - bP_0) e^{-at} \right)^2}$$

Hence $dP/dt > 0$ whenever $0 < a - bP_0$. That is, $P(t)$ is an increasing function whenever $P_0 < a/b$. Moreover,

$$\frac{d^2 P}{dt^2} = \frac{d}{dt} \left(\frac{dP}{dt} \right)$$

$$= \frac{d}{dt} (aP - bP^2)$$

$$= a\frac{dP}{dt} - 2bP\frac{dP}{dt}$$

$$= (a - 2bP)\frac{dP}{dt}$$

and we see that

$$\frac{d^2P}{dt^2} < 0 \quad \text{if } \frac{a}{2b} < P$$

$$\frac{d^2P}{dt} > 0 \quad \text{if } P < \frac{a}{2b}$$

whenever $dP/dt > 0$. Thus the graph of P is concave down whenever $P < a/2b$ ~~$a/2b < P$~~ and concave up whenever ~~$a/2b < P$~~. If $P_0 < a/b$ there are two cases to consider: $P_0 < a/2b$ and $a/2b \leq P_0$. The graph of P in each of these cases is shown in Figure 1.1. The case $a/b \leq P_0$ is considered in Exercise 18.

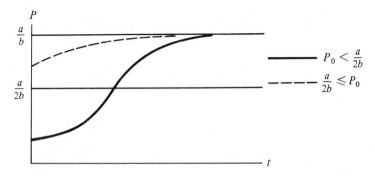

FIGURE 1.1

It is reported [M. Braun, *Differential Equations and Their Applications*, Springer-Verlag, Berlin, 1975, pages 50–51] that in 1845 Verholst used the values $a = .03134$, $b = (1.5887)10^{-10}$, and $P(0) = (3.9)10^6$. This value of $P(0)$ is the approximate population in 1790. With these values in equation (1.24) we will use

$$\frac{dP}{dt} = .03134P - (1.5887)10^{-10}P^2, \quad P(0) = (3.9)10^6 \qquad \textbf{(1.28)}$$

to model the population of the United States. Table 1.2 compares the population at ten-year intervals with the predictions of equation (1.28).

For the years between 1810 and 1950, equation (1.28) gives estimates of the population that are accurate to within four percent. After 1950 the difference between the population and the predicted population grows until by 1970 the predicted population is approximately 83% of the actual population. Since $\lim_{t \to \infty} P(t) = a/b$, we might modify the choices of a and b to obtain better accuracy in the later years. Exercise 18 contains another example of the use of the logistic equation to model the population of the United States.

Table 1.2 ──

Year	Population (in millions)	Predicted Population (in millions)
1810	7.2	7.2
1820	9.6	9.7
1830	12.9	13.0
1840	17.1	17.4
1850	23.2	23.0
1860	31.4	30.2
1870	39.8	39.1
1880	50.2	49.9
1890	63.0	62.4
1900	76.2	76.5
1910	92.2	91.6
1920	106.0	107.0
1930	123.2	122.0
1940	132.2	136.0
1950	151.3	148.4
1960	179.3	159.0
1970	203.2	167.8

In many applications the derivative is interpreted as the rate of change of a quantity with respect to time. For example, suppose that a substance X flows at a specified rate into a tank containing a liquid and that the mixture is kept homogeneous by constant stirring. Further suppose that the mixture is allowed to leave the tank at a specified rate that is not necessarily the same rate at which X enters the tank. The problem is to determine the amount of X in the tank at time t. To solve this problem we begin by letting $x(t)$ denote the amount of X in the tank at time t. Then dx/dt denotes the rate of change (at time t) of the amount of X in the tank. The rate of change of the amount of X in the tank is the rate at which X flows into the tank minus the rate at which X flows out of the tank. That is, x satisfies the differential equation

$$\frac{dx}{dt} = (\text{rate of inflow of X}) - (\text{rate of outflow of X}) \qquad \textbf{(1.29)}$$

▶ **Example 2**   Suppose that a tank initially contains 20 gallons of pure water. At a given instant (taken to be $t=0$) a salt solution containing 1.5 pounds of salt per gallon flows into the tank at a rate of 2 gallons per

2 gal/min

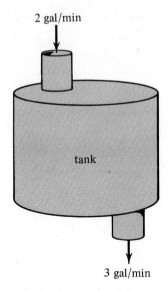

tank

3 gal/min

FIGURE 1.2

minute. The liquid in the tank is kept homogeneous by constant stirring. Also at $t=0$, liquid is allowed to flow from the tank at the rate of 3 gallons per minute. See Figure 1.2. We will find the amount of salt in the tank, $x(t)$, at time t.

We will use equation (1.29) to obtain a differential equation for x. In order to do this, we will first compute the rate at which salt enters the tank. The salt solution enters the tank at the rate of 2 gal/min and contains 1.5 lb/gal of salt. Therefore

$$\text{rate of inflow} = (1.5 \text{ lb/gal})(2 \text{ gal/min})$$

$$= 3 \text{ lb/min} \qquad \text{(1.30)}$$

Next we compute the rate at which salt leaves the tank. This rate (at time t) is the product of the concentration of salt in the tank at time t and the rate at which liquid leaves the tank. The concentration of salt at time t is the quotient of the amount of salt in the tank and the volume of liquid in the tank at time t. After t minutes, $2t$ gallons of solution have entered the tank and $3t$ gallons of liquid have left the tank. Recalling that the tank initially held 20 gallons of water, the volume of liquid in the tank, $v(t)$, at time t is given by

$$v(t) = 20 + 2t - 3t$$

$$= (20 - t)\text{gal}$$

Notice that $v(t) \geq 0$ only if $t \leq 20$. The concentration $C(t)$ of salt in the tank at time t is given by

$$C(t) = \frac{x}{v(t)}$$

$$= \frac{x}{20 - t} \text{ lb/gal}$$

Thus

$$\text{rate of outflow} = C(t)(3 \text{ gal/min})$$

$$= \frac{3x}{20 - t} \text{ lb/min} \qquad (1.31)$$

Using (1.30) and (1.31) in (1.29), we obtain a linear differential equation

$$\frac{dx}{dt} + \frac{3x}{20 - t} = 3$$

for the amount of salt in the tank $x(t)$ at time t. At time $t = 0$ there was no salt in the tank. Hence $x(0) = 0$ is the initial condition. Solving this initial-value problem with the use of equation (1.12), we find that

$$x(t) = \tfrac{3}{2}(20 - t) - \tfrac{3}{800}(20 - t)^3 \qquad (1.32)$$

Since the volume of liquid in the tank $v(t)$ is nonnegative only if $t \leq 20$, the formula for x in (1.32) is valid only for $0 \leq t \leq 20$. In particular, for $t = 10$ there are $x(10) = 11.25$ lb of salt in the tank. For $20 < t$ no liquid accumulates in the tank, and the salt solution flows out of the tank as fast as it enters. ◀

Many other applications of first-order equations are presented in the exercises.

Exercises   SECTION 1.3

1. Suppose that a piece of charcoal from an excavation registers an average of 4.83 counts per minute. How many years ago was the tree cut down?

2. Charcoal from an excavation at Jarmo, Iraq, was dated in 1970 at 5150 B.C. How many counts per minute did the charcoal register? [W. F. Libby, *Radiocarbon Dating*, 2nd ed., University of Chicago Press, Chicago, 1955, page 81.]

3. Deep ocean water has measurably less radioactivity due to carbon-14 than surface ocean water. This seems to indicate that the rate of turnover of the ocean must take place on a scale of thousands of years. A sample of water taken at the surface averaged 6.72 counts per minute, while samples taken at depths of 1829 meters and 2743 meters had average readings of 5.45 and 5.34 counts per minute. Find the apparent ages of the latter two samples. [J. L. Kulp, L. E. Tryin, W. R. Eckelman, W. A. Snell, Lamont natural radiocarbon measurements, II, Science, **116** (1952), 409–414.]

4. Experiments show that the reaction $N_2O_5 \rightarrow 2NO_2 + \frac{1}{2}O_2$ is first-order with respect to the concentration, $[N_2O_2]$, of N_2O_5; i.e.,

$$\frac{d}{dt}[N_2O_5] = -k[N_2O_5]$$

Let $C(t)$ denote the concentration of N_2O_5 at time t. Suppose that $C(t)$ is measured at various times $t_1, t_2, \ldots, t_n$, and that the points $(\ln C(t_1), t_1), (\ln C(t_2), t_2), \ldots, (\ln C(t_n), t_n)$ lie on a straight line with slope k. This would lead one to hypothesize that the point $(\ln C(t), t)$ lies on the line for every $t \geq 0$. Assuming that $(\ln C(t), t)$ is on the line for each $t \geq 0$, show that $dC/dt = kC$. If this method is used to evaluate k, care must be taken because k is quite small; e.g., at 25°C, $k = (3.38)10^{-5}$. [C. W. Castellan, *Physical Chemistry*, 2nd ed., Addison-Wesley, Reading, Mass., 1971, page 736.]

5. A column of a fluid having a cross-sectional area A is subjected to a gravitational field acting downward to give a particle an acceleration g. The vertical component z is measured upward from the ground level where $z = 0$. The pressure $P(z)$ at height z in the column is determined by the total weight of the fluid in the column above that height:

$$\frac{dP}{dz} = kP$$

where k is a constant that depends upon g, A and the fluid. If $P(0) = P_0$, find $P(z)$. [C. W. Castellan, *Physical Chemistry*, 2nd ed., Addison-Wesley, Reading, Mass., 1971, page 22.]

6. If $N(t)$ is the speed of a crankshaft in revolutions per minute for a diesel engine driving a generator, then

$$I\frac{dN}{dt} + cN = A$$

where I is the moment of inertia of the crankshaft, c is a constant such that $-cN$ represents the resisting torque of the generator, and A represents the torque applied to the crankshaft by the gases in the engine. If $N(0)=N_0$, find $N(t)$ in terms of I, c, A, and N_0. [R. Oldenburger, *Mathematical Engineering Analysis*, Macmillan, New York, 1950, page 50.]

7. Consider the discharging of a capacitor through a resistor as shown in Figure 1.3.

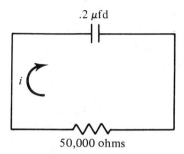

.2 μfd

50,000 ohms

FIGURE 1.3

The instantaneous values of the voltage and current will be denoted by e and i, respectively. If the initial value of e is 110 volts, then it can be shown that

$$e=110+\left[(.2)10^{-6}\right]^{-1}\int_0^t i\,dt$$

$$=110+5(10)^6\int_0^t i\,dt$$

Ohm's law states that the voltage across a resistor equals the product of the resistance and the current. We have $e=-50{,}000\,i$, so that

$$5(10)^4 i+5(10)^6\int_0^t i\,dt+110=0 \qquad\qquad (1.33)$$

Find a formula for the instantaneous values of the current i. Hint: Take the derivative of each side of (1.33) to obtain a first-order linear differential equation.

8. Assume that a nerve contains two substances (or two groups of substances) whose concentrations we will denote by x and y respectively. The ratio x/y determines excitation of the nerve. Whenever x/y becomes larger than a certain constant h, excitation occurs. For simplicity we will take $h=1$. Thus excitation occurs whenever $x \geq y$. Let x_0 and y_0 be the concentrations of x and y in a resting nerve. Since the nerve is at rest, it follows that

$x_0 < y_0$. It can be shown that

$$\frac{dx}{dt} = A - a(x - x_0)$$

$$\frac{dy}{dt} = B - b(y - y_0)$$

for some constants A, a, B, and b. Find x and y as functions of t. Find y as a function of x. [N. Rashevsky, *Mathematical Biophysics*, Vol. I, Dover Publications, New York, 1960, page 390.]

9. Suppose that glucose is infused into the blood of a patient at a constant rate and that blood samples are taken at regular intervals. Let $C(t)$ denote the concentration of glucose in the blood at time t, V the volume of distribution, p the rate of infusion, and k the velocity constant of elimination. If glucose is eliminated at a rate proportional to the amount present in the blood, then the concentration C of glucose satisfies

$$\frac{dC}{dt} = \frac{p}{V} - kC$$

Find a formula for $C(t)$. What happens to $C(t)$ as $t \to \infty$? It was found in one case that the experimental observations could be represented by $k = .0519$ and $p(kV)^{-1} = 53.8$. [S. G. Jokipii, Journal of Clinical Investigation, **34** (1954), 331, 452, 458; repeated in J. G. Defares, I. N. Sneddon, *An Introduction to the Mathematics of Medicine and Biology*, North-Holland Publishing Company, Amsterdam, 1961, page 529.]

10. Let x denote the equity capital of a company. Then under certain assumptions x satisfies the equation

$$\frac{dx}{dt} = (1 - N)rx + s$$

where

$$N = \text{dividend payout ratio}$$

$$r = \text{rate of return on equity}$$

$$s = \text{rate of new stock financing}$$

Suppose that the units have been chosen so that $x(0) = 1$. Find a formula for x. [J. L. Lebowitz, C. O. Lee, P. B. Linhart, Some effects of inflation on a firm with original cost depreciation, The Bell Journal of Economics, **7** (1976), 463–477.]

11. A tank contains 100 gallons of water in which 20 pounds of salt is dissolved. At $t=0$ a salt solution containing 2 pounds of salt per gallon flows into the tank at the rate of 3 gallons per minute. The liquid in the tank is kept homogeneous by constant stirring. Also at $t=0$, liquid begins to flow from the tank at the rate of 2 gallons per minute.

  (a) How much salt is in the tank at the end of 5 minutes?
  (b) If the tank holds 150 gallons, what is the concentration of salt in the tank at the instant the tank overflows?

12. A tank contains 100 gallons of water in which 30 pounds of salt has been dissolved. At $t=0$ pure water flows into the tank at a rate of 5 gallons per minute. The liquid in the tank is kept homogeneous by constant stirring. Also at $t=0$, liquid begins to flow from the tank at the rate of 5 gallons per minute. How long will it take to reduce the amount of salt in the tank to 10 pounds?

13. Newton's law of cooling states that the rate of change of the temperature of an object is proportional to the difference between the object's temperature and the temperature of the environment (assumed to be constant). If $T(t)$ denotes the object's temperature at time t and T_1 denotes the constant temperature of the environment, find a differential equation that represents Newton's law of cooling.

14. Suppose that a thermally insulated container filled with molten iron is exposed to the air. Since the temperature of the iron varies little from point to point, we will assume that it is uniform. The air temperature will also be assumed to be uniform at 500°C. Use the previous exercise to determine how long it takes molten iron at 1700°C to cool to its melting point (1525°C) given that the temperature after one hour is 1650°C. In fact, Newton's law of cooling is not applicable, but it does give a rough description of this cooling process. A more exact description is given in Exercise 15.

15. Suppose that a thermally insulated container filled with molten iron is exposed to the air. Since the temperature of the iron varies little from point to point, we will assume that it is uniform. The air temperature will also be assumed to be uniform at T_0. Then for various constants a and b, the temperature $T(t)$ of the iron at time t satisfies

$$\frac{dT}{dt} + aT^4 + b(T-T_0) = 0$$

[R. Oldenburger, *Mathematical Engineering Analysis*, Macmillan, New York, 1950, page 257.] In the special case $T_0 = 0$, this differential is the Bernoulli equation. Show that if $T_0 = 0$ and if the above differential equation is used to model the temperature of the iron for all $0<t$, then the coefficient b must be ≥ 0. Hint: Argue that $\lim_{t\to\infty} T(t) = T_0$.

16. The equation

$$\frac{dx}{dt} = ax - \frac{a}{b}(1 + e^{c-dt})x^2$$

arises in one model of the growth process of ownership of a consumer durable. Determine x for cases $a \neq d$ and $a = d$. [W. J. Oomens, *The Demand for Consumer Durables*, Tilburg University Press, The Netherlands, 1976, page 150.]

17. In certain cases a steady state diffusion process can be represented by a differential equation

$$C\frac{d^2 z}{dx^2} - V\frac{dz}{dx} = 0$$

where C is a diffusion constant and V represents the velocity of the process. Find a general solution of this differential equation.

18. Sketch the graph of $P(t)$ in equation (1.27) when $a/b \leq P_0$.

19. In 1920 the logistic equation was used to model the population of the United States. From this model it was determined that

$$P_1(t) = \frac{197.27}{1 + 67.32e^{-0.0313t}}$$

where time t is measured in years from 1780 and the predicted population $P_1(t)$ at time t is measured in millions. In 1940 this model was modified to another logistic equation that yielded

$$P_2(t) = \frac{184.00}{1 + 66.69e^{-0.0322t}}$$

where time t is measured in years and the predicted population $P_2(t)$ at time t is measured in years from 1780. Using the data in Table 1.1 (p. 19), determine which model more accurately predicted the population in 1850, 1910, and 1970. [The logistic curve and the census of 1940, Science, **92** (1940), 486–488.]

20. Take 1910 to be the time $t=0$. Find an equation similar to equation (1.23) and predict the population for the years 1930, 1940, 1950, 1960, and 1970.

21. When diatoms, a type of microscopic algae, are grown in a "batch culture," a phase of the growth cycle satisfies the differential equation

$dN/dt = \mu N$ where N is the cell concentration and time t is measured in days. Let N_0 denote the initial cell concentration. Different species have different values for μ at the optimum growing temperatures. These values have been tabulated for various species. [*The Biology of Diatoms*, edited by D. Werner, University of California Press, Berkeley, 1977, page 32.] For the species tabulated, μ varies from .57 to 6, with 22 of the 27 entries lying between 1 and 3. For example, *Asterionella formosa* has $\mu = 2.1$. How long does it take for an initial concentration of *Asterionella* to increase by 50%?

22. Let an annulus, i.e., the area between two concentric circles, be described in polar coordinates by $0 < a \leqslant r \leqslant b$. If the inner and outer boundaries are held at constant temperatures T_1 and T_2, respectively, for a long period of time, then the temperature T at a point in the annulus will depend only upon the distance r of the point from the origin. In such a case the temperature distribution T satisfies the equation

$$\frac{d^2T}{dr^2} + \frac{1}{r}\frac{dT}{dr} = 0$$

Find the temperature distribution.

1.4 Separable Differential Equations

A biomolecular chemical reaction can be represented by

$$A + B \rightarrow \text{products}$$

If equal portions of the chemicals A and B are combined to form the products, the decrease in the concentrations $x(t)$ is the same for either chemical. Thus, if a and b denote the initial concentrations of A and B, respectively, then the concentrations of A and B at time t are $a - x(t)$ and $b - x(t)$, respectively. At any time t after the reaction has begun, a portion of A and B have been converted into the products. Since the reaction is not reversible (the products do not decompose back to A and B), we must have $x(t) > 0$. Moreover, as time passes, A and B continue to be converted into the products. Hence $x(t)$ is an increasing function, i.e., $dx/dt > 0$. One way to model such a reaction is to suppose that the rate of change dx/dt of the decrease in concentration (of either A or B) is proportional to the concentrations of A and B. That is,

$$\frac{dx}{dt} = k(a-x)(b-x)$$

$$= k\left(ab - (a+b)x + x^2\right) \tag{1.34}$$

where k is a positive constant. The reaction of isobutyl bromide and sodium ethoxide in dry ethyl alcohol is an example of such a reaction.

Equation (1.34) is a nonlinear equation since the dependent variable x occurs with an exponent other than zero or one. This equation is also an example of a class of differential equations that can be written in the form

$$\frac{dx}{dt} = \frac{f(t)}{g(x)} \tag{1.35}$$

A differential equation of this form is called **separable**. The function of x has been written as $1/g$ for algebraic convenience. Equation (1.35) can be rewritten as

$$g(x)\frac{dx}{dt} = f(t) \tag{1.36}$$

whenever $g(x) \neq 0$. Upon integrating each side of equation (1.36) with respect to t, we obtain

$$\int g(x)\frac{dx}{dt}\,dt = \int f(t)\,dt \tag{1.37}$$

or, equivalently,

$$\int g(x)\,dx = \int f(t)\,dt$$

If G and F are antiderivatives of g and f, respectively, then

$$G(x) = F(t) + c \tag{1.38}$$

where c is an arbitrary constant. If an initial condition $x(t_0) = x_0$ is given, the definite integral instead of the indefinite integral may be used. In such a case equation (1.37) becomes

$$\int_{t_0}^{t} g(x)\frac{dx}{ds}\,ds = \int_{t_0}^{t} f(s)\,ds$$

or, equivalently,

$$\int_{x_0}^{x} g(u)\,du = \int_{t_0}^{t} f(s)\,ds$$

Hence

$$G(x) = F(t) - F(t_0) + G(x_0) \tag{1.39}$$

Equations (1.38) and (1.39) give implicit solutions to the differential equation (1.35) in all of the cases we will consider in this text. If an inverse function G^{-1} exists for G, then x may be found as a function of t. Otherwise, the implicit relationship between x and t in equation (1.39) is the best that can be achieved.

▶ **Example 1**   Consider the equation

$$\frac{dx}{dt} = \frac{t}{x} \tag{1.40}$$

Separating variables and integrating, we have

$$\int x\,dx = \int t\,dt$$

so that

$$\frac{x^2}{2} = \frac{t^2}{2} + c_1$$

where c_1 is an arbitrary constant. The relation $x^2 - t^2 = c$, where $c = 2c_1$ is an arbitrary constant, is an implicit solution for equation (1.40). Note that t^2 cannot equal $-c$ for any t since zero is not a permitted value of x in the original equation.   ◀

▶ **Example 2**   Consider the equation

$$\frac{dx}{dt} = x^2 t \tag{1.41}$$

Separating variables and integrating, we have

$$\int x^{-2}\,dx = \int t\,dt$$

whenever $x \neq 0$. The evaluation of each of these integrals yields

$$-x^{-1} = \frac{t^2}{2} + c_1 \qquad (x \neq 0)$$

where c_1 is an arbitrary constant. Hence

$$x(t) = \frac{-2}{t^2 + c} \qquad (x(t) \neq 0) \tag{1.42}$$

where $c=-2c_1$ is an arbitrary constant. We now need to consider the exceptional case $x(t)=0$ more carefully. It is easy to see that the solutions given in equation (1.42) are never zero, regardless of the choice of c. We must ask ourselves if it is possible to have a solution x_1 which is zero for some value of t. If $x_1(t_1)\neq0$ for some t_1, then our calculations show that x_1 would have the form given in equation (1.42), and $x_1(t)\neq0$ for every t since the solutions given in equation (1.42) are never zero. Hence, if x_1 is zero for one value of t, it must be zero for all values of t. That is, if x_1 is zero for one value of t, then $x_1(t)\equiv0$. It is easy to show that the constant function $x_1(t)\equiv0$ is in fact a solution of equation (1.41). We have shown that the solutions of equation (1.41) are

$$x(t)\equiv0 \quad \text{or} \quad x(t)=\frac{-2}{t^2+c}$$

where c is an arbitrary constant. ◀

▶ **Example 3**   Consider the equation

$$\frac{dx}{dt}=\frac{x^3+1}{3x^2} \tag{1.43}$$

Separating variables and integrating, we have

$$\int\frac{3x^2}{x^3+1}\,dx=\int dt$$

whenever $x^3+1\neq0$. Integrating each side of this equation, we obtain $\ln|x^3+1|=t+c$, so that the relation

$$|x^3+1|=e^{t+c} \tag{1.44}$$

gives an implicit solution of equation (1.43) whenever $x^3+1\neq0$. Setting $c_1=e^c$, we have

$$x^3+1=\pm c_1 e^t$$

Since c is an arbitrary constant, c_1 is an arbitrary positive constant and $\pm c_1$ is an arbitrary constant we will denote by C. With this change in notation we now solve for x to obtain

$$x(t)=(Ce^t-1)^{1/3}$$

Since $e^z \neq 0$ for any number z, the solutions that satisfy the relationship in (1.44) do not equal -1 for any value of t. Note that $x_1(t) \equiv -1$ is a solution of equation (1.43). Hence, if x is a solution of equation (1.43) such that $x(t_0) = -1$ for some number t_0, then $x(t) \equiv -1$. We have shown that if x is a solution of equation (1.43) other than $x(t) \equiv -1$, then $x(t) = (Ce^t - 1)^{1/3}$ for some choice of the constant C.   ◀

▶ **Example 4**   Consider the equation

$$\frac{dx}{dt} = k(a-x)(b-x), \quad x(0)=0$$

which describes the chemical reaction discussed at the beginning of this section. At least for small values of t, neither A nor B is exhausted. That is, for small values of t we have $a - x(t) \neq 0$ and $b - x(t) \neq 0$. In fact, as we shall see, $a - x(t) \neq 0$ or $b - x(t) \neq 0$ for any $t \geq 0$. Separating variables, we have

$$\int_0^x \frac{du}{(a-u)(b-u)} = \int_0^t k\, dt$$

Using the method of partial fractions to evaluate the left-hand integral, we find that

$$\int_0^x \frac{du}{(a-u)(b-u)} = \frac{1}{b-a} \int_0^x \left(\frac{1}{a-u} - \frac{1}{b-u} \right) du$$

$$= \frac{1}{b-a} \ln \frac{a(b-x)}{b(a-x)}$$

whenever $a \neq b$. Hence

$$\frac{1}{b-a} \ln \frac{a(b-x)}{b(a-x)} = kt \qquad (1.45)$$

whenever $a \neq b$, and

$$\frac{1}{a-x} - \frac{1}{a} = kt \qquad (1.46)$$

whenever $a = b$. The relations in (1.45) and (1.46) give implicit solutions of the differential equation. If we solve these relations for x, we obtain the solutions

$$x(t) = ab \frac{e^{akt} - e^{bkt}}{ae^{akt} - be^{bkt}} \qquad (1.47)$$

if $a \neq b$, and

$$x(t) = a - \frac{a}{kta+1} \qquad (1.48)$$

if $a = b$.

We will now determine the concentration, $a - x(t)$, of A at time t. Using equation (1.47) we have, for the case $a \neq b$,

$$a - x(t) = a - ab \frac{e^{akt} - e^{bkt}}{ae^{akt} - be^{bkt}}$$

$$= \frac{a}{ae^{akt} - be^{bkt}} \left[ae^{akt} - be^{bkt} - b(e^{akt} - e^{bkt}) \right]$$

$$= \frac{a(a-b)e^{akt}}{ae^{akt} - be^{bkt}} \qquad (1.49)$$

Since $a \neq b$, we have $a - x(t) \neq 0$ for any $t \geq 0$. Similarly,

$$b - x(t) = \frac{b(a-b)e^{bkt}}{ae^{akt} - be^{bkt}} \neq 0 \qquad (1.50)$$

for any $t \geq 0$. If $a = b$, then from equation (1.48) we find that

$$a - x(t) = b - x(t) = \frac{a}{kta+1} \neq 0 \qquad (1.51)$$

for any $t \geq 0$. Thus in either case the concentrations of A and B are never zero. However, in the case $a = b$, we see from equation (1.51) that

$$\lim_{t \to \infty} \left[a - x(t) \right] = \lim_{t \to \infty} \left[b - x(t) \right] = 0$$

In this case the concentrations of A and B are identical, and their graph resembles that in Figure 1.4.

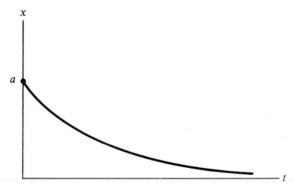

FIGURE 1.4

If $a \neq b$, the situation is slightly more complicated. Suppose that $a > b$. Then from equation (1.49) we have

$$\lim_{t \to \infty} [a - x(t)] = \lim_{t \to \infty} \frac{a(a-b)e^{akt}}{ae^{akt} - be^{bkt}}$$

$$= \lim_{t \to \infty} \frac{a(a-b)}{a - be^{(b-a)kt}}$$

$$= a - b$$

and

$$\lim_{t \to \infty} [b - x(t)] = \lim_{t \to \infty} \frac{b(a-b)e^{bkt}}{ae^{akt} - be^{bkt}}$$

$$= \lim_{t \to \infty} \frac{b(a-b)e^{(b-a)kt}}{a - be^{(b-a)kt}}$$

$$= 0$$

If $a > b$, then the graphs of the concentrations of A and B resemble those in Figure 1.5.

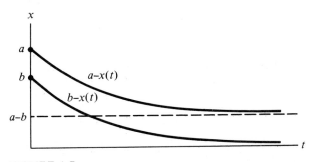

FIGURE 1.5

The case $b > a$ is identical with that done above if a and b are interchanged. ◀

Exercises   SECTION 1.4

In Exercises 1–12 find all solutions of the given equation.

1. $\dfrac{dx}{dt} = \dfrac{t}{x^2}$

2. $\dfrac{dx}{dt} = (x+1)t(t^2+1)^{1/2}$

3. $\dfrac{dx}{dt} = (x^2+1)(t^3+5)$

4. $\dfrac{dx}{dt} = (x^2-1)te^{-t^2}$

5. $\dfrac{dx}{dt} = \dfrac{\sin t}{\cos x}$

6. $\dfrac{dx}{dt} = \dfrac{x^2-1}{t^2+1}$

7. $\dfrac{dx}{dt} = e^{x+t}$

8. $\dfrac{dx}{dt} = \dfrac{(x-t)^2}{t^2} + 1$

Hint: Set $u = x - t$.

9. $\dfrac{dx}{dt} = \dfrac{x^4}{t^3}$

10. $\dfrac{dx}{dt} = \cos(x-t) + \cos(x+t)$

11. $\dfrac{dx}{dt} = \dfrac{tx+x}{x^2+1}$

12. $\dfrac{dx}{dt} = \dfrac{xt-t-x+1}{t-1}$

In Exercises 13–22 find the solution of the given initial-value problem.

13. $\dfrac{dx}{dt} = e^t \sec x, \quad x(0) = \pi$

14. $\dfrac{dx}{dt} = \dfrac{t+1}{x^2+1}, \quad x(1) = 0$

15. $\dfrac{dx}{dt} = x \tan t, \quad x(0) = 2$

16. $\dfrac{dx}{dt} = \dfrac{(1+x^2)^{1/2}}{xt+x}, \quad x(0) = 4$

17. $\dfrac{dx}{dt} = e^{-t+x}, \quad x(1) = e$

18. $\dfrac{dx}{dt} = 2(x^2+1)t, \quad x(\sqrt{\pi/4}) = \sqrt{3}$

19. $\dfrac{dx}{dt} = te^{t-x}, \quad x(0) = 5$

20. $\dfrac{dx}{dt} = \dfrac{\sqrt{1-x^2}}{t}, \quad x(e^\pi) = \tfrac{1}{2}$

21. $\dfrac{dx}{dt} = \sin(x-t) - \sin(x+t), \quad x(0) = 0$

22. $\dfrac{dx}{dt} = t^2 x^2 - t^2 + x^2 - 1, \quad x(1) = 0$

23. Show that the method of separation of variables can be used to solve the initial-value problem

$$\frac{dy}{dx} + p(x)y = 0, \quad y(0) = y_0$$

and that the solution obtained by this method is identical with the solution found in Example 3 of Section 1.2.

24. Suppose that a granular substance is to be dissolved in a liquid. If each grain is a cube of length L on each edge and if we assume that the grain remains cubical as it dissolves, then

$$\frac{dL}{dt} = -\rho^{-1} D^{2/3} v^{-1/6} (u/L)^{1/2} (c_s - C)$$

where

$$\rho = \text{density of the grains}$$

$$D = \text{coefficient of diffusion}$$

$$v = \text{kinematic viscoscity}$$

$$u = \text{flow velocity}$$

$$C = \text{concentration of the solution}$$

$$c_s = \text{saturation concentration}$$

Assuming that all of the listed quantities are constants, find a formula for the length of a side of a grain at time t. How long will it take for the length of a side to be half its initial length? [D. Marsal, Mathematical models for solution rates of different-sized particles in liquids, in *Mathematical Models of Sedimentary Processes*, edited by D. F. Merriam, Plenum Press, New York, 1972, pages 191–201.]

25. The Cobb-Douglas neoclassical growth model in economics simplifies to

$$\frac{dk}{dt} = a e^{bt} [k(t)]^\alpha$$

where $k(t)$ is the aggregate input of capital stock at time t, and a, b, α are constants depending upon the output elasticity of capital, the input of

labor, the growth of labor and technology, and the ratio of full employ-
ment saving to the gross national product. Find $k(t)$ if $k(0)=k_0$. [D.
Hamberg, *Models of Economic Growth*, Harper & Row, New York, 1971,
page 44.]

26. In considering a case of slow selection involving two genes, the equation

$$\frac{dy}{dx} = \frac{y^2(1-y)(x^2-a^2)}{x^2(1-x)(y^2-b^2)}$$

arises, where a and b are constants. Find an implicit relationship between y
and x. The exact meanings of x and y are technical; the interested reader is
referred to page 192 of *Causes of Evolution* by J. B. S. Haldane, Cornell
University Press, 1966.

27. Let $V(t)$ denote the volume of a tumor at time t. The growth of many solid
tumors has been shown to obey the Gompertzian equation

$$\frac{dV}{dt} = ae^{-bt}V$$

where a and b are positive constants, rather than the simple exponential
equation $dV/dt=aV$. [A. C. Burton, Rate of growth of solid tumors as a
problem of diffusion, Growth, **30** (1966), 157–176.] Show that in such a
case the volume of the tumor is monotonically increasing with respect to
time and approaches a finite limit as $t \to \infty$. Determine this limit.

28. A man starts at a point A and his dog at a point B in a large field. The
man begins to walk at a constant velocity v along a path perpendicular to
the line passing through the points A and B, while the dog runs at a
constant velocity $2v$ always in the direction of the man. We will determine
the path along which the dog runs and the distance the man travels before
he is overtaken by the dog.

    We take the x-axis to be the line passing through the points A and B,
and the y-axis to be the line along which the man walks. (See Figure 1.6.)
For simplicity we choose the units so that the distance from A to B is one
unit. Then A and B are the points $(0,0)$ and $(1,0)$, respectively. After a
time t the dog has moved to a point $P=(x, y)$, and the man has moved to
a point $Q=(0, a)$ on the y-axis. If the path taken by the dog is the graph of
a twice differentiable function f, then the line passing through the points P
and Q is tangent to the graph of f at the point P (since the dog always
moves toward the man).

    (a) Show that $f'(x)=-a-f(x)/x$ so that $a=f(x)-xf'(x)$. Hint: Express
    the slope of the tangent line in two ways.

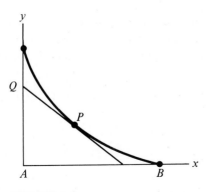

FIGURE 1.6

(b) Let s denote the length of the arc AP along the dog's path. Show that $s=2a=2(f(x)-xf'(x))$ so that

$$\frac{ds}{dx}=-2xf''(x)$$

(c) In most calculus courses it is shown that $(ds/dx)^2=1+[f'(x)]^2$. Argue that

$$\frac{ds}{dx}=-\left(1+[f'(x)]^2\right)^{1/2}$$

and obtain the equation

$$2xf''(x)=\left[1+f'(x)\right]^{1/2}$$

for the dog's path.
(d) Set $p=f'(x)$ so that the equation in (c) becomes

$$2x\frac{dp}{dx}=\left(1+p^2\right)^{1/2}$$

Show that

$$p+\left(1+p^2\right)^{1/2}=cx^{1/2}$$

where c is a constant. Hint: $\int dp/(1+p^2)^{1/2}=\ln(p+(1+p^2)^{1/2})+$ constant.
(e) Use the first equation in (b) to show that $p(1)=f'(1)=0$. Use this to show that $x^{1/2}=p+(1+p^2)^{1/2}$ and $x^{-1/2}=(1+p^2)^{1/2}-p$.

(f) Show that $f'(x)=\frac{1}{2}(x^{1/2}-x^{-1/2})$ and that $f(x)=\frac{1}{3}x^{3/2}-x^{1/2}+$ constant.

(g) Show that $f(1)=0$ and that the dog's path is given by $f(x)=\frac{1}{3}x^{3/2}-x^{1/2}+\frac{2}{3}$.

(h) Determine the distance traveled by the man before he is overtaken by the dog.

1.5 Homogeneous Differential Equations

There are equations that are not separable, but that can be made separable by an appropriate change of variables. We will consider one such type of equation in this section.

A function $F(x,t)$ is called homogeneous if $F(kx,kt)=F(x,t)$ for every number k. For example, the functions

$$G(x,t)=\frac{x^2+xt+t^2}{x^2+t^2} \quad \text{and} \quad H(x,t)=\cos\left(\frac{x}{x+t}\right)$$

are homogeneous, while the functions

$$P(x,y)=\frac{x+1}{x+t} \quad \text{and} \quad Q(x,y)=x-t$$

are not homogeneous. If we set $k=1/t$, then

$$F(x,t)=F\left(\frac{x}{t},1\right)$$

Thus the function $F(x,t)$ is a function of the single variable x/t.

The differential equation

$$\frac{dx}{dt}=F(x,t)$$

is called **homogeneous** if the function $F(x,t)$ is homogeneous. Letting $f(z)=F(z,1)$, we see that a homogeneous differential equation can be written in the form

$$\frac{dx}{dt}=f\left(\frac{x}{t}\right) \tag{1.52}$$

Unfortunately, the term homogeneous has two unrelated meanings with respect to differential equations. One meaning is the one just given. The other occurs when considering the linear differential equation in (1.6), where homogeneous means that $r(x)=0$ for every x in I. The importance of the latter meaning will be discussed in Chapter 2.

If we change variables according to $v=x/t$ in equation (1.52), then

$$\frac{dx}{dt}=v+t\frac{dv}{dt}$$

and equation (1.52) can be rewritten as

$$\frac{dv}{dt}=\frac{f(v)-v}{t} \tag{1.53}$$

which is a separable differential equation that may be solved by the method given in the previous section.

▶ **Example 1**   Consider the equation

$$\frac{dx}{dt}=\frac{x-t}{x+t}$$

which can be rewritten in the form

$$\frac{dx}{dt}=\frac{\dfrac{x}{t}-1}{\dfrac{x}{t}+1} \tag{1.54}$$

The equation now has the form of equation (1.52). Setting $v=x/t$ and using (1.53), equation (1.54) becomes

$$\frac{dv}{dt}=\frac{1}{t}\left(\frac{v-1}{v+1}-v\right) \tag{1.55}$$

$$=-\frac{1}{t}\left(\frac{v^2+1}{v+1}\right)$$

so that

$$\frac{v+1}{v^2+1}\frac{dv}{dt}=-\frac{1}{t}$$

or, equivalently,

$$\left(\frac{v}{v^2+1}+\frac{1}{v^2+1}\right)\frac{dv}{dt}=-\frac{1}{t}$$

Upon integrating each side of this equation with respect to t, we obtain an implicit solution to equation (1.55):

$$\tfrac{1}{2}\ln(v^2+1)+\tan^{-1}v=-\ln t+c$$

where c is any constant. Since $v=x/t$,

$$\tfrac{1}{2}\ln\left[\left(\frac{x}{t}\right)^2+1\right]+\tan^{-1}\frac{x}{t}=-\ln t+c$$

is an implicit solution of equation (1.55) for any constant c.   ◄

Other equations that may be transformed into separable equations by an appropriate change of variables are considered in Exercises 5–15.

Exercises   SECTION 1.5

In Exercises 1–4 find all solutions of the given equation.

1. $\dfrac{dx}{dt}=\dfrac{x}{t}+\left(\dfrac{x}{t}\right)^2$

2. $\dfrac{dx}{dt}=\dfrac{x}{x+t}$

3. $\dfrac{dx}{dt}=\dfrac{x^2+2xt}{t^2}$

4. $\dfrac{dx}{dt}=e^{x/t}+\dfrac{x}{t}$

5. Show that the change of variables $v=ax+bt$ transforms an equation of the form

$$\frac{dx}{dt}=f(ax+bt)$$

into an equation which is separable.

In Exercises 6–9 use Exercise 5 to find all solutions of the given equation.

6. $\dfrac{dx}{dt}=(x+t)^2$

7. $\dfrac{dx}{dt}=\tan^2(x+t)$

8. $\dfrac{dx}{dt}=(2x+3t)^2-4(2x+3t)-\tfrac{3}{2}$

9. $\dfrac{dx}{dt}=\dfrac{x+t-1}{x+t+1}$

10. Show that an equation of the form

$$\frac{dx}{dt} = f\left(\frac{ax+bt}{cx+dt}\right) \qquad (ad-bc \neq 0)$$

is homogeneous.

11. Show that there are numbers h and k such that the change of variables $X = x-h$, $T = t-k$ transforms an equation of the form

$$\frac{dx}{dt} = f\left(\frac{ax+bt+r}{cx+dt+s}\right) \qquad (ad-bc \neq 0)$$

into the equation

$$\frac{dX}{dT} = f\left(\frac{aX+bT}{cX+dT}\right)$$

which was considered in Exercise 10.

In Exercises 12–15 use Exercises 10 and 11 to find all solutions of the given equation.

12. $\dfrac{dx}{dt} = \dfrac{x-t+1}{x+3t+1}$

13. $\dfrac{dx}{dt} = \dfrac{t+3}{2x+t+5}$

14. $\dfrac{dx}{dt} = \dfrac{x+2}{x-t+3}$

15. $\dfrac{dx}{dt} = \dfrac{x-1}{x+t}$

1.6 Exact Differential Equations (OPTIONAL)

This section should be omitted by readers who have not studied partial differentiation. It is not necessary for the rest of the text.

In this section we will consider a special class of first-order differential equations of the form

$$\frac{dx}{dt} = -\frac{M(x,t)}{N(x,t)} \tag{1.56}$$

(The minus sign has been included for algebraic convenience.) Equation (1.56) can be rewritten as

$$M(x,t)+N(x,t)\frac{dx}{dt}=0 \tag{1.57}$$

The reader should now recall that if x is a function of t and if F is a function of x and t, then by the chain rule (where F_x and F_t denote the partial derivatives of F with respect to x and t),

$$\frac{d}{dt}F(x,t)=F_t(x,t)+F_x(x,t)\frac{dx}{dt}$$

whenever the derivatives F_t, F_x, and dx/dt exist. If F happens to be a function such that

$$F_t(x,t)=M(x,t) \quad \text{and} \quad F_x(x,t)=N(x,t) \tag{1.58}$$

then equation (1.57) may be rewritten as

$$\frac{d}{dt}F(x,t)=0$$

Hence, if c is a constant such that the relation $F(x,t)=c$ determines x as a function of t on an open interval I, then $F(x,t)=c$ is an implicit solution of equation (1.57) on I. This shows that we are able to find solutions of equation (1.57) whenever we are able to find a function F that satisfies the equations in (1.58). The following theorem gives necessary and sufficient conditions for the existence of such a function F whenever N and M have continuous derivatives.

THEOREM 1.2   Let R denote the rectangle $a<x<b$, $c<t<d$. Let M_x, M_t, N_x, and N_t be continuous functions on the rectangle R. Then there is a function F such that $F_t(x,t)=M(x,t)$ and $F_x(x,t)=N(x,t)$ on R if and only if $M_x(x,t)=N_t(x,t)$ on R. In particular, if $M_x(x,t)=N_t(x,t)$ on R, then there is a function F such that

$$\frac{d}{dt}F(x,t)=M(x,t)+N(x,t)\frac{dx}{dt}$$

on R.

Partial Proof   Suppose that there exists a function F such that $F_t(x,t)=M(x,t)$ and $F_x(x,t)=N(x,t)$ on R. Then $F_{tx}(x,t)=M_x(x,t)$ and $F_{xt}(x,t)=N_t(x,t)$. Since M_x and N_t are continuous, the functions F_{tx} and F_{xt} are continuous and, therefore, identical. Thus $M_x(x,t)=N_t(x,t)$ on R. The proof of the converse is more complicated and will be omitted. The interested reader is referred to page 496 of *Advanced Calculus*, second edition, by Angus E. Taylor and W. Robert Mann, Xerox College Publishing Company, 1972. ∎

For our purposes, the important aspect of Theorem 1.2 is that it assures the existence of implicit solutions

$$F(x, t) = \text{constant}$$

of equation (1.57) whenever M_x and N_t are continuous and equal.

> **DEFINITION 1.1** A differential equation $M(x, t) + N(x, t)dx/dt = 0$ is called **exact** on a rectangle R, $a < x < b$, $c < t < d$, if M_x, M_t, N_x, and N_t are continuous on R and $M_x(x, t) = N_t(x, t)$ at each point (x, t) of R.

A procedure to solve our exact differential equation can be described as follows.

(1) Verify that the equation $M(x, t) + N(x, t)dx/dt = 0$ is exact by showing that $M_x(x, t) = N_t(x, t)$.

(2) Supposing that the equation is exact, there is a function F such that $F_t(x, t) = M(x, t)$ and $F_x(x, t) = N(x, t)$. Obtain a representation for F by holding x fixed and integrating each side of the equation $F_t(x, t) = M(x, t)$ with respect to t. This yields $F(x, t) = G(x, t) + f(x)$, where f is a function that we must determine and G is a known function.

(3) If we take the representation for F obtained in step (2) and insert it into the relation $F_x(x, t) = N(x, t)$, we obtain an equation of the form $f'(x) = g(x)$, where g is a known function. Integrating, we find that $f(x) = \int g(x) \, dx$.

(4) We have found $F(x, t) = G(x, t) + f(x)$.

This process for solving an exact differential equation will be illustrated in the following two examples.

▶ **Example 1** Consider the equation

$$2tx + (t^2 - x)\frac{dx}{dt} = 0 \qquad \text{(1.59)}$$

In terms of our previous notation, $M(x, t) = 2tx$ and $N(x, t) = t^2 - x$. Since M and N are polynomials in x and t, their derivatives exist and are continuous on any rectangle. Moreover,

$$M_x(x, t) = 2t = N_t(x, t)$$

Therefore there is a function F such that

$$F_t(x,t)=2tx \quad \text{and} \quad F_x(x,t)=t^2-x \tag{1.60}$$

When F_t is computed, the variable x is held fixed. Hence, if we wish to obtain F from F_t, we must hold x fixed and integrate with respect to t. Doing this, we find that

$$F(x,t)=t^2x+f(x) \tag{1.61}$$

where f is a function that will be determined subsequently. The "constant" of integration is a function of x because functions of x act as though they were constants when the partial derivative with respect to t is taken. The function f may now be determined by using the relation $F_x(x,t)=t^2-x$. From (1.60) and (1.61) we obtain

$$t^2-x=F_x(x,t)=t^2+f'(x)$$

so that

$$f'(x)=-x$$

and

$$f(x)=-\frac{x^2}{2}+c_1$$

where c_1 is an arbitrary constant. Therefore

$$F(x,t)=t^2x-\frac{x^2}{2}+c_1$$

and the relation

$$t^2x-\frac{x^2}{2}=c$$

determines implicit solutions of equation (1.59).   ◄

▶ **Example 2**   Consider the equation

$$x\cos t+\frac{1}{x}+\left(\sin t-\frac{t}{x^2}+\frac{1}{x}\right)\frac{dx}{dt}=0 \tag{1.62}$$

Maintaining our previous notation, we have $M(x, t) = x \cos t + 1/x$ and $N(x, t) = \sin t - (t/x^2) + 1/x$. It is easy to show that the partial derivatives are continuous on any rectangle that does not contain points (x, t) with $x = 0$. Moreover,

$$M_x(x, t) = \cos t - \frac{1}{x^2} = N_t(x, t)$$

Therefore there is a function F such that

$$F_t(x, t) = x \cos t + \frac{1}{x} \quad \text{and} \quad F_x(x, t) = \sin t - \frac{t}{x^2} + \frac{1}{x}$$

If we hold x fixed and integrate F_t, we obtain

$$F(x, t) = x \sin t + \frac{t}{x} + f(x)$$

Taking the partial derivative with respect to x of this form for F and using the relation $F_x(x, t) = \sin t - (t/x^2) + 1/x$ yields

$$\sin t - \frac{t}{x^2} + \frac{1}{x} = F_x(x, t) = \sin t - \frac{t}{x^2} + f'(x)$$

Hence

$$f'(x) = \frac{1}{x}$$

and

$$f(x) = \ln|x| + c_1$$

where c_1 is an arbitrary constant. Therefore

$$F(x, t) = x \sin t + \frac{t}{x} + \ln|x| + c_1$$

and the relation

$$x \sin t + \frac{t}{x} + \ln|x| = c$$

determines implicit solutions of equation (1.62).   ◀

Exercises   SECTION 1.6

Determine which of the following equations are exact. If the equation is exact, find its solutions.

1. $(2x^3t+5xt^4)+(3x^2t^2+t^5+1)\dfrac{dx}{dt}=0$

2. $x\cos xt+1+t\cos xt\dfrac{dx}{dt}=0$

3. $1+tx(1+t^2)^{-1/2}+(1+t^2)^{1/2}\dfrac{dx}{dt}=0$

4. $tx(1+t^2)^{-1/2}+(1+t^2)^{1/2}\dfrac{dx}{dt}=0$

5. $(x^2+t)+(t^2+x)\dfrac{dx}{dt}=0$

6. $(2t+x+2)+(3x^2+t-3)\dfrac{dx}{dt}=0$

7. $(e^{x+t}-2)+(e^{x+t}+3)\dfrac{dx}{dt}=0$

8. $\left(xe^{xt}-\dfrac{x}{(t+x)^2}\right)+\left(te^{tx}+\dfrac{x}{(t+x)^2}\right)\dfrac{dx}{dt}=0$

9. $\dfrac{x}{t}+(\ln t)\dfrac{dx}{dt}=0$

10. $\left(x+\dfrac{x}{t}\right)+(t+\ln t)\dfrac{dx}{dt}=0$

11. $(5x^3+9xt^2+4x^2t+4x)+(2x+15x^2t+3t^3+4xt^2+4t)\dfrac{dx}{dt}=0$

12. $(x^2+t)+(2tx+x^7)\dfrac{dx}{dt}=0$

13. Show that a separable differential equation can be rewritten as an exact differential equation. Is the converse true?

1.7 Integrating Factors (OPTIONAL)

This section should be omitted by readers who have not studied partial differentiation. It is not necessary for the rest of the text.

Consider the first-order linear differential equation

$$\frac{dx}{dt}+p(t)x-q(t)=0$$

on an open interval I. Since

$$\frac{\partial}{\partial x}\left(p(t)x-q(t)\right)=p(t) \quad \text{and} \quad \frac{\partial}{\partial t}(1)=0$$

this equation is not exact unless $p(t)\equiv 0$ on the interval I, in which case the equation may be solved merely by integrating. In Section 1.2 we found that this equation can be solved by multiplying each side by $e^{P(t)}$, where $P(t)$ is any antiderivative on I of $p(t)$. Doing this yields the equation

$$e^{P(t)}\frac{dx}{dt}+p(t)e^{P(t)}x-q(t)e^{P(t)}=0 \tag{1.63}$$

Now

$$\frac{\partial}{\partial x}\left(p(t)e^{P(t)}x-q(t)e^{P(t)}\right)=p(t)e^{P(t)}$$

$$=P'(t)e^{P(t)}$$

$$=\frac{\partial}{\partial t}\left(e^{P(t)}\right)$$

so that equation (1.63) is exact. Thus in certain cases it is possible to transform a nonexact equation into an exact equation by multiplying each side by an appropriate function. This technique is precisely what enabled us to solve first-order linear differential equations by the method presented in Section 1.2.

DEFINITION 1.2   A function f such that

$$f(x,t)M(x,t)+f(x,t)N(x,t)\frac{dx}{dt}=0$$

is an exact differential equation is called an **integrating factor** for the differential equation

$$M(x,t)+N(x,t)\frac{dx}{dt}=0$$

If $M(x,t)+N(x,t)dx/dt=0$ is not an exact differential equation, there is no general method of finding an integrating factor if in fact one exists. However, there are two cases in which an integrating factor can be computed. We will discuss one such case and present the other in Exercise 5.
   Suppose there is an integrating factor f for

$$M(x,t)+N(x,t)\frac{dx}{dt}=0 \tag{1.64}$$

Then

$$f(x,t)M(x,t)+f(x,t)N(x,t)\frac{dx}{dt}=0$$

is an exact differential equation. This means that

$$\frac{\partial}{\partial x}\left(f(x,t)M(x,t)\right)=\frac{\partial}{\partial t}\left(f(x,t)N(x,t)\right)$$

or, equivalently,

$$f_x(x,t)M(x,t)+f(x,t)M_x(x,t)=f_t(x,t)N(x,t)+f(x,t)N_t(x,t)$$

(1.65)

Any integrating factor of equation (1.64) must satisfy the partial differential equation (1.65). In general, it is not possible to solve equation (1.65). If we make the simplifying assumption that there is an integrating factor which is only a function of t, then equation (1.65) can be rewritten as the separable differential equation

$$\frac{df}{dt}=\frac{M_x(x,t)-N_t(x,t)}{N(x,t)}f$$

(1.66)

In order for our assumption that f is a function of the single variable t to be meaningful, the expression $(M_x-N_t)/N$ must be a function of the single variable t. If this is the case, equation (1.66) can be solved for the integrating factor f:

$$f(t)=e^{\int \frac{M_x-N_t}{N}dt}$$

▶ **Example 1**   Consider the equation

$$x^2+t^2+2t+2x\frac{dx}{dt}=0$$

(1.67)

Setting

$$M(x,t)=x^2+t^2+2t \quad \text{and} \quad N(x,t)=2x$$

we have

$$M_x(x,t)=2x\neq 0=N_t(x,t)$$

Equation (1.67) is not exact, but

$$\frac{M_x(x,t)-N_t(x,t)}{N(x,t)}=\frac{2x}{2x}=1$$

so that equation (1.66) becomes

$$\frac{df}{dt}=f$$

which is a separable differential equation having $f(t)=e^t$ as a solution. Hence the function $f(t)=e^t$ is an integrating factor for equation (1.67). ◀

▶ **Example 2**   Consider the equation

$$4xt+6(t^2+1)+(t^2+1)\frac{dx}{dt}=0 \tag{1.68}$$

and set

$$M(x,t)=4xt+6(t^2+1) \qquad N(x,t)=t^2+1$$

Since

$$M_x(x,t)=4t\neq2t=N_t(x,t)$$

the equation is not exact. However,

$$\frac{M_x(x,t)-N_t(x,t)}{N(x,t)}=\frac{2t}{t^2+1}$$

so that there is an integrating factor f which is a solution of the separable differential equation

$$\frac{df}{dt}=\frac{2t}{t^2+1}f \tag{1.69}$$

Upon separating variables and integrating, we have

$$\int\frac{1}{f}df=\int\frac{2t}{t^2+1}dt$$

Evaluating each of these integrals shows that

$$\ln|f|=\ln(t^2+1)+C$$

Since we are seeking any solution of equation (1.69) other than $f(x)\equiv0$, we will choose C to be 0. Then

$$\ln|f|=\ln(t^2+1)$$

so that

$$|f(t)|=t^2+1$$

or, equivalently,

$$f(t)=t^2+1 \quad \text{or} \quad f(t)=-(t^2+1)$$

Hence $f(t)=t^2+1$ is an integrating factor for equation (1.68).   ◀

Exercises   SECTION 1.7

In Exercises 1–4 use an integrating factor to find solutions of the given equation.

1. $2(1+x)+3t\dfrac{dx}{dt}=0$

2. $\dfrac{x^3}{3}+xt^2+2xt+(x^2+t^2)\dfrac{dx}{dt}=0$

3. $x^{-1}-t^{-1}+t^2-x^{-2}\dfrac{dx}{dt}=0$

4. $\tfrac{1}{2}tx^2+xt^2+x+(x+t)\dfrac{dx}{dt}=0$

5. Show that if $(N_t-M_x)M$ is a function of the single variable x, then any function f such that

$$\frac{df}{dx}=\frac{N_t(x,t)-M_x(x,t)}{M(x,t)}f$$

is an integrating factor for the equation

$$M(x,t)+N(x,t)\frac{dx}{dt}=0$$

In Exercises 6–9 use Exercise 5 to find solutions of the given equation.

6. $tx+(1+t^2)\dfrac{dx}{dt}=0$

7. $\sin x+(2t\cos x)\dfrac{dx}{dt}=0$

8. $x^2 + t + (x^3t + \frac{1}{2}xt^2 + 2xt)\frac{dx}{dt} = 0$

9. $x + t(x+1)\frac{dx}{dt} = 0$

10. Find an integrating factor for the equation

$$\frac{dy}{dx} + p(x)y = q(x)$$

and find its solution. Compare this solution with the one given in equation (1.11).

11. Bernoulli's equation,

$$\frac{dy}{dx} + p(x)y = q(x)y^n$$

can be rewritten in the form

$$y^{-n}\frac{dy}{dx} + p(x)y^{1-n} - q(x) = 0$$

Find the solution of this equation by using an integrating factor. Compare your answer with that obtained using the method described in Section 1.2.

1.8 Numerical Methods I: Euler's Method

When a sloping field is irrigated, water is let onto the field at the uphill end and then flows downhill. Suppose that water at a constant depth is introduced at the uphill end of the field. At time $t=0$ the water is cut off. Since the water is both flowing down the field and being absorbed by the soil, a portion of the field near the uphill end is no longer covered by water shortly after the water is cut off. Let $x(t)$ denote the distance at time t from the uphill end of the field to the portion of the field covered by water. A cross-sectional view of the field at time t is shown in Figure 1.7.

FIGURE 1.7

We would like to determine the function x. Under certain conditions it can be shown that [V. P. Singh and R. C. McCann, Mathematical modeling of hydraulics of irrigation recession, in Proceedings of the Second International Conference of Mathematical Modeling, 1979]

$$\frac{dx}{dt} = A + Bx^k, \quad x(0) = 0 \tag{1.70}$$

for some positive numbers A, B, and k (k is not necessarily an integer). This is a separable first-order differential equation. However,

$$\int \frac{dx}{A + Bx^k} \tag{1.71}$$

cannot be expressed in terms of elementary functions for all values of k. Even if we could evaluate the integral for a particular value of k, the method of separation of variables would yield t as a function of x: $t = f(x)$. In order to find x as a function of t, we would now need to determine f^{-1} (assuming that f^{-1} exists) so that $x = f^{-1}(t)$. Unfortunately, the computation of an inverse function is frequently a difficult if not impossible task. Hence even if we can evaluate the integral in (1.71) in terms of elementary functions, we still may not be able to determine x as a function of t, except theoretically.

In practice, we will not need to know $x(t)$ exactly. A "good" approximation will suffice. One of the simplest methods for computing approximations to the solutions of the initial-value problem,

$$\frac{dx}{dt} = f(t, x), \quad x(t_0) = x_0 \tag{1.72}$$

is called **Euler's method**. In this method we take a finite sequence of numbers $\{t_i\}_{i=0}^n$ beginning with t_0 such that $t_0 < t_1 < t_2 < \cdots < t_n$ and compute numbers $x_1, x_2, \ldots, x_n$ so that $|x_i - x(t_i)|$ is small for each i. We begin by recalling the definition of derivative:

$$\frac{dx}{dt} = \lim_{h \to 0} \frac{x(t+h) - x(t)}{h}$$

Thus for small values of h we have

$$\frac{dx}{dt} \simeq \frac{x(t+h) - x(t)}{h}$$

or, equivalently,

$$x(t+h) \simeq x(t) + hx'(t) = x(t) + hf(t, x(t)) \tag{1.73}$$

Given a small number $h>0$, we will take the finite sequence $\{t_i\}_{i=0}^n$ to be equally spaced by a distance h. That is,

$$t_{i+1}=t_i+h$$

for $i=0,1,2,\ldots,n-1$. Equivalently,

$$t_i=t_0+ih \tag{1.74}$$

for $i=0,1,2,\ldots,n$. If we take $t=t_i$ in (1.73), we have

$$x(t_{i+1})=x(t_i+h)\simeq x(t_i)+hf(t,x(t_i))$$

This suggests that we define the finite sequence $\{x_i\}_{i=0}^n$ iteratively by the rule

$$x_{i+1}=x_i+hf(t_i,x_i) \tag{1.75}$$

for $i=0,1,2,\ldots,n-1$. According to this iterative scheme, we have

$$x_1=x_0+hf(t_0,x_0)$$
$$x_2=x_1+hf(t_1,x_1)$$
$$x_3=x_2+hf(t_2,x_2)$$
$$\cdots\cdots\cdots\cdots\cdots$$
$$x_n=x_n+hf(t_{n-1},x_{n-1})$$

Notice that in order to compute x_1 we need t_0 and x_0, which are given as the initial data. In order to compute x_{i+1} we need t_i, which is given by (1.74), and x_i, which was computed in the previous step. We hope that x_i is a good approximation to $x(t_i)$ whenever h is "small." The purpose here is to describe the procedure and not to give details concerning the errors in these approximations. The interested reader should consult a text on numerical analysis such as *A First Course in Numerical Analysis* (McGraw-Hill, New York, 1965) by A. Ralston. Roughly speaking, the smaller the choice of h, the better the approximation. This statement is not precisely true due to errors in computation, such as roundoff and truncation errors. Nonetheless, the approximations tend to get better as h gets smaller. In fact it can be shown that, neglecting errors in computation, if we restrict our attention to an interval $[t_0,T]$, then there is a constant M such that

$$|x(t_i)-x_i|\leq hM$$

for every $t_i=t_0+ih$ in $[t_0,T]$.

▶ **Example 1**   Consider the initial-value problem

$$\frac{dx}{dt} = x + t, \quad x(0) = 1$$

which has $x(t) = 2e^t - t - 1$ as its solution. We will use Euler's method to approximate the solution of the differential equation and then compare the approximation with the true solution. Here $f(t, x) = x + t$, so that Euler's method is

$$x_{i+1} = x_i + hf(x_i, t_i)$$

$$= x_i + h(x_i + t_i)$$

$$= (1 + h)x_i + ht_i \qquad \text{(1.76)}$$

We will approximate the solution on the interval $[0, 1]$. We begin by dividing the interval $[0, 1]$ into n equal subintervals each of length $h = 1/n$. Then

$$t_i = ih = \frac{i}{n}$$

for $i = 0, 1, 2, \ldots, n$, and the equation in (1.76) becomes

$$x_{i+1} = \left(1 + \frac{1}{n}\right)x_i + \frac{i}{n^2}$$

When $n = 10$ we have

$$x_1 = (1 + .1)x_0$$

$$= (1.1)(1) = 1.1$$

$$x_2 = (1 + .1)x_1 + \frac{1}{100}$$

$$= (1.1)(1.1) + .01 = 1.22$$

Continuing in this fashion, we have

i	t_i	x_i	$x(t_i)$
0	.0	1.00000	1.00000
1	.1	1.10000	1.11034
2	.2	1.22000	1.24281
3	.3	1.36200	1.39972
4	.4	1.52820	1.58365
5	.5	1.72102	1.79744
6	.6	1.94312	2.04424
7	.7	2.19743	2.32751
8	.8	2.48718	2.65108
9	.9	2.81590	3.01921
10	1.0	3.18749	3.43656

When $n = 100$ and $h = 1/100$

i	t_i	x_i	$x(t_i)$
0	.00	1.00000	1.00000
1	.01	1.01000	1.01010
2	.02	1.02020	1.02040
3	.03	1.03060	1.03091
4	.04	1.04121	1.04162
5	.05	1.05202	1.05254
6	.06	1.06304	1.06367
7	.07	1.07427	1.07502
8	.08	1.08571	1.08657
9	.09	1.09737	1.09835
10	.10	1.10924	1.11034
20	.20	1.24038	1.24281
30	.30	1.39570	1.39972
40	.40	1.57773	1.58365
50	.50	1.78926	1.79744
60	.60	2.03339	2.04424
70	.70	2.31353	2.32751
80	.80	2.63343	2.65108
90	.90	2.99726	3.01921
100	1.00	3.40962	3.43656

Two very important things should be noted in these tables. The smaller value of h gives better approximations to the true values of the solution, and the larger the value of t_i, the poorer tends to be the approximation. Both of these are typical occurrences when using numerical methods to approximate solutions to an initial-value problem. ◀

▶ **Example 2**   Let us consider the initial-value problem in equation (1.70) given at the beginning of this section for the special case $A=B=1$ and $k=1.2$. Then Euler's method for this equation can be written as

$$x_0 = 0$$
$$x_{i+1} = hf(x_i, t_i)$$
$$= x_i + h(1 + x_i^{1.2})$$

We will approximate the solution on the interval $[0, 1]$ using $n = 100$ and $h = 1/100$. Then

$$x_1 = x_0 + .01(1 + x_0^{1.2}) = .01000$$
$$x_2 = x_1 + .01(1 + x_1^{1.2})$$
$$= .01 + .01(1 + (.01)^{1.2}) = .02004$$

Continuing in this fashion, we have

i	t_i	x_i
0	.0	.00000
10	.1	.10260
20	.2	.21305
30	.3	.33353
40	.4	.46608
50	.5	.61289
60	.6	.77643
70	.7	.95958
80	.8	1.16568
90	.9	1.39871
100	1.0	1.66338

◀

Euler's method has an elementary geometrical interpretation. Notice that $f(t_i, x_i)$ is the slope of the solution to the initial-value problem, and consider the line determined by

$$p(s) = (s - t_i)f(t_i, x_i) + x_i$$

Clearly the point (t_i, x_i) is on the line. Also, using equation (1.74),

$$p(t_{i+1}) = (t_{i+1} - t_i)f(t_i, x_i) + x_i$$
$$= hf(t_i, x_i) + x_i$$
$$= x_{i+1}$$

This shows that the point (t_{i+1}, x_{i+1}) is on the line through the point (t_i, x_i) having slope $f(t_i, x_i)$. In essence, the graph of the solution x is being approximated by a continuous, piecewise-linear curve (Figure 1.8).

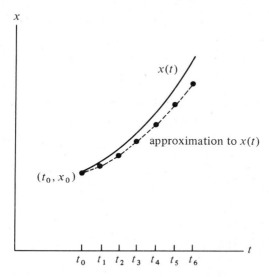

FIGURE 1.8

We now give an algorithm for approximating the solution of the initial-value problem

$$\frac{dx}{dt} = f(t, x), \quad x(t_0) = x_0$$

at $n+1$ equally spaced points in the interval $[t_0, T]$ by Euler's method.

1. Set $h = (T - t_0)/n$ and $i = 0$.
2. Set $t_i = t_0 + ih$ and $x_{i+1} = x_i + hf(t_i, x_i)$.
3. If $i < n - 1$, add one to i and return to Step 2.
4. If $i = n - 1$, the procedure is finished and x_i approximates $x(t_i)$ for $i = 1, 2, \ldots, n$.

Exercises   SECTION 1.8

In Exercises 1–5 use Euler's method with $n = 10$ and $h = 1/n$ to approximate the solution of the given initial-value problem. Compare your results with the true solution.

1. $\dfrac{dx}{dt} = x - t, \quad x(0) = 2$

2. $\dfrac{dx}{dt} = tx, \quad x(0) = 1$

3. $\dfrac{dx}{dt} = \dfrac{x^2}{t+1}, \quad x(0) = 1$

4. $\dfrac{dx}{dt} = \dfrac{x}{t}, \quad x(1) = 3$

5. $\dfrac{dx}{dt} = x + e^{-t}, \quad x(1) = 0$

In Exercises 6–10 use Euler's method to approximate the solutions of the given initial-value problems at the numbers $i/10$, where $i = 1, 2, \ldots, 10$. Begin with $h = 1/10$. Then use $h = 1/100$, $h = 1/1000, \ldots$, until the first three digits of the approximations to $x(i/10)$, $i = 1, 2, \ldots, 10$ remain unchanged.

6. $\dfrac{dx}{dt} = x^2 + t, \quad x(0) = 0$

7. $\dfrac{dx}{dt} = \cos(x + t), \quad x(0) = 0$

8. $\dfrac{dx}{dt} = x^{1/2} + 1, \quad x(0) = 0$

9. $\dfrac{dx}{dt} = \dfrac{x}{x^2 + t^2}, \quad x(0) = 1$

10. $\dfrac{dx}{dt} = (x^2 + t^2)^{-1/2}, \quad x(0) = 4$

11. Let n be a positive integer and set $h = 1/n$. Let $x_{i,h}$ denote the Euler's method approximation to the solution of the initial-value problem $dx/dt = x$, $x(0) = 1$, at $t = ih$. Prove that $\lim_{h \to 0} x_{n,h} = x(1) = e$.

1.9 Numerical Methods II: The Runge-Kutta Method

Though conceptually simple, Euler's method has the disadvantage that it usually takes a relatively small value of h to achieve a reasonable degree of accuracy. For this reason Euler's method is usually not used. We would expect to get more accurate approximations for a given value of h if we used a better approximation to $x(t+h)$. Such an approximation can be obtained using Taylor's formula. Assuming that the fifth derivative of x is continuous, we have

$$x(t+h) = x(t) + hx'(t) + \frac{h^2}{2!}x''(t) + \frac{h^3}{3!}x'''(t) + \frac{h^4}{4!}x^{(4)}(t) + \frac{h^5}{5!}x^{(5)}(w(t))$$

where $w(t)$ is some number between t and $t+h$. Thus for small values of h we have

$$x(t+h)\simeq x(t)+hx'(t)+\frac{h^2}{2!}x''(t)+\frac{h^3}{3!}x'''(t)+\frac{h^4}{4!}x^{(4)}(t) \qquad \textbf{(1.77)}$$

As we saw earlier, $x'(t)=f(t,x(t))$. Moreover,

$$x''(t)=\{x'(t)\}'$$

$$=\{f(t,x(t))\}'$$

$$=f_t(t,x(t))+f_x(t,x(t))x'(t)$$

$$=f_t(t,x(t))+f_x(t,x(t))f(t,x(t))$$

The computation of $x'''(t)$ and $x^{(4)}(t)$ is even more complicated. Due to these complications, a numerical scheme based upon equation (1.77) is rarely used. Instead, the terms on the right-hand side of equation (1.77) are approximated by terms that are more easily computed. The resulting numerical scheme, called a (fourth order) **Runge-Kutta method**, is given by

$$x_{i+1}=x_i+\frac{h}{6}(k_{1i}+2k_{2i}+2k_{3i}+k_{4i})$$

where

$$k_{1i}=hf(t_i,x_i)$$

$$k_{2i}=hf(t_i+h/2,x_i+k_{1i}/2)$$

$$k_{3i}=hf(t_i+h/2,x_i+k_{2i}/2)$$

$$k_{4i}=hf(t_i+h,x_i+k_{3i})$$

This formula involves a weighted average of values of $f(t,x)$ taken at four different values of (t,x). Since $x'(t)=f(t,x)$, this amounts to taking a weighted average of four values of the slope of the graph of the solution x. Hence the term $(k_{1i}+2k_{2i}+2k_{3i}+k_{4i})/6$ can be interpreted as an average slope. It can be shown that if we restrict our attention to an interval $[t_0,T]$, then there is a constant N such that

$$|x(t_i)-x_i|\leqslant h^4N$$

for every $t_i = t_0 + ih$ in $[t_0, T]$. Comparing this inequality with the error estimate $|x(t_i) - x_i| \leq hM$ given for Euler's method in Section 1.8, we see that the Runge-Kutta method is more accurate than Euler's method for sufficiently small values of h.

We now give an algorithm for approximating the solution of the initial-value problem

$$\frac{dx}{dt} = f(t, x), \quad x(t_0) = x_0$$

at $n+1$ equally spaced points in the interval $[(t_0, T]$ by the Runge-Kutta method.

1. Set $h = (T - t_0)/n$ and $i = 0$.

2. Set $t_i = a + ih$

$$k_{1i} = hf(t_i, x_i)$$

$$k_{2i} = hf(t_i + h/2, x_i + k_{1i}/2)$$

$$k_{3i} = hf(t_i + h/2, x_i + k_{2i}/2)$$

$$k_{4i} = hf(t_i + h, x_i + k_{3i})$$

$$x_{i+1} = x_i + (k_{1i} + 2k_{2i} + 2k_{3i} + k_{4i})/6$$

3. If $i < n-1$, add one to i and return to Step 2.

4. If $i = n-1$, the procedure is complete and x_i approximates $x(t_i)$ for $i = 1, 2, \ldots, n$.

▶ **Example 1**   Consider the initial-value problem

$$\frac{dx}{dt} = x + t, \quad x(0) = 1$$

which we considered in Example 1 of Section 1.8. As in this earlier example, we will take $n = 10$ and $h = 1/10$. Here $f(t, x) = t + x$. Then

$$k_{11} = hf(0, 1) = .1$$

$$k_{21} = hf(.05, 1 + .05) = .11$$

$$k_{31} = hf(.05, 1 + .055) = .1105$$

$$k_{41} = hf(.1, 1 + .1105) = .12105$$

so that

$$x_1 = 1 + \tfrac{1}{6}\big(.1 + 2(.11) + 2(.1105) + .12105\big)$$

$$= 1.11034$$

Continuing in this fashion, we have

i	t_i	x_i	$x(t_i)$
0	.0	1.00000	1.00000
1	.1	1.11034	1.11034
2	.2	1.24281	1.24281
3	.3	1.39972	1.39972
4	.4	1.58365	1.58365
5	.5	1.79744	1.79744
6	.6	2.04424	2.04424
7	.7	2.32750	2.32751
8	.8	2.65108	2.65108
9	.9	3.01920	3.01921
10	1.0	3.43656	3.43656	◀

Notice that the Runge-Kutta method with $h = .1$ gives four-place accuracy, which is considerably better than Euler's method with $h = .01$. It is almost always the case that for a given value of h, the Runge-Kutta method gives much better accuracy than Euler's method.

▶ **Example 2**   Let us reconsider the initial-value problem

$$\frac{dx}{dt} = 1 + x^{1.2}, \quad x(0) = 0$$

on which we used Euler's method in Example 2 of Section 1.8. We will begin by taking $n = 10$ and $h = 1/10$. Here $f(t, x) = 1 + x^{1.2}$. Then

$$k_{11} = hf(0,0) = .1$$

$$k_{21} = hf(.05, .05) = (.1)\big(1 + (.05)^{1.2}\big) = .102746$$

$$k_{31} = hf(.05, .051373) = (.1)\big(1 + (.051373)^{1.2}\big) = .102837$$

$$k_{41} = hf(.1, .100179) = (.1)\big(1 + (.102873)^{1.2}\big) = .106525$$

so that

$$x_1 = 0 + \tfrac{1}{6}\big(.1 + 2(.102746) + 2(.102837) + .106525\big) = .102949$$

Continuing in this manner, we have

i	t_i	x_i
0	.0	.00000
1	.1	.10295
2	.2	.21390
3	.3	.33505
4	.4	.46845
5	.5	.61634
6	.6	.78125
7	.7	.96612
8	.8	1.17439
9	.9	1.41015
10	1.0	1.67826	◀

By now the reader may be wondering how to choose h to assure a given degree of accuracy. There are some theorems which, in certain circumstances, give criteria to choose h. However, these theorems are not always easy to apply. One simple method is to decrease the size of h until the computed results no longer differ in the number of decimal places required. For example, suppose we wished to approximate the solution of

$$\frac{dx}{dt} = x + t, \quad x(0) = 1$$

at $t = .1, .2, \ldots, 1.0$. We could begin with $h = 10^{-1}$. If we then use $h = 10^{-2}$, every tenth value computed would approximate x at one of the desired values of t. With $h = 10^{-3}$, every hundredth value computed would approximate x at one of the desired values of t. This process is continued until there are no changes in the desired number of decimal places. The accuracy of the computed values is *not* guaranteed using this process. Nonetheless, this process usually does give values to the desired accuracy.

Exercises   SECTION 1.9

1–10. Do Exercises 1–10 of Section 1.8 using the Runge-Kutta method instead of Euler's method. Compare the results with those obtained using Euler's method.

1.10 Existence of Solutions

In Sections 1.2–1.7 we have discussed methods of finding solutions of differential equations which have special forms, such as first-order linear, separable, and exact. We have not shown that these methods yield all of the solutions of these equations, although a careful analysis of these methods shows this to be the case. Neither have we discussed the problem of determining whether a general first-order differential equation

$$\frac{dy}{dx} = f(x, y) \tag{1.78}$$

has a solution. In applications in which equation (1.78) arises, the problem is usually not just to find any solution, but to find one particular solution. Frequently this particular solution satisfies an initial condition $y(a) = b$. Moreover, in almost all applications we would like there to be only one such solution. This is not possible with the definition of solution given in Section 1.1. For example, the functions

$$y_1(x) = x \quad \text{on } (-1, 2)$$

and

$$y_2(x) = x \quad \text{on } (-3, 1)$$

are both solutions of the initial-value problem

$$\frac{dy}{dx} = 1, \quad y(0) = 0 \tag{1.79}$$

It is important to notice that $y_1(x) = y_2(x)$ at every x that their domains have in common. It is possible to refine the definition of solution so that $y(x) = x$ on $(-\infty, \infty)$ is the only solution of the initial-value problem in (1.79), but this is a subtlety that is beyond the scope of this text. Instead, we will state a theorem which shows that our example concerning the initial-value problem in (1.79) is typical of almost all initial-value problems that arise in applications.

In order to shorten the statement of this theorem, we introduce the following definition. A subset Y of the xy-plane is called a **region** if

1. Y contains none of its boundary points.
2. Any two points in Y can be connected by a curve that lies entirely in Y.

THEOREM 1.3   If f and f_y are continuous on a region Y of the xy-plane, then for any point (a, b) in Y there is a solution of the initial-value problem

$$\frac{dy}{dx} = f(x, y), \quad y(a) = b$$

on some interval $(a - c, a + c)$, $c > 0$. Moreover, if y_1 and y_2 are any two solutions of this initial-value problem whose graphs lie in Y, then $y_1(x) = y_2(x)$ for every x that their domains have in common.

Before proceeding, it needs to be emphasized that almost all differential equations of the form $dy/dx = f(x, y)$ that arise in applications satisfy the hypotheses of Theorem 1.3.

▶ **Example 1**   Consider the initial-value problem

$$\frac{dy}{dx} = \sin xy, \quad y(1) = 2 \tag{1.80}$$

The differential equation in (1.80) is not one of the special forms studied earlier in this chapter. Nonetheless, Theorem 1.3 assures the existence of a solution because $\sin(xy)$ and $(\sin(xy))_y$ are continuous for all values of x and y. If this initial-value problem arose in an application, a numerical procedure, such as the Runge-Kutta method, could be used to obtain approximations to a solution evaluated at various values of x.   ◀

The conclusions of Theorem 1.3 need not be valid if f does not satisfy the stated hypotheses. For example, consider the initial-value problem

$$\frac{dy}{dx} = \frac{2}{3}y^{1/3}, \quad y(0) = 0 \tag{1.81}$$

Note that $(\partial/\partial y)(\frac{2}{3}y^{1/3}) = \frac{2}{9}y^{-2/3}$ is not continuous on any rectangle containing the point $(0,0)$. The functions

$$y_1(x) = t^{3/2}$$

and

$$y_2(x) = \begin{cases} (t-1)^{3/2} & \text{if } t \geqslant 1 \\ 0 & \text{if } t < 0 \end{cases}$$

both having domain $(-\infty, \infty)$, are solutions of the initial-value problem (1.81). If we consider any open interval I containing zero, the functions y_1 and y_2 do not agree on I. Thus the conclusion of Theorem 1.3 need not be valid if the hypotheses are not satisfied.

Notice that although Theorem 1.3 gives conditions that assure the existence of solutions of equation (1.78), it does not provide a method of finding solutions. One method of proving Theorem 1.3 consists of constructing a sequence of functions $\{y_n(x)\}_{n=1}^{\infty}$ that converges to a solution of the initial-value problem. Thus this method of proof constructs a solution, but only theoretically. Even though a solution may not be obtained explicitly, it is possible to obtain sketches of the graphs of solutions of the differential equation $dy/dx = f(x, y)$. This is done by interpreting dy/dx as the slope of a solution. If (a, b)

is a point in the plane at which f is defined, then $f(a, b)$ gives the slope of the graph of the solution that passes through the point (a, b); i.e., the slope of the graph of the solution to the initial-value problem $dy/dx=f(x, y)$, $y(a)=b$ at the point (a, b). An approximation to the graph of this solution near (a, b) may be obtained by sketching a short line segment with slope $f(a, b)$ and midpoint (a, b). If this process is repeated for a large number of points, the line segments can be pieced together to obtain an approximation to a larger portion of the graphs of the solutions. This process may be simplified by considering curves, called **isoclines**, on which solutions have constant slope. That is, an isocline is determined by the equation

$$f(x, y)=c$$

for some constant c.

▶ **Example 2**   Consider the differential equation

$$\frac{dy}{dx}=x+y$$

The isoclines are determined by the equations of the form $x+y=c$, where c is a constant. At each point (a, b) of the line determined by $y=-x+c$, the slope of the graph of the solution at (a, b) has slope c. The graphs of some typical isoclines and short line segments approximating portions of the graphs of solutions are given in Figure 1.9. If we piece

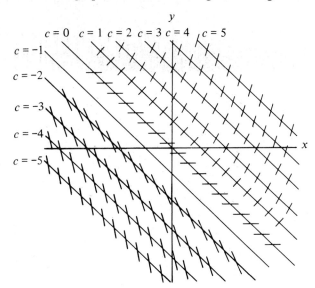

FIGURE 1.9

together the line segments, the graphs of the solutions resemble the curves in Figure 1.10.   ◄

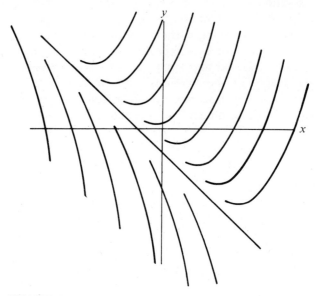

FIGURE 1.10

Exercises   SECTION 1.10

For which of the following initial-value problems (Exercises 1–6) does Theorem 1.3 guarantee the existence of a solution?

1. $\dfrac{dy}{dx} = y^2 + x, \quad y(2) = 3$

2. $\dfrac{dy}{dx} = y^{1/3} + 3, \quad y(0) = 1$

3. $\dfrac{dy}{dx} = y^{1/3} + 3, \quad y(1) = 0$

4. $\dfrac{dy}{dx} = \dfrac{x+y}{x-y}, \quad y(2) = 2$

5. $\dfrac{dy}{dx} = x^2 - y^{-2}, \quad y(0) = 0$

6. $\dfrac{dy}{dx} = \dfrac{x}{y} - y, \quad y(2) = 1$

7. For which points (a, b) does Theorem 1.3 guarantee the existence of a solution to

$$\frac{dy}{dx} = \frac{y}{x(x+y)}, \quad y(a) = b$$

In Exercises 8–13 sketch a few isoclines and then sketch the graphs of a few solutions of each equation.

8. $\dfrac{dy}{dx} = -\dfrac{x}{y}$

9. $\dfrac{dy}{dx} = \dfrac{x}{y}$

10. $\dfrac{dy}{dx} = -\dfrac{y}{x}$

11. $\dfrac{dy}{dx} = \dfrac{y}{x}$

12. $\dfrac{dy}{dx} = y^2$

13. $\dfrac{dy}{dx} = x^2 + y^2$

2 Second-Order Linear Differential Equations

2.1 Introduction

Let a_0, a_1, and r be functions defined on an open interval I. A differential equation of the form

$$\frac{d^2y}{dx^2}+a_1(x)\frac{dy}{dx}+a_0(x)y=r(x) \tag{2.1}$$

on I is called a second-order linear differential equation. If $r(x)\equiv0$ on I, the equation is said to be **homogeneous**. Otherwise, the equation is said to be **nonhomogeneous**. A solution to equation (2.1) on I is a function u such that u and its first two derivatives are defined on I and

$$\frac{d^2u}{dx^2}+a_1(x)\frac{du}{dx}+a_0(x)u(x)=r(x)$$

on I.

As we will see throughout the remainder of this chapter, many problems from physics, engineering, biology, and economics may be expressed in terms of second-order linear differential equations.

For simplicity of notation we will denote d^2y/dx^2 and dy/dx by y'' and y', respectively. In this notation, equation (2.1) may be rewritten as

$$y''+a_1(x)y'+a_0(x)y=r(x)$$

One of the simplest second-order linear differential equations arises when analyzing the motion of a falling object. Suppose that a ball of mass m is held stationary and then dropped from a tower. Let $y(x)$ denote the distance the ball travels in x seconds after it is released. If gravity is the only force acting on the ball, then, by Newton's second law of motion (where the acceleration a is given by $a=y''$ and g is the constant acceleration due to the force of gravity),

$$my''=ma=\text{force}=mg$$

so that

$$y''=g$$

This second-order linear differential equation is easily solved by integrating twice:

$$y(x)=\tfrac{1}{2}gx^2+c_1x+c_2$$

where c_1 and c_2 are arbitrary constants. Since $y(x)$ is the distance the ball travels in x seconds after being released, we must have $y(0)=0$. We must also have $y'(0)=0$ because the ball is held stationary before being released. Thus

$$0=y(0)=c_2$$

$$0=y'(0)=c_1$$

and we have

$$y(x)=\tfrac{1}{2}gx^2$$

Notice that the initial conditions $y(0)=0$ and $y'(0)=0$ enabled us to determine a unique solution on the interval $[0, \infty)$.

The following theorem assures the existence of solutions to equation (2.1) provided the functions a_0, a_1, and r are continuous. A proof may be found in almost any advanced text of ordinary differential equations.

THEOREM 2.1   Let a_0, a_1, and r be continuous functions on an open interval I. Let x_0 be any number in I and let y_0 and y_1 be arbitrary real numbers. Then there is a unique function y such that

$$y''+a_1(x)y'+a_0(x)y=r(x)$$

on the interval I and

$$y(x_0)=y_0 \quad \text{and} \quad y'(x_0)=y_1$$

The problem of finding the solution of the differential equation

$$y''+a_1(x)y'+a_0(x)y=r(x)$$

that satisfies

$$y(x_0)=y_0 \quad \text{and} \quad y'(x_0)=y_1$$

is called the **initial-value problem.**

Theorem 2.1 assures that under the stated hypotheses equation (2.1) has infinitely many solutions: one solution for each choice of the constants y_0 and y_1. Even though there are infinitely many solutions of equation (2.1), the set of all solutions possesses properties that will enable us to find all of the solutions in many cases of interest. One of the basic properties of a homogeneous linear differential equation is given in the following theorem. The remaining properties in which we are interested will be presented in the following section.

> **THEOREM 2.2**   Let a_0 and a_1 be functions defined on an open interval I. If y_1 and y_2 are solutions on I of the homogeneous linear differential equation
>
> $$y''+a_1(x)y'+a_2(x)y=0 \tag{2.2}$$
>
> then $c_1 y_1+c_2 y_2$ is also a solution on I for every choice of the constants c_1 and c_2.

Proof   Since the derivative of a sum is the sum of the derivatives, we have

$$(c_1 y_1+c_2 y_2)''+a_1(x)(c_1 y_1+c_2 y_2)'+a_0(x)(c_1 y_1+c_2 y_2)$$
$$=c_1 y_1''+c_2 y_2''+c_1 a_1(x)y_1'+c_2 a_1(x)y_2'+c_1 a_0(x)y_1+c_2 a_0(x)y_2$$
$$=c_1\left(y_1''+a_1(x)y_1'+a_0(x)y_1\right)+c_2\left(y_2''+a_1(x)y_2'+a_0(x)y_2\right)$$
$$=c_1\cdot 0+c_2\cdot 0$$
$$=0$$

Hence $c_1 y_1+c_2 y_2$ is also a solution on I of equation (2.2)   ■

▶ **Example 1**   The functions $y_1(x)=e^x$ and $y_2=e^{-x}$ are solutions on $(-\infty,\infty)$ of $y''-y=0$ since $(e^x)''=e^x$ and $(e^{-x})''=e^{-x}$. Then for any choice of the numbers c_1 and c_2, we have

$$(c_1 y_1(x)+c_2 y_2(x))''=(c_1 e^x+c_2 e^{-x})''$$
$$=c_1(e^x)''+c_2(e^{-x})''$$
$$=c_1 e^x+c_2 e^{-x}$$

so that $c_1 y_1+c_2 y_2$ is a solution on $(-\infty,\infty)$ of $y''-y=0$, as is assured by Theorem 2.2.   ◀

▶ **Example 2** The functions $y_1(x)=\cos x$ and $y_2(x)=\sin x$ are solutions on $(-\infty,\infty)$ of $y''+y=0$ since $(\cos x)''=-\cos x$ and $(\sin x)''=-\sin x$. Let c_1 and c_2 be real numbers, and consider the function $u(x)=c_1\cos x+c_2\sin x$. By Theorem 2.2 the function u is a solution of $y''+y=0$ since y_1 and y_2 are solutions. Moreover,

$$u(0)=c_1\cos 0+c_2\sin 0=c_1$$

and

$$u'(0)=-c_1\sin 0+c_2\cos 0=c_2$$

Hence u is the unique solution on $(-\infty,\infty)$ which satisfies $y(0)=c_1$, $y'(0)=c_2$. ◀

Exercises   SECTION 2.1

1. Let a_1 and a_0 be functions defined on an interval I. Show that the function $u(x)\equiv 0$ on I is a solution on I of the second-order linear differential equation

$$y''+a_1(x)y'+a_0(x)y=0$$

2. (a) Show that $y_1(x)=e^{2x}$ and $y_2(x)=e^{-3x}$ are solutions on $(-\infty,\infty)$ of $y''+y'-6y=0$.
   (b) By Theorem 2.1 the function $u(x)=c_1e^{2x}+c_2e^{-3x}$ is a solution of $y''+y'-6y=0$ for every choice of the constants c_1 and c_2. Choose the constants c_1 and c_2 so that the function u is a solution of the initial-value problem

$$y''+y'-6y=0$$

$$y(0)=1, \quad y'(0)=0$$

3. Find all possible values of r so that $u(x)=e^{rx}$ is a solution of $y''+4y'+3y=0$.

4. (a) Find two values, a and b, for r so that $u(x)=e^{rx}$ is a solution of $y''-5y'-14y=0$.
   (b) In part (a) we found two distinct solutions: $y_1(x)=e^{ax}$ and $y_2(x)=e^{bx}$ of $y''-5y'-14y=0$. Does this contradict Theorem 2.1? Why or why not?

5. Let u be a solution on an interval I of

$$y''+a_1(x)y'+a_0(x)y=0$$

and let z be a solution on I of

$$y''+a_1(x)y'+a_0(x)y=f(x)$$

Show that $u+z$ is a solution on I of

$$y''+a_1(x)y'+a_0(x)y=f(x)$$

6. (a) Determine the number A so that the function $u(x)=Ae^{-x}$ is a solution of $y''+y'-6y=e^{-x}$.
   (b) By Exercise 2(a) and Theorem 2.1, the function $y_h(x)=c_1e^{2x}+c_2e^{-3x}$ is a solution of $y''+y'-6y=0$ for every choice of the constants c_1 and c_2. Use Exercise 5 to find the solution of the initial-value problem

$$y''+y'-6y=e^{-x}$$

$$y(0)=0, \quad y'(0)=0$$

7. Let a_1 and a_0 be continuous functions defined on an open interval I, and let x_0 be any number in I. Find the solution on I of the initial-value problem

$$y''+a_1(x)y'+a_0(x)y=0$$

$$y(x_0)=0, \quad y'(x_0)=0$$

Hint: Consider the function $u(x)\equiv0$ on I.

2.2 General Solutions

In order to discuss further the nature of the solutions of the second-order linear differential equation

$$y''+a_1(x)y'+a_0(x)y=0 \tag{2.3}$$

we need a basic concept of linear algebra.

DEFINITION 2.1   Let f_1 and f_2 be two functions defined on an interval I. The functions f_1 and f_2 are said to be **linearly dependent** on I if one of the functions is a constant multiple of the other. If f_1 and f_2 are not linearly dependent, they are said to be **linearly independent.**

▶ **Example 1**   Consider the functions $f_1(x)=\cos x+\sin x$ and $f_2(x)=\cos x - \sin x$ on the interval $(-\infty,\infty)$. If there is a constant c such that

$f_1(x)=cf_2(x)$, then

$$1=f_1(0)=cf_2(0)=c$$

so that

$$\cos x+\sin x=\cos x-\sin x$$

for every x. This is obviously impossible. Therefore there is no constant c such that $f_1(x)=cf_2(x)$ for every x. A similar argument shows that there is no constant d such that $f_2(x)=df_1(x)$ for every x. Thus neither function is a constant multiple of the other. The functions f_1 and f_2 are linearly independent on $(-\infty,\infty)$. In fact, f_1 and f_2 are linearly independent on any interval. ◀

▶ **Example 2** Consider the function $f_1(x)\equiv0$ on an interval I. If f_2 is any function on I, then

$$f_1(x)=0\cdot f_2(x)$$

for all x in I. Thus f_1 and f_2 are linearly dependent on I. ◀

▶ **Example 3** Let a and b be distinct numbers. Consider the functions $f_1(x)=e^{ax}$ and $f_2(x)=e^{bx}$ on the interval $(-\infty,\infty)$. If f_1 and f_2 are linearly dependent, then there is a constant c such that either $e^{ax}=ce^{bx}$ or $e^{bx}=ce^{ax}$ for every x. Clearly we cannot have $c=0$. Therefore in either case $e^{(b-a)x}$ must be a constant function. Since $a\neq b$, this is obviously impossible. Therefore the functions f_1 and f_2 cannot be linearly dependent. The functions $f_1(x)=e^{ax}$ and $f_2(x)=e^{bx}$ are linearly independent on $(-\infty,\infty)$. In fact, these functions are linearly independent on any interval. ◀

It is now possible to state a theorem which will serve as a basis for determining all solutions of equation (2.3). A proof of this theorem can be based on Theorem 2.1, but will be postponed until the following chapter, where it will be an immediate consequence of a similar theorem for an nth-order linear differential equation.

THEOREM 2.3 Let a_0 and a_1 be continuous functions on an open interval I. Then there are two linearly independent solutions on I of

$$y''+a_1(x)y'+a_0(x)y=0$$

Moreover, if y_1 and y_2 are any two linearly independent solutions on I, and if u is any solution on I, then there are numbers c_1 and c_2 such that

$$u(x)=c_1 y_1(x)+c_2 y_2(x)$$

for every x in I.

Theorem 2.3 tells us that if we are able to find two linearly independent solutions y_1 and y_2 of equation (2.3) on I, then we can find all of its solutions on I by considering sums of the form

$$c_1 y_1(x)+c_2 y_2(x)$$

This leads us to the following definition.

DEFINITION 2.2 Let a_0 and a_1 be continuous functions on an open interval I, and let y_1 and y_2 be linearly independent solutions on I of

$$y''+a_1(x)y+a_2(x)y=0$$

The expression

$$y(x)=c_1 y_1(x)+c_2 y_2(x)$$

where c_1 and c_2 are constants, is called a **general solution** on I of the differential equation.

▶ **Example 4** Consider the differential equation $y''-y=0$ and its solutions $y_1(x)=e^x$ and $y_2(x)=e^{-x}$ on $(-\infty,\infty)$. (See Example 1 of Section 2.1.) Since neither y_1 nor y_2 is a constant multiple of the other, they are linearly independent. Hence $y(x)=c_1 e^x+c_2 e^{-x}$ is a general solution of $y''-y=0$ on $(-\infty,\infty)$. ◀

▶ **Example 5** Consider the differential equation $y''+y=0$ and its solutions $y_1(x)=\cos x$ and $y_2(x)=\sin x$. (See Example 2 of Section 2.1.) Clearly these solutions are linearly independent. Hence $y(x)=c_1\cos x+c_2\sin x$ is a general solution of $y''+y=0$ on $(-\infty,\infty)$. Consider the functions $y_3(x)=\cos x+\sin x$ and $y_4(x)=\cos x-\sin x$ on $(-\infty,\infty)$. By Theorem 2.2, the functions y_3 and y_4 are also solutions of $y''+y=0$. In Example 1 we showed that y_3 and y_4 are linearly independent. Hence

$$z(x)=d_1(\cos x-\sin x)+d_2(\cos x+\sin x)$$

where d_1 and d_2 are arbitrary constants, is also a general solution on $(-\infty,\infty)$ of $y''+y=0$. In general, a homogeneous linear differential equation has infinitely many general solutions. Note that

$$z(x) = d_1(\cos x - \sin x) + d_2(\cos x + \sin x)$$

$$= (d_1 + d_2)\cos x + (d_2 - d_1)\sin x$$

If we set $c_1 = d_1 + d_2$ and $c_2 = d_2 - d_1$, then the general solution z can be written in the same form as the general solution y. This is an example of a general property: if z_1 and z_2 are general solutions on an open interval I of a homogeneous linear differential equation, then for an appropriate choice of the arbitrary constants in z_2, the general solutions z_1 and z_2 have the same form (after possibly rearranging terms, using trigonometric identities, etc.). ◀

We now consider properties of the nonhomogeneous second-order linear differential equation.

THEOREM 2.4 Let a_0, a_1, and r be functions on an open interval I with a_0 and a_1 continuous, let y_1 and y_2 be linearly independent solutions of

$$y'' + a_1(x)y' + a_0(x)y = 0$$

on I, and let z be any solution of

$$y'' + a_1(x)y' + a_0(x)y = r(x) \qquad (2.4)$$

on I. If w is any solution of equation (2.4), then there exist constants c_1 and c_2 such that

$$w(x) = c_1 y_1(x) + c_2 y_2(x) + z(x)$$

on I.

Proof We begin by considering the difference $u = w - z$. Then

$$u'' + a_1(x)u' + a_0(x)u = (w - z)'' + a_1(x)(w - z)' + a_0(w - z)$$

$$= (w'' + a_1(x)w' + a_0(x)w) - (z'' + a_1(x)z' + a_0(x)z)$$

$$= r(x) - r(x)$$

$$= 0$$

Thus $w - z$ is a solution of the homogeneous equation. By Theorem 2.3 there exist constants c_1 and c_2 such that

$$w(x) - z(x) = c_1 y_1(x) + c_2 y_2(x)$$

so that

$$w(x) = c_1 y_1(x) + c_2 y_2(x) + z(x) \quad \blacksquare$$

This theorem leads us to the following definition.

DEFINITION 2.3   Let a_0, a_1, and r be defined on an open interval I with a_0 and a_1 continuous. Let y_h be a general solution on I of the homogeneous differential equation

$$y'' + a_1(x)y' + a_0(x)y = 0$$

and let z be any solution on I (referred to as a **particular solution**) of the nonhomogeneous differential equation

$$y'' + a_1(x)y' + a_0(x)y = r(x)$$

The expression

$$y(x) = y_h(x) + z(x)$$

is called a **general solution** on I of the nonhomogeneous differential equation.

▶ **Example 6**   Consider the equation $y'' - y = x$. The function $z(x) = -x$ is a particular solution of this equation since $(-x)'' - (-x) = x$. In Example 4 we found that $y_h(x) = c_1 e^x + c_2 e^{-x}$ is a general solution of the homogeneous equation $y'' - y = 0$. Therefore $y(x) = c_1 e^x + c_2 e^{-x} - x$ is a general solution of the nonhomogeneous equation $y'' - y = x$.   ◀

▶ **Example 7**   Consider the equation $y'' + y = e^x$. The function $z(x) = \frac{1}{2}e^x$ is a particular solution of this equation since $(\frac{1}{2}e^x)'' + \frac{1}{2}e^x = e^x$. In Example 5 we found that $y_h(x) = c_1 \cos x + c_2 \sin x$ is a general solution of the homogeneous equation $y'' + y = 0$. Therefore $y(x) = c_1 \cos x + c_2 \sin x + \frac{1}{2}e^x$ is a general solution of the nonhomogeneous equation $y'' + y = e^x$.   ◀

THEOREM 2.5   (Principle of Superposition) Let a_0, a_1, f, and g be defined on an open interval I with a_0 and a_1 continuous. Let y_1 and y_2 be solutions on I of

$$y''+a_1(x)y'+a_0(x)y=f(x)$$

and

$$y''+a_1(x)y'+a_0(x)y=g(x)$$

respectively. Then y_1+y_2 is a solution on I of

$$y''+a_1(x)y'+a_0(x)y=f(x)+g(x)$$

Proof

$$(y_1+y_2)''+a_1(x)(y_1+y_2)'+a_0(x)(y_1+y_2)$$
$$=(y_1''+a_1(x)y_1'+a_0(x)y_1)+(y_2''+a_1(x)y_2'+a_0(x)y_2)$$
$$=f(x)+g(x) \quad\blacksquare$$

The purpose of superposition is to decompose the problem of finding a solution of the differential equation

$$y''+a_1(x)y'+a_0(x)y=f(x)+g(x)$$

into the two separate problems of finding solutions of the differential equations

$$y''+a_1(x)y'+a_0(x)y=f(x)$$
$$y''+a_1(x)y'+a_0(x)y=g(x)$$

▶ **Example 8** Consider the problem of finding a solution of

$$y''+2y'+y=x^2+2x-2+4e^x \qquad \text{(2.5)}$$

Superposition allows us to construct a solution of this equation by taking a sum of solutions of the equations

$$y''+2y'+y=x^2+2x-2 \qquad \text{(2.6)}$$

$$y''+2y'+y=4e^x \qquad \text{(2.7)}$$

The functions $y_1(x)=x^2-2x$ and $y_2(x)=e^x$ are easily shown to be particular solutions of equations (2.6) and (2.7), respectively. (The reader is urged to verify that these functions are in fact solutions of the given differential equations.) By superposition the function $z(x)=x^2-2x+e^x$ is a particular solution of equation (2.5).   ◀

Exercises   SECTION 2.2

1. Verify that $y_1(x)=e^{2x}$ and $y_2(x)=e^{-2x}$ are linearly independent solutions of $y''-4y=0$ on $(-\infty,\infty)$. Find a general solution. Find the solution that satisfies $y(0)=1$, $y'(0)=2$.

2. Verify that $y_1(x)=e^{2x}-e^{-3x}$ and $y_2(x)=e^{2x}+e^{-3x}$ are linearly independent solutions of $y''+y-6y=0$ on $(-\infty,\infty)$. Find a general solution. Find a solution that satisfies $y(0)=1$, $y'(0)=0$.

3. (a) Show that $y''+3y'+2y=0$ has solutions of the form $y(x)=e^{kx}$. Find a general solution.
   (b) Show that $y''+3y'+2y=x$ has a solution of the form $z(x)=ax+b$.
   (c) Find a general solution of the nonhomogeneous equation in (b).
   (d) Find the solution of the nonhomogeneous equation in (b) which satisfies $y(0)=1$, $y'(0)=1$.

4. (a) Show that $y''-2y'-15y=0$ has solutions of the form $y(x)=e^{kx}$. Find a general solution.
   (b) Show that $y''-2y'-15y=e^x$ has a solution of the form $z(x)=ce^x$.
   (c) Find a general solution of the nonhomogeneous equation in (b).
   (d) Find the solution of the nonhomogeneous equation in (b) which satisfies $y(1)=-\frac{1}{16}e$, $y'(1)=-\frac{1}{16}e$.

5. (a) Show that the equation $y''+4x^{-1}y'+2x^{-2}y=0$ has solutions of the form $y(x)=x^k$. Find a general solution on $(0,\infty)$.
   (b) Show that $y''+4x^{-1}y'+2x^{-2}y=9$ has a solution of the form $z(x)=ax^2+bx+c$.
   (c) Find a general solution of the nonhomogeneous equation in (b).
   (d) Find the solution of the nonhomogeneous equation in (b) which satisfies $y(1)=3$, $y'(1)=-1$.

6. Let f_1 and f_2 be functions on an interval I. Show that f_1 and f_2 are linearly dependent if and only if there are constants c_1 and c_2, at least one of which is nonzero, such that $c_1f_1(x)+c_2f_2(x)=0$ for every x in I.

7. Let a_0, a_1, and r be continuous functions on an open interval I containing the number x_0. If y_1 is the unique solution of the initial-value problem

$$y''+a_1(x)y'+a_0(x)y=r(x)$$

$$y(x_0)=y_0,\quad y'(x_0)=y_1$$

show that cy_1 is the unique solution of the initial-value problem

$$y''+a_1(x)y'+a_0(x)y=cr(x)$$

$$y(x_0)=cy_0, \quad y'(x_0)=cy_1$$

8. Let a_0, a_1, r_1, and r_2 be continuous functions on an open interval I containing x_0. Let y_1 and y_2 be the solutions of the initial-value problem

$$y''+a_1(x)y'+a_0(x)y=r_1(x)$$

$$y(x_0)=c_1, \quad y'(x_0)=c_2$$

and

$$y''+a_1(x)y'+a_0(x)y=r_2(x)$$

$$y(x_0)=c_3, \quad y'(x_0)=c_4$$

Show that y_1+y_2 is the unique solution of

$$y''+a_1(x)y'+a_0(x)y=r_1(x)+r_2(x)$$

$$y(x_0)=c_1+c_3, \quad y'(x_0)=c_2+c_4$$

9. An equation of the form

$$y''+a_1x^{-1}y'+a_0x^{-2}y=0$$

is called a **Cauchy-Euler** equation. Show that $y(x)=x^r$ is a solution of this equation if and only if $r^2+(a_1-1)r+a_0=0$.

In Exercises 10–13 use Exercise 9 to find a general solution of the given differential equation.

10. $y''+4x^{-1}y'+2x^{-2}y=0$
11. $y''-2x^{-1}y'+2x^{-2}y=0$
12. $y''-4x^{-1}y'-6x^{-2}y=0$
13. $y''+x^{-1}y'-x^{-2}y=0$

14. (a) Use Euler's formula, $e^{i\theta}=\cos\theta+i\sin\theta$, and the identity $x^{bi}=e^{ib\ln x}$ to show that $x^{a+bi}=x^a[\cos(b\ln x)+i\sin(b\ln x)]$.
    (b) Let $r=a+bi$ be a root of $r^2+(a_1-1)r+a_0=0$. Show that $y_1(x)=x^a\cos(b\ln x)$ and $y_2(x)=x^a\sin(b\ln x)$ are linearly independent solutions of the Cauchy-Euler equation $y''+a_1x^{-1}y'+a_0x^{-2}y=0$.

In Exercises 15–18 use Exercise 14 to find a general solution of the given differential equation.

15. $y'' + x^{-1}y' + x^{-2}y = 0$

16. $y'' + x^{-1}y' + 4x^{-2}y = 0$

17. $y'' + 2x^{-1}y' + 5x^{-2}y = 0$

18. $y'' + 2x^{-1}y' + 3x^{-2}y = 0$

2.3 Linear Differential Equations with Constant Coefficients, I

In the next four sections we will consider linear differential equations of the form

$$y'' + a_1 y' + a_0 y = g(x) \tag{2.8}$$

where the coefficients a_1 and a_0 are constants. It will be convenient to introduce yet another notation for the derivative:

$$D = \frac{d}{dx}$$

With this notation,

$$D^2 = DD = \frac{d}{dx}\frac{d}{dx} = \frac{d^2}{dx^2}$$

If we set

$$f(z) = z^2 + a_1 z + a_0$$

we are able to rewrite equation (2.8) as

$$g(x) = y'' + a_1 y' + a_0 y$$

$$= D^2 y + a_1 Dy + a_0 y$$

$$= (D^2 + a_1 D + a_0) y$$

$$= f(D) y$$

By Theorem 2.4 we know that a general solution of the equation $f(D)y = g(x)$ is the sum of a general solution of the homogeneous equation $f(D)y = 0$ and any particular solution of $f(D)y = g(x)$. This section and Section 2.5 will be devoted to finding solutions of $f(D)y=0$, while Sections 2.7, 2.8, and 2.11 will be devoted to finding solutions of $f(D)y=g(x)$.

One of the basic techniques of mathematics is to transform a problem one would like to solve into a problem one knows how to solve (at least in theory). We will change the problem of solving $f(D)y=0$ into the problem of finding the roots of the polynomial f. We begin by noting that

$$De^{rx}=re^{rx} \quad \text{and} \quad D^2e^{rx}=r^2e^{rx}$$

so that

$$f(D)e^{rx}=D^2e^{rx}+a_1De^{rx}+a_0e^{rx}$$

$$=r^2e^{rx}+a_1re^{rx}+a_0e^{rx}$$

$$=(r^2+a_1r+a_0)e^{rx}$$

$$=f(r)e^{rx}$$

Thus $f(D)e^{rx}=0$ if and only if r is a root of the polynomial f. This means that $y(x)=e^{rx}$ is a solution of $f(D)y=0$ whenever $f(r)=0$. The equation

$$f(r)=0$$

is called the **auxiliary equation** for $f(D)y=0$.

If the auxiliary equation $f(r)=0$ has distinct roots p and q, then $y_1(x)=e^{px}$ and $y_2(x)=e^{qx}$ are solutions of $f(D)y=0$. Since p and q are different, y_1 and y_2 are linearly independent. (See Example 3 of Section 2.2.) Therefore,

$$y(x)=c_1e^{px}+c_2e^{qx}$$

is a general solution of $f(D)y=0$.

▶ **Example 1**   Consider the differential equation $(D^2+5D+6)y=0$. The roots of the auxiliary equation $r^2+5r+6=0$ are $r=-2$ and $r=-3$. Hence $y_1(x)=e^{-2x}$ and $y_2(x)=e^{-3x}$ are linearly independent solutions. A general solution of the differential equation is

$$y(x)=c_1e^{-2x}+c_2e^{-3x} \qquad\qquad ◀$$

▶ **Example 2**   Consider the differential equation $(D^2-1)y=0$. The roots of the auxiliary equation $r^2-1=0$ are $r=1$ and $r=-1$. Therefore

$$y(x)=c_1e^x+c_2e^{-x}$$

is a general solution of the differential equation.   ◀

Next we will consider the case in which a real number p is a double root of the auxiliary equation $f(r)=0$. In such a case

$$f(D)y=(D-p)^2 y=0$$

We know that $y_1(x)=e^{px}$ is one solution of this equation, and now we show that $y_2(x)=xe^{px}$ is another solution:

$$f(D)(xe^{px})=(D-p)^2(xe^{px})$$

$$=(D-p)((D-p)(xe^{px}))$$

$$=(D-p)(e^{px})$$

$$=0$$

Evidently $y_1(x)=e^{px}$ and $y_2(x)=xe^{px}$ are linearly independent since neither function is a constant multiple of the other. Thus a general solution of $(D-p)^2 y=0$ is

$$y(x)=c_1 e^{px}+c_2 xe^{px}$$

▶ **Example 3** A general solution of $(D^2-2D+1)y=0$ is

$$y(x)=c_1 e^x+c_2 xe^x$$

since $r=1$ is a root of multiplicity two of the auxiliary equation $r^2-2r+1=0$. ◀

▶ **Example 4** A general solution of $(D^2+2\sqrt{3}\,D+3)y=0$ is

$$y(x)=c_1 e^{-\sqrt{3}x}+c_2 xe^{-\sqrt{3}x}$$

since $r=-\sqrt{3}$ is a root of multiplicity two of the auxiliary equation $r^2+2\sqrt{3}\,r+3=0$. ◀

The remaining case, in which the auxiliary equation has complex roots, will be postponed until Section 2.5.

Many physical phenomena can be modeled by differential equations that can be solved by the method described above. We will conclude this section by determining the shape of a hanging cable that is loaded in a special manner. Some other applications are indicated in the exercises and in the following section.

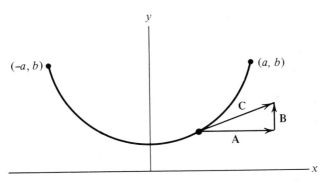

FIGURE 2.1

▶ **Example 5**   Consider a cable fixed at each end, loaded by its own weight or in some other manner, and hanging in a vertical plane. For simplicity we choose the rectangular coordinate system so that the ends of the cable are at the points $(-a, b)$ and (a, b) with $b > 0$ (Figure 2.1). We will make the assumption that the position of the cable in the plane can be represented by the graph of a function y whose domain is the interval $[-a, a]$. Hence points of the cable correspond to points $(x, y(x))$ in the plane for $-a \leqslant x \leqslant a$. Let $T = T(x)$ denote the magnitude of the tension in the cable at the point $(x, y(x))$, and let $X = X(x)$, $Y = Y(x)$ denote the horizontal and vertical components of T, so that the vectors **A**, **B**, and **C** in Figure 2.1 have lengths X, Y, and T, respectively. Since T acts in a direction tangent to the graph of y, we have

$$\frac{dy}{dx} = \frac{Y}{X} \qquad\qquad (2.9)$$

We will now make the additional assumption that the cable is acted upon by no force other than the vertically acting force due to gravity, except at the end points. It can be shown by the methods of statics that in such a case the horizontal component X of the tension is constant and that $dY(x)/dx$ is the vertical force acting at the point $(x, y(x))$ of the cable.

For example, suppose that a curtain consisting of a collection of thin homogeneous strips with constant linear density k extends from the cable to the ground (the x-axis) and that the weight of the cable is negligible. Then the vertical force acting at a point $(x, y(x))$ of the cable is the weight, $ky(x)$, of the strip suspended at the point $(x, y(x))$. Therefore $dY/dx = ky(x)$. If we differentiate each side of equation (2.9), we find that, remembering that X is constant,

$$\frac{d^2y}{dx^2} = \frac{1}{X}\frac{dY}{dx}$$

$$= \frac{k}{X}y$$

Thus y satisfies the differential equation

$$\frac{d^2y}{dx^2} - c^2y = 0$$

where we have set $c = (k/X)^{1/2}$ for ease of notation. The auxiliary equation $r^2 - c^2 = 0$ has $r = c$ and $r = -c$ as its roots. Hence

$$y(x) = c_1e^{cx} + c_2e^{-cx}$$

is a general solution. In order for the graph of this function to represent the hanging cable, we must choose the constants c_1 and c_2 so that the graph of y contains the points $(-a, b)$ and (a, b). That is, we must choose the constants c_1 and c_2 so that

$$b = y(-a) = c_1e^{-ca} + c_2e^{ca}$$

$$b = y(a) = c_1e^{ca} + c_2e^{-ca}$$

Thus we have two linear equations in the two unknowns c_1 and c_2. A short calculation shows that

$$c_1 = c_2 = \frac{b}{e^{ca} + e^{-ca}}$$

Hence

$$y(x) = b\frac{e^{cx} + e^{-cx}}{e^{ca} + e^{-ca}} = b\frac{\cosh cx}{\cosh ca}$$

is the function whose graph gives the shape of the hanging cable suspending a curtain of thin homogeneous strips. The graph of y is called a catenary. ◀

Exercises   SECTION 2.3

Find a general solution of each of the following equations (where $D \equiv d/dx$).

1. $(D^2 - 4)y = 0$
2. $(D^2 + 4D + 3)y = 0$
3. $(D^2 + D - 6)y = 0$
4. $(D^2 - 6D + 9)y = 0$
5. $(D^2 + 5D + 3)y = 0$

6. $(D^2+6D+8)y=0$

7. $(D^2-6D+8)y=0$

8. $(D^2-2\sqrt{3}\,D+3)y=0$

9. $(D^2-16D)y=0$

10. $(D-4)(D+7)y=0$

11. $(D^2-2D+1)y=0$

12. $(D^2+D)y=0$

13. $(D+5)^2y=0$

14. $(D^2+3D+1)y=0$

15. $(D^2+4D+2)y=0$

16. $(D^2+16D+64)y=0$

Find the solution of each of the following problems.

17. $(D^2+2D+1)y=0, \quad y(0)=0, \quad y'(0)=1$

18. $(D^2+2D+1)y=0, \quad y(0)=1, \quad y'(0)=0$

19. $(D^2+3D+2)y=0, \quad y(0)=1, \quad y'(0)=2$

20. $(D^2-3D-10)y=0, \quad y(0)=4, \quad y'(0)=2$

21. **(a)** An equation of the form

$$\frac{dy}{dx}=P(x)y^2+Q(x)y+R(x)$$

is called **Riccati's equation.** Show that the change of variable

$$y=-\frac{1}{P(x)z}\frac{dz}{dx}$$

transforms Riccati's equation into the second-order linear differential equation

$$\frac{d^2z}{dx^2}-\left(Q(x)+\frac{1}{P(x)}\frac{dP}{dx}\right)\frac{dz}{dx}+P(x)R(x)z=0$$

(b) Use part (a) to find a solution of

$$\frac{dy}{dx}=xy^2+\left(2-\frac{1}{x}\right)y-\frac{3}{x} \qquad (0<x)$$

22. Consider the differential equation $(D-a)(D-b)y=0$, $a\neq b$.

(a) Explain why the function

$$f_b(t)=\frac{e^{bt}-e^{at}}{b-a}$$

is a solution.

(b) Show that the function $g(t)=\lim_{b\to a}f_b(t)$ is a solution of the differential equation $(D-a)^2y=0$.

23. In certain cases a steady state diffusion process can be represented by a differential equation,

$$C\frac{d^2z}{dx^2}-V\frac{dz}{dx}=0$$

where C is a diffusion constant and V represents the velocity of the process. Find a general solution of this equation. Compare your answer with that to Exercise 17 of Section 1.3.

24. In its simplest form, a fixed-bed reactor consists of a cylindrical tube packed with a catalyst. Reactants are passed through the catalyst and transformed into products at a rate depending on various properties of the system, such as rate of flow, temperature, length of the reactor, and concentrations of the reactants. In the study of a one-dimensional reactor, the following differential equation is encountered:

$$E\frac{d^2C}{dz^2}-\bar{U}\frac{dC}{dz}-kC=0$$

where

$C=$ reactant concentration

$z=$ axial coordinate

$E=$ axial dispersion coefficient

$\bar{U}=$ average velocity of the reaction mixture

$k=$ reaction constant

Each of these quantities (except C) is a positive constant. [E. Petersen, *Chemical Reaction Analysis*, Prentice-Hall, Englewood Cliffs, N.J., 1965, pages 195–196.] Find a general solution of this differential equation.

25. The equation of motion of a damped pendulum consisting of a particle of mass m at the end of a massless rod of length L is

$$\frac{d^2\theta}{dt^2} + \frac{b}{mL}\frac{d\theta}{dt} + \frac{g}{L}\sin\theta = 0$$

where θ is the angular displacement of the rod from the vertical, g is the gravitational acceleration, and $b > 0$ is a constant representing frictional retardation. For small values of θ, $\sin\theta \approx \theta$, so that the differential equation may be approximated by

$$\frac{d^2\theta}{dt^2} + \frac{b}{mL}\frac{d\theta}{dt} + \frac{g}{L}\theta = 0 \qquad\qquad \textbf{(2.10)}$$

Suppose that the pendulum is displaced by a small angle θ_0, held stationary, and then released. In the special case $b = 2m\sqrt{gL}$, use equation (2.10) to determine the approximate motion of the pendulum. Describe the motion of the pendulum.

2.4 An Example from Oceanography
(OPTIONAL)

This section contains an example illustrating how homogeneous second-order linear differential equations may be used to model the physical and chemical changes occurring in sea sediments between the time of deposit and solidification.

Samples from various depths of the sediment in a muddy zone along the Belgian North Sea coast show that the concentration of silica varies according to depth as shown in Figure 2.2. Apparently there is a discontinuity in the derivative of the concentration at a depth of approximately 3.5 cm. Recall that a differentiable function is continuous, so that any function that has a second derivative has a continuous first derivative. Therefore, if we want a model that maintains the discontinuity of the derivative of the concentration, it must consist of more than a single second-order differential equation. One proposed model [J. Vanderborght, R. Wollast, and G. Billen, Kinetic models of diagenesis in disturbed sediments, Part I: Mass transfer properties and silica diagenesis, Limnology and Oceanography **22** (1977), 787–793] assumes that the concentration of silica at a depth z varies differently according to whether $0 \leqslant z \leqslant 3.5$ or $3.5 \leqslant z$. It is assumed that the first 3.5 cm of the sediment is kept relatively agitated by the movement of the overlying water, while the lower

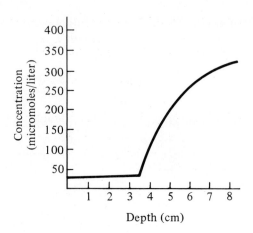

FIGURE 2.2

layer is compacted. It is further assumed that the only processes affecting the rate of change dS/dz of the concentration S at depth z are diffusion and precipitation. The contribution of diffusion is proportional to d^2S/dz^2, while it is assumed that the contribution of precipitation is proportional to the difference between S and an equilibrium concentration S_∞. These quantities are related by two second-order linear differential equations, one for the layer $0 \leqslant z \leqslant 3.5$ and the other for the layer $3.5 \leqslant z$. These equations are

$$D_1 \frac{d^2S}{dz^2} - W \frac{dS}{dz} + k(S_\infty - S) = 0 \qquad \text{(2.11)}$$

for $0 \leqslant z \leqslant 3.5$, and

$$D_2 \frac{d^2S}{dz^2} - W \frac{dS}{dz} + k(S_\infty - S) = 0 \qquad \text{(2.12)}$$

for $3.5 < z$, where D_1, D_2, W, and k are constants. It is assumed that:

(1) The concentration of silica at the surface of the sediment equals the concentration of silica in the overlying water.

(2) The concentration of silica tends to a finite value as z tends to infinity.

(3) The concentration varies continuously.

(4) The flux across the interface of the two layers is conservative; that is,

$$\lim_{z \to 3.5^-} D_1 \frac{dS}{dz} = \lim_{z \to 3.5^+} D_2 \frac{dS}{dz}$$

The referenced article gives following values for the constants:

$$D_1 = 10^{-4} \text{cm}^2/\text{sec}, \quad D_2 = 10^{-6} \text{cm}^2/\text{sec}$$

$$W = 3 \text{cm}/100 \text{ years} \approx 9.5(10)^{-8} \text{cm}/\text{sec}$$

$$k = 2(10)^{-7}/\text{sec}, \quad S_\infty = 400 \text{ micromoles}/\text{liter}$$

We will also suppose that the concentration of silica in the water is 25 micromoles/liter. With these constants, and making the change of variable $T = S - 400$, equations (2.11) and (2.12) become

$$10^{-4} \frac{d^2 T}{dz^2} - 9.5(10)^{-8} \frac{dT}{dz} - 2(10)^{-7} T = 0$$

for $0 \leqslant z \leqslant 3.5$, and

$$10^{-6} \frac{d^2 T}{dz^2} - 9.5(10)^{-8} \frac{dT}{dz} - 2(10)^{-7} T = 0$$

for $3.5 < z$, or, equivalently,

$$\frac{d^2 T}{dz^2} - 9.5(10)^{-4} \frac{dT}{dz} - 2(10)^{-3} T = 0 \qquad \text{(2.13)}$$

for $0 \leqslant z \leqslant 3.5$ and

$$\frac{d^2 T}{dz^2} - 9.5(10)^{-2} \frac{dT}{dz} - 2(10)^{-1} T = 0 \qquad \text{(2.14)}$$

for $3.5 < z$. The roots of the auxiliary equation $r^2 - 9.5(10)^{-4} r - 2(10)^{-3} = 0$ for equation (2.13) are

$$\frac{9.5(10)^{-4} \pm \sqrt{(9.5)^2 (10)^{-8} + 8(10)^{-3}}}{2} \approx \frac{9.5(10)^{-4} \pm 8.9(10)^{-2}}{2}$$

That is, the roots are approximately .045 and $-.044$. Equation (2.13) has a general solution of the form

$$T_1(x) = c_1 e^{.045z} + c_2 e^{-.044z}$$

Since $T = S - 400$, any solution of equation (2.11) has the form

$$S_1(z) = c_1 e^{.045z} + c_2 e^{-.044z} + 400$$

The roots of the auxiliary equation $r^2 - 9.5(10)^{-2}r - 2(10)^{-1} = 0$ for equation (2.14) are

$$\frac{9.5(10)^{-2} \pm \sqrt{(9.5)^2(10)^{-4} + 8(10)^{-1}}}{2} \simeq \frac{9.5(10)^{-2} \pm 9.0(10)^{-1}}{2}$$

That is, the roots are approximately .50 and $-.40$. Equation (2.14) has a general solution of the form

$$T_2(x) = d_1 e^{.5z} + d_2 e^{-.4z}$$

Since $T = S - 400$, any solution of equation (2.12) has the form

$$S_2(z) = d_1 e^{.5z} + d_2 e^{-.4z} + 400$$

Assumption (2) means that $\lim_{z \to \infty} S_2(z)$ exists and is finite. For this to be the case, we must have $d_1 = 0$. Conditions (1), (3), and (4) give

$$S_1(0) = 25$$

$$S_1(3.5) = S_2(3.5)$$

$$D_2 S_1'(3.5) = D_2 S_2'(3.5)$$

Hence

$$c_1 + c_2 + 400 = 25$$

$$c_1 e^{.045(3.5)} + c_2 e^{-.044(3.5)} + 400 = d_2 e^{-.4(3.5)} + 400$$

$$10^{-4}\left[(.045)c_1 e^{.045(3.5)} - (.044)c_2 e^{-.044(3.5)}\right] = -10^{-6}(.4)d_2 e^{-.4(3.5)}$$

To three decimal places these equations can be rewritten as

$$c_1 + c_2 = -375$$

$$1.171c_1 + .857c_2 - .247d_2 = 0$$

$$.053c_1 - .038c_2 + .001d_2 = 0 \qquad \text{(2.15)}$$

The first equation, which gives $c_2 = -375 - c_1$, enables us to rewrite the other two equations as

$$.314c_1 - .247d_2 = 321.375 \qquad \text{(2.16)}$$

$$.091c_1 + .001d_2 = -14.25 \qquad \text{(2.17)}$$

Equation (2.16) gives (to three decimal places) $c_1 = 1024 + .787d_2$. Inserting this value for c_1 into equation (2.17) gives

$$.091(1024 + .787d_2) + .001d_2 = -14.3$$

so that $d_2 = -1480$ and $c_1 = -141$. Since $c_1 + c_2 = -375$, we must have $c_2 = -234$. The concentration of silica is given by

$$S(z) = \begin{cases} -141e^{.045z} - 234e^{-.044z} + 400 & \text{if } 0 \leqslant z \leqslant 3.5 \qquad \textbf{(2.18)} \\ -1480e^{-.4z} + 400 & \text{if } 3.5 < z \qquad\qquad \textbf{(2.19)} \end{cases}$$

If we evaluate (2.18) and (2.19) at $z=3.5$, we get 34.346 and 35.036, respectively. This slight discontinuity in S at $z=3.5$ is caused by the approximations for exponential terms used to obtain the equations in (2.15) and is not inherent in the phenomenon being modeled. In fact, assumption (3) requires that the concentration S be continuous. In general, numerical approximations that are made to simplify calculations may lead to "solutions" that violate known properties of whatever is being modeled. The graph of S is a close fit to the empirical plot in Figure 2.2.

Exercise   SECTION 2.4

In the sequel to the paper referred to in this section [J. Vanderborght, R. Wollast, and G. Billen, Kinetic models of diagenesis in disturbed sediments, Part II: Nitrogen diagenesis, Limnology and Oceanography **22** (1977), 794–803], the authors model the concentration of various quantities in the sediments of a muddy zone of the North Sea coast of Belgium. In particular, it is shown that the concentration of sulfate $C(z)$ at a depth z satisfies the differential equations

$$D_1 \frac{d^2C}{dz^2} - W\frac{dC}{dz} = 0 \qquad \text{if } z \leqslant 3.5\,\text{cm}$$

$$D_2 \frac{d^2C}{dz^2} - W\frac{dC}{dz} - kC = 0 \quad \text{if } z \geqslant 3.5\,\text{cm}$$

where $D_1 = 10^{-4}\text{cm}^2/\text{sec}$, $D_2 = 10^{-6}\text{cm}^2/\text{sec}$, $W = 9.5(10)^{-8}\text{cm}/\text{sec}$, and $k = 2.5(10)^{-8}/\text{sec}$. As in the previous discussion, it is assumed that the concentration at the surface of the sediment equals the concentration in the overlying water, 28 millimoles/liter; the concentration tends to a finite value as z tends to infinity; the concentration varies continuously; and the flux across the interface of the two layers is conservative; that is,

$$\lim_{z \to 3.5^-} D_1 \frac{dC}{dz} = \lim_{z \to 3.5^+} D_2 \frac{dC}{dz}$$

Determine the concentration $C(z)$ at a depth z. Sketch the graph of $C(z)$.

2.5 Linear Differential Equations with Constant Coefficients, II

In Section 2.3 we considered the equation

$$f(D)y = (D^2 + a_1 D + a_0)y = 0 \tag{2.20}$$

for the case where the roots of the auxiliary equation $f(r)=0$ are real.

It remains to consider the case where the auxiliary equation has complex roots. Suppose that $r=a+bi$, $b \neq 0$, is a root of the auxiliary equation $f(r)=0$. Recall that if $a+bi$ is a root of a polynomial with real coefficients, then $a-bi$ is also a root of the same polynomial. Hence $y_1(x)=e^{(a+bi)x}$ and $y_2(x)=e^{(a-bi)x}$ are complex-valued solutions of $f(D)y=0$. By Euler's formula (see Appendix 1)

$$e^{(a+bi)x} = e^{ax}e^{bxi}$$

$$= e^{ax}(\cos bx + i \sin bx)$$

$$= e^{ax}\cos bx + ie^{ax}\sin bx$$

Then

$$0 = f(D)e^{(a+bi)x}$$

$$= f(D)(e^{ax}\cos bx + ie^{ax}\sin bx)$$

$$= f(D)(e^{ax}\cos bx) + if(D)(e^{ax}\sin bx)$$

Since a complex number is zero if and only if both the real and imaginary parts of the number are zero, we have

$$f(D)(e^{ax}\cos bx) = 0 \quad \text{and} \quad f(D)(e^{ax}\sin bx) = 0$$

Thus $y_3(x)=e^{ax}\cos bx$ and $y_4(x)=e^{ax}\sin bx$ are real solutions of $f(D)y=0$. In fact, it can be verified directly that y_3 and y_4 are solutions of $f(D)y=0$. We will do this for the function y_3. To begin, we have

$$f(D)(e^{ax}\cos bx) = D^2(e^{ax}\cos bx) + a_1 D(e^{ax}\cos bx) + a_0 e^{ax}\cos bx$$

$$= \left[(a^2 - b^2)e^{ax}\cos bx - 2abe^{ax}\cos bx\right]$$

$$+ a_1(ae^{ax}\cos bx - be^{ax}\sin bx) + a_0 e^{ax}\cos bx$$

$$= (a^2 - b^2 + a_1 a + a_0)e^{ax}\cos bx - (2a + a_1)be^{ax}\cos bx$$

$$\tag{2.21}$$

Since $a+bi$ is a root of $f(r)=0$, we have

$$0=f(a+bi)$$
$$=(a+bi)^2+a_1(a+bi)+a_0$$
$$=(a^2-b^2+a_1a+a_0)+ib(2a+a_1)$$

Again, since a complex number is zero if and only if both its real and imaginary parts are zero, $a^2-b^2+a_1a+a_0=0$ and $b(2a+a_1)=0$. Using these identities, equation (2.21) becomes

$$f(D)(e^{ax}\cos bx)=0$$

This proves that $y_3(x)=e^{ax}\cos bx$ is a solution of equation (2.20) whenever $a\pm bi$ are the roots of the auxiliary equation. A similar argument shows that $y_4(x)=e^{ax}\sin bx$ is also a solution of equation (2.20). Since neither solution is a constant multiple of the other, these two solutions are linearly independent. Therefore

$$y(x)=c_1e^{ax}\cos bx+c_2e^{ax}\sin bx$$

is a general solution of equation (2.20) whenever $a\pm bi$ are the roots of the auxiliary equation.

▶ **Example 1** A general solution of $(D^2+D+1)y=0$ is

$$y(x)=c_1e^{-x/2}\cos\frac{\sqrt{3}}{2}x+c_2e^{-x/2}\sin\frac{\sqrt{3}}{2}x$$

since the roots of the auxiliary equation $r^2+r+1=0$ are $(-1+\sqrt{3}\,i)/2$ and $(-1-\sqrt{3}\,i)/2$. ◀

▶ **Example 2** A general solution of $(D^2+1)y=0$ is

$$y(x)=c_1\cos x+c_2\sin x$$

since the roots of the auxiliary equation $r^2+1=0$ are i and $-i$. ◀

▶ **Example 3** A general solution of $(D^2+6D+25)y=0$ is

$$y(x)=c_1e^{-3x}\cos 4x+c_2e^{-3x}\sin 4x$$

since the roots of the auxiliary equation $r^2+6r+25=0$ are $-3+4i$ and $-3-4i$. ◀

Exercises   SECTION 2.5

Find the general solution of each of the following differential equations (where $D \equiv d/dx$).

1. $(D^2 + 9)y = 0$
2. $(D^2 + 2D + 3)y = 0$
3. $(D^2 + D + 1)y = 0$
4. $(D^2 + 16)y = 0$
5. $(D^2 - 2D + 3)y = 0$
6. $(D^2 + 4D + 8)y = 0$
7. $(D^2 - D + 1)y = 0$
8. $(D^2 + 6D + 34)y = 0$
9. $(D^2 + 10D + 34)y = 0$
10. $(D^2 - 9D + 22)y = 0$
11. $(D^2 + 2D + 4)y = 0$
12. $(D^2 + D + 4)y = 0$

13. In applications it is not uncommon to require that a solution of a differential equation $f(D)y = 0$ satisfy $y(a) = 0$ and $y(b) = 0$ where $a \neq b$. Such a problem is called a **boundary-value problem**. For example,

$$(D^2 + 1)y = 0, \quad y(0) = 0, \quad y(\pi) = 0$$

is a boundary-value problem. Note that the existence theorem (Theorem 2.1) does not apply to such a problem. In fact, there is in general no guarantee that there is a solution other than $y(x) \equiv 0$ that satisfies the prescribed conditions. The following examples show that, depending on the choice of a,

$$(D^2 + 1)y = 0, \quad y(0) = 0, \quad y(a) = 0$$

may or may not have a solution other than $y(x) \equiv 0$.

(a) Show that if $a = \pi/2$, then there is no solution other than $y(x) \equiv 0$.
(b) Show that if $a = \pi$, then there are solutions other than $y(x) \equiv 0$. Find all of these solutions.

14. It can be shown that the deflection w of a uniform column of length L under a constant axial load p satisfies

$$EI\frac{d^2 w}{dx^2} + pw = 0, \quad w(0) = w(L) = 0$$

where E is Young's modulus, and I is the moment of inertia of the column. For which values of p does this problem have a solution other than $w(x) \equiv 0$. The smallest value of p is the upper limit for the stability of the undeflected equilibrium position of the column.

15. Suppose that an electron is confined so that it can move only along an interval $[0, L]$. We would like to know the position of the electron at any given time, but, according to quantum mechanics, the best we can hope to know is its probable position. The one-dimensional, time-independent Schrödinger equation is

$$\frac{h^2}{2m} \frac{d^2 \Psi}{dx^2} + (E - V)\Psi = 0$$

where m is the mass of the electron, h is Planck's constant, V is the potential energy, and E is the total energy. From a knowledge of Ψ, the probability of finding the electron at a given point of the interval can be determined. Since the electron cannot leave the interval, the appropriate boundary conditions are $\Psi(0) = \Psi(L) = 0$. Assuming that $E - V$ is constant, determine the values of L for which this problem has a solution.

2.6 Harmonic Motion

Consider the simple mechanical system consisting of a mass m connected to a spring as shown in Figure 2.3. If the mass is displaced from its equilibrium position by a small distance x, then (according to Hooke's law) a force proportional to x is exerted on the mass by the spring. Initially we assume that this is the only force that acts upon the mass. Newton's second law of motion enables us to obtain a relation between the acceleration $a = d^2x/dt^2$ and the position x of the mass:

$$m\frac{dx^2}{dt^2} = ma = \text{force} = -kx$$

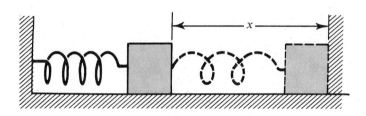

FIGURE 2.3

or

$$\frac{d^2x}{dt^2} + \frac{k}{m}x = 0 \tag{2.22}$$

where k is a positive constant. Any mechanical system described by equation (2.22) is called a **harmonic oscillator**. For convenience of notation we set

$$w = \sqrt{\frac{k}{m}}$$

A general solution of equation (2.22) may be written in the form

$$x(t) = c_1 \cos wt + c_2 \sin wt$$

$$= \left(c_1^2 + c_2^2\right)^{1/2} \left[\frac{c_1}{\left(c_1^2 + c_2^2\right)^{1/2}} \cos wt + \frac{c_2}{\left(c_1^2 + c_2^2\right)^{1/2}} \sin wt \right] \tag{2.23}$$

Note that

$$\left[\frac{c_1}{\left(c_1^2 + c_2^2\right)^{1/2}} \right]^2 + \left[\frac{c_2}{\left(c_1^2 + c_2^2\right)^{1/2}} \right]^2 = 1$$

Hence there is an angle α, $-\pi < \alpha \leq \pi$, such that

$$\sin \alpha = \frac{c_1}{\left(c_1^2 + c_2^2\right)^{1/2}}, \quad \cos \alpha = \frac{c_2}{\left(c_1^2 + c_2^2\right)^{1/2}}$$

The general solution in (2.23) may now be rewritten in the form (where $A = (c_1^2 + c_2^2)^{1/2}$)

$$x(t) = A(\cos wt \sin \alpha + \sin wt \cos \alpha)$$

$$= A \sin(wt + \alpha)$$

$$= A \sin\left(\sqrt{\frac{k}{m}}\, t + \alpha\right) \tag{2.24}$$

The constants A and α have important physical interpretations. The function $\sin(\sqrt{k/m}\, t + \alpha)$ varies between -1 and 1 as t varies from 0 to ∞. Therefore the maximum and minimum values of x are A and $-A$, respectively. The constant A is called the **amplitude** of the oscillation. The argument, $\sqrt{k/m}\, t + \alpha$, of the sine function is called the **phase angle**. When $t = 0$ the phase angle is α. Hence α is called the **initial phase**.

It should be noted that the motion described by the function x in (2.23) is periodic, with period $T=2\pi\sqrt{m/k}$; that is,

$$x(t+T)=A\sin\left(\sqrt{\frac{k}{m}}\,(T+t)+\alpha\right)$$

$$=A\sin\left(\sqrt{\frac{k}{m}}\,t+2\pi+\alpha\right)$$

$$=A\sin\left(\sqrt{\frac{k}{m}}\,t+\alpha\right)$$

$$=x(t)$$

▶ **Example 1** A mass of .5 kilogram is attached to a spring with constant $k=4.5$ Newtons/meter. At time $t=0$ the mass is displaced a distance $x=.2$ meter, held stationary, and released. We will find the position of the mass at time t. In this case, equation (2.22) becomes

$$\frac{d^2x}{dt^2}+\frac{4.5}{.5}x=0$$

which has, by equation (2.24), a general solution

$$x(t)=A\sin(3t+\alpha)$$

Initially the velocity of the mass is 0 and the displacement of the mass is .2, so that

$$.2=x(0)=A\sin\alpha \qquad\qquad\qquad \textbf{(2.25)}$$

$$0=x'(0)=3A\cos\alpha \qquad\qquad\qquad \textbf{(2.26)}$$

Recalling that $A=(c_1^2+c_2^2)^{1/2}$, we see that A must be positive. From equation (2.25) we observe that $\sin\alpha$ must also be positive. Hence $0<\alpha<\pi$. In order for equation (2.26) to be valid with $0<\alpha<\pi$, we must have $\alpha=\pi/2$. Equation (2.25) now becomes

$$.2=A\sin\frac{\pi}{2}=A$$

The position $x(t)$ of the mass at time t is given by

$$x(t)=.2\sin\left(3t+\frac{\pi}{2}\right) \qquad\qquad ◀$$

A more realistic model for this harmonic oscillator would include a frictional force. When the mass moves slowly, the air resistance is proportional to the velocity dv/dt of the mass, but oppositely directed. This is in contrast to friction between solid bodies, which is independent of speed within wide limits. We will assume that the frictional force is due entirely to air resistance. This modifies equation (2.22) to

$$m\frac{d^2x}{dt^2} + a\frac{dx}{dt} + kx = 0 \qquad (2.27)$$

where a and k are positive constants. Any mechanical system described by equation (2.27) is called a **damped harmonic oscillator**. For convenience of notation we set

$$p = \sqrt{\frac{k}{m}} \quad \text{and} \quad 2b = \frac{a}{m}$$

In this notation, equation (2.27) may be written as

$$\frac{d^2x}{dt^2} + 2b\frac{dx}{dt} + p^2x = 0 \qquad (2.28)$$

which has

$$r^2 + 2br + p^2 = 0 \qquad (2.29)$$

as its auxiliary equation. It is important to note that $b > 0$.

The roots of equation (2.29) are $-b \pm \sqrt{b^2 - p^2}$. There are three cases to consider: (1) $b^2 - p^2 < 0$; (2) $b^2 - p^2 = 0$; and (3) $b^2 - p^2 > 0$.

Case 1: $b^2 - p^2 < 0$. If $b^2 - p^2 < 0$, then the roots of the auxiliary equation (2.29) are the distinct complex numbers $-b + \sqrt{p^2 - b^2}\,i$ and $-b - \sqrt{p^2 - b^2}\,i$. In this case a general solution of equation (2.28) is

$$x(t) = e^{-bt}\left[c_1\cos\left(\sqrt{p^2 - b^2}\,t\right) + c_2\sin\left(\sqrt{p^2 - b^2}\,t\right) \right]$$

$$= \left(c_1^2 + c_2^2\right)^{1/2} e^{-bt}\left[\frac{c_1}{\left(c_1^2 + c_2^2\right)^{1/2}}\cos\left(\sqrt{p^2 - b^2}\,t\right) + \frac{c_2}{\left(c_1^2 + c_2^2\right)^{1/2}}\sin\left(\sqrt{p^2 - b^2}\,t\right) \right]$$

$$= Ae^{-bt}\left[\sin\beta\cos\left(\sqrt{p^2 - b^2}\,t\right) + \cos\beta\sin\left(\sqrt{p^2 - b^2}\,t\right) \right]$$

where $A = \left(c_1^2 + c_2^2 \right)^{1/2}$, $\sin \beta = \dfrac{c_1}{\left(c_1^2 + c_2^2 \right)^{1/2}}$, and $\cos \beta = \dfrac{c_2}{\left(c_1^2 + c_2^2 \right)^{1/2}}$

Thus

$$x(t) = A e^{-bt} \sin\left(\sqrt{p^2 - b^2} \, t + \beta \right) \tag{2.30}$$

Since $|\sin(\sqrt{p^2 - b^2} \, t + \beta)| \leqslant 1$ and $e^{-bt} \to 0$ as $t \to \infty$ (remember $b > 0$), we have $x(t) \to 0$ as $t \to \infty$. Moreover, $\sin(\sqrt{p^2 - b^2} \, t + \beta)$ varies between -1 and 1 as $t \to \infty$, so $x(t)$ oscillates between positive and negative values as $t \to \infty$. A sketch of the graph of x is indicated in Figure 2.4. The motion described by x in (2.30) is called **damped harmonic motion**.

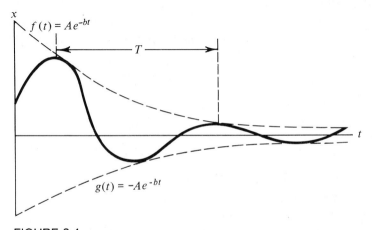

FIGURE 2.4

Case 2: $b^2 - p^2 = 0$. If $b^2 - p^2 = 0$, then $-b$ is a root of multiplicity two of the auxiliary equation (2.29). In this case a general solution of equation (2.28) is

$$x(t) = e^{-bt}(c_1 + c_2 t)$$

Since $e^{-bt} > 0$, we have that $x(t) = 0$ if and only if $c_1 + c_2 t = 0$. Thus $x(t) = 0$ for at most one value of t. Since x is a continuous function of t, we conclude that $x(t)$ has constant sign for t sufficiently large. The motion is not oscillatory. Moreover, by L'Hospital's rule,

$$\lim_{t \to \infty} x(t) = \lim_{t \to \infty} \frac{c_1 + c_2 t}{e^{bt}} = \lim_{t \to \infty} \frac{c_2}{b e^{bt}} = 0$$

In this case the motion is called **critically damped**.

If we let x_0 and x_1 denote the initial displacement and initial velocity of the mass, respectively, then

$$x_0 = x(0) = c_1$$

$$x_1 = x'(0) = c_2 - bc_1 = c_2 - bx_0$$

so that $c_1 = x_0$ and $c_2 = x_1 + bx_0$. We now have

$$x(t) = e^{-bt}\left(x_0 + (x_1 + bx_0)t\right)$$

If the initial displacement x_0 is positive, there are two cases to consider: $x_1 + bx_0 \geqslant 0$ and $x_1 + bx_0 < 0$.

(i) If $x_1 + bx_0 \geqslant 0$, then $x(t) > 0$ for all positive values of t, and the mass does not pass through its equilibrium position. The graph of x for such a case is shown in Figure 2.5.

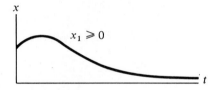
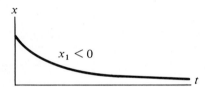

FIGURE 2.5

(ii) If $x_1 + bx_0 < 0$, then $x(t) = 0$ when $t = -x_0(x_1 + bx_0)^{-1}$. In this case the mass passes through its equilibrium position. The graph of x is shown in Figure 2.6.

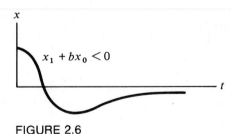

FIGURE 2.6

Case 3: $b^2 - p^2 > 0$.   If $b^2 - p^2 > 0$, then the auxiliary equation (2.29) has two distinct real roots:

$$r_1 = -b + \sqrt{b^2 - p^2} \quad \text{and} \quad r_2 = -b - \sqrt{b^2 - p^2}$$

In this case a general solution of equation (2.28) is

$$x(t) = c_1 e^{r_1 t} + c_2 e^{r_2 t}$$

Since $p > 0$, we have

$$b^2 - p^2 < b^2$$

so that

$$\sqrt{b^2 - p^2} < b$$

or, equivalently,

$$r_1 = -b + \sqrt{b^2 - p^2} < 0$$

It is clear that

$$r_2 = -b - \sqrt{b^2 - p^2} < 0$$

because $b > 0$. It now follows that

$$\lim_{t \to \infty} x(t) = 0$$

The solution $x(t)$ can be rewritten as

$$x(t) = e^{r_1 t}\left(c_1 + c_2 e^{(r_2 - r_1)t} \right)$$

$$= e^{r_1 t}\left(c_1 + c_2 e^{-2\sqrt{b^2 - p^2}\, t} \right) \tag{2.31}$$

Thus $x(t) = 0$ if and only if

$$c_1 + c_2 e^{-2\sqrt{b^2 - p^2}\, t} = 0$$

This equation has at most one solution. Since x is a continuous function of t, we conclude that $x(t)$ has constant sign for all t sufficiently large. Again the motion is not oscillatory. In this case the motion is called **overdamped**. If we let x_0 and x_1 denote the initial position and velocity of the mass, then from (2.31) we have

$$x_0 = x(0) = c_1 + c_2$$

$$x_1 = x'(0) = r_1(c_1 + c_2) + (r_2 - r_1)c_2$$

$$= r_1 x_0 + (r_2 - r_1)c_2$$

Solving the second equation for c_2 and then using this value in the first equation to determine c_1, we have

$$c_2 = \frac{x_1 - r_1 x_0}{r_2 - r_1}$$

$$c_1 = x_0 - c_2$$

$$= x_0 - \frac{x_1 - r_1 x_0}{r_2 - r_1}$$

$$= \frac{r_2 x_0 - x_1}{r_2 - r_1}$$

The graph of x, depending upon the relative sizes of x_0 and x_1, is similar to one of the graphs given in Case 2.

Exercises   SECTION 2.6

1. A mass of .5 kilogram is attached to a spring with constant $k = 4.5$ Newtons/meter. Suppose there is a frictional force of numerical value $2 dx/dt$ opposing the motion of this mass. Determine the motion of the mass if:

   (a) The mass is displaced a distance $x = .2$ meter, held stationary, and released.
   (b) The mass is stationary at its equilibrium $(x = 0)$ and then given a push that gives it an initial velocity of .3 meter/second.
   (c) The mass is displaced a distance $x = .2$ meter and then given an initial velocity of .3 meter/second.

2. The undamped harmonic oscillator may be used to model the vibrations of a car without shock absorbers. We will suppose that the car weighs 2000 pounds and that the spring has a constant of 520 pounds per inch. The car body will begin 1 inch above equilibrium with a downward velocity of 1 foot per second. Converting weight in pounds to mass in slugs and ft/sec to in/sec, we find that the numerical value of the mass of the car is $2000/(32)(12) \approx 5.2$. The equation for the displacement y of the car body is

$$5.2 y'' + 520 y = 0$$

$$y(0) = 1, \quad y'(0) = -12$$

Find the displacement $y(t)$ of the car body at time t.

3. Using the previous problem, we add shock absorbers that produce a force of 83.2 times the velocity of the car body, but oppositely directed. In this case the equation for the displacement y of the car body is

$$5.2\,y'' + 83.2\,y' + 520\,y = 0$$

$$y(0) = 1, \quad y'(0) = -12$$

Find the displacement $y(t)$ of the car body at time t.

4. Consider the simple electrical circuit shown in Figure 2.7, where R is a resistor measured in ohms, C a capacitor measured in farads, L an inductor

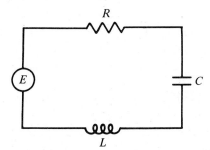

FIGURE 2.7

measured in henries, and E an electromotive force measured in volts. It can be shown by Kirchhoff's laws that the charge q on the capacitor satisfies

$$L\frac{d^2q}{dt^2} + R\frac{dq}{dt} + \frac{1}{C}q = E$$

For the special case $E = 0$, what relationship must L, R, and C have in order for q to be a damped harmonic motion?

2.7 The Method of Undetermined Coefficients, I

The method (or modifications of it) described in this and the following sections is frequently called the **method of undetermined coefficients.**
In Section 2.2 we showed that a general solution of

$$\left(D^2 + a_1 D + a_0\right)y = g(x)$$

is the sum of any particular solution of this equation and any general solution of the homogeneous equation

$$(D^2 + a_1 D + a_0)y = 0$$

In Sections 2.3 and 2.5 we saw how to find a general solution of the homogeneous equation. We will now discuss a method of determining a solution of the nonhomogeneous equation whenever the function g has one of several special forms.

To begin, we consider the equation

$$(D^2 + 1)y = x^2 \qquad\qquad \textbf{(2.32)}$$

and try to guess a particular solution. A particular solution is a function that, when added to its second derivative, yields the polynomial $g(x) = x^2$. If P is a nonconstant polynomial, then $D^2 P$ is a polynomial of degree less than the degree of P. Therefore $(D^2 + 1)P$ is a polynomial of the same degree as P. This suggests that we try to find a second-degree polynomial

$$P(x) = d_0 + d_1 x + d_2 x^2$$

that is a particular solution of equation (2.32). Inserting this guess of a particular solution into equation (2.32) yields

$$x^2 = (D^2 + 1)(d_0 + d_1 x + d_2 x^2)$$

$$= d_0 + 2d_2 + d_1 x + d_2 x^2$$

If we now equate the coefficients of like powers of x on each side of this equation, we find that

$$d_0 + 2d_2 = 0 \qquad d_1 = 0 \qquad d_2 = 1$$

so that

$$d_0 = -2 \qquad d_1 = 0 \qquad d_2 = 1$$

With these choices for the coefficients,

$$P(x) = x^2 - 2$$

is a particular solution of equation (2.32).

An argument similar to that given above shows that if P is a polynomial, then $(D^2+a_1D+a_0)P$ is a polynomial whose degree is equal to:

(i) The degree of P if $a_0\neq0$; i.e., if zero is not a root of the auxiliary equation.

(ii) One less than the degree of P if $a_0=0$ and $a_1\neq0$; i.e., if zero is a root of multiplicity one of the auxiliary equation.

(iii) Two less than the degree of P if $a_0=0$ and $a_1=0$; i.e., if zero is a root of multiplicity two of the auxiliary equation.

This leads us to speculate that if zero is a root of multiplicity k of the equation

$$r^2+a_1r+a_0=0$$

($k=0$ if zero is not a root), then the differential equation

$$\left(D^2+a_1D+a_0\right)y=b_0+b_1x+\cdots+b_nx^n$$

has a solution of the form

$$y_1(x)=x^k\left(d_0+d_1x+\cdots+d_nx^n\right)$$

In fact, this can be shown to be the case.

▶ **Example 1**   Consider the differential equation

$$(D^2-1)y=3x^2+2x+4 \tag{2.33}$$

Since the roots of the auxiliary equation $r^2-1=0$ are nonzero, there is a particular solution of equation (2.33) having the form

$$z(x)=d_0+d_1x+d_2x^2$$

where the coefficients d_0, d_1, and d_2 will be chosen so that

$$3x^2+2x+4=(D^2-1)\left(d_2x^2+d_1x+d_0\right)$$

$$=-d_2x^2-d_1x+(2d_2-d_0)$$

In order to determine d_2, d_1, and d_0, we equate the coefficients of like powers of x to obtain

$$3=-d_2 \qquad 2=-d_1 \qquad 4=2d_2-d_0$$

Hence

$$d_2 = -3 \qquad d_1 = -2 \qquad d_0 = -10$$

so that

$$z(x) = -3x^2 - 2x - 10$$

is a particular solution of equation (2.33). The roots of the auxiliary equation are 1 and -1. Therefore

$$y(x) = c_1 e^x + c_2 e^{-x} - 3x^2 - 2x - 10$$

is a general solution of equation (2.33).   ◀

▶ **Example 2**   Consider the differential equation

$$(D^2 + 4D + 20)y = 10x^3 - 1 \qquad\qquad \textbf{(2.34)}$$

Since the roots of the auxiliary equation $r^2 + 4r + 20 = 0$ are nonzero, there is a particular solution of equation (2.34) having the form

$$z(x) = d_0 + d_1 x + d_2 x^2 + d_3 x^3$$

where the coefficients d_0, d_1, d_2, and d_3 will be chosen so that

$$10x^3 - 1 = (D^2 + 4D + 20)(d_3 x^3 + d_2 x^2 + d_1 x + d_0)$$

$$= 20d_3 x^3 + (12d_3 + 20d_2)x^2 + (6d_3 + 8d_2 + 20d_1)x$$

$$+ (2d_2 + 4d_1 + 20d_0)$$

In order to determine d_3, d_2, d_1, and d_0, we equate the coefficients of like powers of x to obtain

(i) $20d_3 = 10$ (iii) $6d_3 + 8d_2 + 20d_1 = 0$

(ii) $12d_3 + 20d_2 = 0$ (iv) $2d_2 + 4d_1 + 20d_0 = -1$

From (i) we have $d_3 = \frac{1}{2}$. Inserting this value into (ii), we find that $d_2 = -\frac{3}{10}$. These values enable us to compute d_1 from (iii): $d_1 = -\frac{3}{10}$. Since we know d_1 and d_2, we can compute d_0 from (iv): $d_0 = -\frac{7}{500}$.

Hence

$$z(x)=\tfrac{1}{2}x^3-\tfrac{3}{10}x^2-\tfrac{3}{100}x-\tfrac{7}{500}$$

is a particular solution of equation (2.34). The roots of the auxiliary equation are $-2\pm4i$. Therefore

$$y(x)=c_1e^{-2x}\cos4x+c_2e^{-2x}\sin4x+\tfrac{1}{2}x^3-\tfrac{3}{10}x^2-\tfrac{3}{100}x-\tfrac{7}{500}$$

is a general solution of equation (2.34).   ◄

▶ **Example 3**   Consider the differential equation

$$(D^2+3D)y=4x^3+13x^2-3x \tag{2.35}$$

Since zero is a root of multiplicity one of the auxiliary equation, there is a particular solution of equation (2.35) having the form

$$z(x)=x(d_0+d_1x+d_2x^2+d_3x^3)$$

where the coefficients d_0, d_1, d_2, and d_3 will be chosen so that

$$4x^3+13x^2-3x=(D^2+3D)(d_0x+d_1x^2+d_2x^3+d_3x^4)$$

$$=12d_3x^3+(12d_3+9d_2)x^2+(6d_2+6d_1)x+(2d_1+3d_0)$$

In order to determine d_0, d_1, d_2, and d_3, we equate the coefficients of like powers of x to obtain

(i) $12d_3=4$ (iii) $6d_2+6d_1=-3$

(ii) $12d_3+9d_2=13$ (iv) $2d_1+3d_0=0$

From (i) we have $d_3=\tfrac{1}{3}$. Inserting this value into (ii), we find that $d_2=1$. These values enable us to compute d_1 from (iii): $d_1=-\tfrac{3}{2}$. The coefficient d_0 is now obtained from (iv): $d_0=1$. Hence

$$z(x)=\tfrac{1}{3}x^4+x^3-\tfrac{3}{2}x^2+x$$

is a particular solution of equation (2.35). The roots of the auxiliary equation are 0 and -3. Therefore

$$y(x)=c_1e^{-3x}+c_2+\tfrac{1}{3}x^4+x^3-\tfrac{3}{2}x^2+x$$

is a general solution of equation (2.35).   ◄

The method just described can be adapted to find a particular solution of

$$(D^2 + a_1 D + a_0)y = e^{ax}(b_0 + b_1 x + \cdots + b_n x^n) \qquad \textbf{(2.36)}$$

If we make the change of variable $y = ve^{ax}$, then

$$Dy = D(ve^{ax})$$

$$= e^{ax}(D + a)v \qquad \textbf{(2.37)}$$

and

$$D^2 y = D^2(ve^{ax})$$

$$= e^{ax}(D^2 + 2aD + a^2)v \qquad \textbf{(2.38)}$$

so that equation (2.36) becomes

$$e^{ax}(b_0 + b_1 x + \cdots + b_n x^n) = (D^2 + a_1 D + a_0)(ve^{ax})$$

$$= e^{ax}\left[D^2 + (2a + a_1)D + (a^2 + aa_1 + a_0)\right]v$$

or, equivalently,

$$\left[D^2 + (2a + a_1)D + (a^2 + aa_1 + a_0)\right]v = b_0 + b_1 x + \cdots + b_n x^n$$

Thus the change of variable $y = ve^{ax}$ transforms equation (2.36) into an equation which we just learned how to solve.

▶ **Example 4**   Consider the differential equation

$$(D^2 + 4)y = e^{3x}(x + 2) \qquad \textbf{(2.39)}$$

If we set $y = e^{3x}v$ and use equations (2.37) and (2.38) with $a = 3$, equation (2.39) becomes

$$e^{3x}(x + 2) = (D^2 + 4)(ve^{3x})$$

$$= D^2(ve^{3x}) + 4ve^{3x}$$

$$= e^{3x}(D^2 + 6D + 9)v + 4ve^{3x}$$

$$= e^{3x}(D^2 + 6D + 13)v$$

so that

$$(D^2+6D+13)v=x+2 \qquad \text{(2.40)}$$

Since 0 is not a root of the auxiliary equation for equation (2.40), there is a particular solution of the form $v(x)=d_1x+d_0$. Inserting this choice of v into equation (2.40) yields

$$x+2=(D^2+6D+13)(d_1x+d_0)$$

$$=13d_1x+6d_1+13d_0$$

It follows that $13d_1=1$ and $6d_1+13d_0=2$, so that

$$d_1=\tfrac{1}{13} \qquad d_0=\tfrac{20}{169}$$

Thus $v(x)=\tfrac{1}{13}x+\tfrac{20}{169}$ is a particular solution of equation (2.40), and

$$e^{3x}v(x)=\left(\tfrac{1}{13}x+\tfrac{20}{169}\right)e^{3x}$$

is a particular solution of equation (2.39). Since the roots of the auxiliary equation for equation (2.39) are $2i$ and $-2i$,

$$y(x)=c_1\cos 2x+c_2\sin 2x+\left(\tfrac{1}{13}x+\tfrac{20}{169}\right)e^{3x}$$

is a general solution of equation (2.39).   ◀

▶ **Example 5**   Consider the differential equation

$$(D^2+4)y=e^x+e^{3x}(x+2) \qquad \text{(2.41)}$$

By the principle of superposition (Theorem 2.5), a particular solution of this equation may be found by adding together particular solutions of

$$(D^2+4)y=e^x \qquad \text{(2.42)}$$

and

$$(D^2+4)y=e^{3x}(x+2)$$

A particular solution of the latter equation was found in Example 4. We will now find a particular solution of equation (2.42). We begin by setting $y=ve^x$. Equation (2.42) can now be written as

$$e^x = (D^2+4)(ve^x)$$

$$=e^x(D^2+2D+5)v$$

so that

$$(D^2+2D+5)v=1 \qquad\qquad \textbf{(2.43)}$$

Since 0 is not a root of the auxiliary equation for equation (2.43), there is a particular solution of equation (2.43) having the form $v(x)=d_0$. Inserting this choice for v into equation (2.43) yields

$$1=(D^2+2D+5)(d_0)$$

$$=5d_0$$

It follows that $d_0=\frac{1}{5}$. Thus $v(x)=\frac{1}{5}$ is a particular solution of equation (2.43) and

$$e^x v(x)=\tfrac{1}{5}e^x$$

is a particular solution of equation (2.42). Using Example 4 and the principle of superposition, we have that

$$z(x)=\tfrac{1}{5}e^x+\left(\tfrac{1}{13}x+\tfrac{20}{169}\right)e^{3x}$$

is a particular solution of equation (2.41). Since the roots of the auxiliary equation for equation (2.41) are $2i$ and $-2i$,

$$y(x)=c_1\cos 2x+c_2\sin 2x+\tfrac{1}{5}e^x+\left(\tfrac{1}{13}x+\tfrac{20}{169}\right)e^{3x}$$

is a general solution of equation (2.41). ◀

Exercises   SECTION 2.7

Find a general solution of each of the following differential equations.

1. $(D^2+1)y=2x+3$
2. $(D^2+4D-5)y=x^2+x$
3. $(D^2+2D+3)y=x^4+5$

4. $(D^2 + 3D + 2)y = 6x^2 + 2x + 5$

5. $(D^2 - D)y = x^2 + 6x + e^{5x}$

6. $(D^2 - 1600)y = x^2 - 4$

7. $(D^2 - 3D + 2)y = e^{-x} + x^2$

8. $(D^2 - 3D + 2)y = e^x + e^{3x}$

9. $(D^2 - 3D + 4)y = 32x^2 e^{-4x} + 3$

10. $(D^2 + 3D - 4)y = xe^{3x}$

11. $(D^2 + 7D)y = e^x + x$

12. $(D^2 + 4D - 45)y = e^{2x}$

13. $(D^2 + 2D + 1)y = e^{-x}$

14. $(D^2 - 5D + 4)y = e^x + e^{-2x}$

15. $(D^2 + 5D + 4)y = e^{3x} + e^{2x} - x$

16. $(D^2 - 5D + 6)y = e^{3x} - e^{2x}$

17. $(D^2 + 5D + 4)y = e^{4x} + e^{2x}$

18. $(D^2 - D - 6)y = x^3 - 4 + e^x - e^{-x}$

19. $(D^2 - 5D)y = xe^x - x + 5$

20. $(D^2 - 16D + 64)y = xe^{8x} + xe^{-x}$

21. Find a solution to the equation in Exercise 1 such that $y(0)=0$, $y'(0)=1$.
22. Find a solution to the equation in Exercise 2 such that $y(0)=1$, $y'(0)=0$.
23. Find a solution to the equation in Exercise 3 such that $y(0)=2$, $y'(0)=3$.

2.8 The Method of Undetermined Coefficients, II

In this section we extend the method developed in the previous section to find particular solutions of differential equations having the form

$$(D^2 + a_1 D + a_0)y = g(x) \tag{2.44}$$

where

$$g(x) = e^{ax}(b_0 + b_1 x + \cdots + b_n x^n)h(x)$$

and the function $h(x)$ is either $\cos bx$ or $\sin bx$. We will show that a particular solution of equation (2.44) for such a function $g(x)$ may be obtained from a particular solution of the differential equation

$$\left(D^2+a_1D+a_0\right)y=e^{(a+bi)x}\left(b_0+b_1x+\cdots+b_nx^n\right)$$

The basis of this method of solving equation (2.44) is contained in the following theorem.

> **THEOREM 2.6**   If $y(x)=u_1(x)+iu_2(x)$ is a complex-valued solution (u_1 and u_2 are real-valued functions) on an open interval I of
>
> $$\left(D^2+a_1D+a_0\right)y=g_1(x)+ig_2(x) \tag{2.45}$$
>
> where a_1, a_0 are real numbers and g_1, g_2 are real-valued functions, then
>
> $$\left(D^2+a_1D+a_0\right)u_1=g_1(x)$$
>
> and
>
> $$\left(D^2+a_1D+a_0\right)u_2=g_2(x)$$

Proof   From equation (2.45) we obtain

$$g_1(x)+ig_2(x)=\left(D^2+a_1D+a_0\right)(u_1+iu_2)$$

$$=\left(D^2+a_1D+a_0\right)u_1+i\left(D^2+a_1D+a_0\right)u_2 \tag{2.46}$$

Equating the real and imaginary parts of each side of equation (2.46) yields $(D^2+a_1D+a_0)u_1=g_1(x)$ and $(D^2+a_1D+a_0)u_2=g_2(x)$, which is the desired result.   ∎

Now let $y(x)=u_1(x)+iu_2(x)$, with u_1 and u_2 real-valued functions, be a solution of

$$\left(D^2+a_1D+a_0\right)y=e^{(a+bi)x}\left(b_kx^k+b_{k-1}x^{x-1}+\cdots+b_1x+b_0\right) \tag{2.47}$$

where a_1 and a_0 are real numbers. In light of Theorem 2.6,

$$\left(D^2+a_1D+a_0\right)u_1=e^{ax}\cos bx\left(b_kx^k+b_{k-1}x^{k-1}+\cdots+b_1x+b_0\right) \tag{2.48}$$

and

$$\left(D^2+a_1D+a_0\right)u_2=e^{ax}\sin bx\left(b_kx^k+b_{k-1}x^{k-1}+\cdots+b_1x+b_0\right) \tag{2.49}$$

Thus if we can find a solution of equation (2.47), we are able to find solutions of equations (2.48) and (2.49) also. Note that equation (2.47) is equation (2.36), provided complex numbers are permitted as exponents. The derivative of e^{ax} is ae^{ax} even when a is a complex number.

▶ **Example 1**   Consider the equation

$$(D^2 - 4D + 4)y = x\cos 2x \tag{2.50}$$

We begin by finding a particular solution of

$$(D^2 - 4D + 4)y = xe^{i2x} \tag{2.51}$$

The change of variable $y = ve^{i2x}$ transforms equation (2.51) into

$$\left(D^2 + (-4 + 4i)D - 8i\right)v = x \tag{2.52}$$

Since zero is not a root of the auxiliary equation $r^2 + (-4 + 4i)r - 8i = 0$, there is a solution of equation (2.52) having the form $v_1(x) = d_0 + d_1 x$. Inserting this choice for v into equation (2.52) yields

$$x = \left(D^2 + (-4 + 4i)D - 8i\right)(d_1 x + d_0)$$

$$= -8id_1 x + (-4 + 4i)d_1 - 8id_0$$

We now equate the coefficients of like powers of x to obtain $-8id_1 = 1$ and $(-4 + 4i)d_1 - 8id_0 = 0$, so that $d_1 = -1/8i = i/8$ and $d_0 = (i-1)/16$. Thus

$$v(x) = \tfrac{1}{8}ix + \tfrac{1}{16}(i - 1)$$

is a particular solution of equation (2.52). Then

$$y(x) = \left(\tfrac{1}{8}ix + \tfrac{1}{16}(i - 1)\right)e^{2ix}$$

$$= \left[-\tfrac{1}{16} + i\left(\tfrac{1}{8}x + \tfrac{1}{16}\right)\right](\cos 2x + i\sin 2x)$$

$$= \left(-\tfrac{1}{16}\cos 2x - \left(\tfrac{1}{8}x + \tfrac{1}{16}\right)\sin 2x + i\left[\left(\tfrac{1}{8}x + \tfrac{1}{16}\right)\cos 2x - \tfrac{1}{16}\sin 2x\right]\right)$$

By Theorem 2.6

$$u_1(x) = -\tfrac{1}{16}\cos 2x - \left(\tfrac{1}{8}x + \tfrac{1}{16}\right)\sin 2x$$

is a particular solution of

$$(D^2-4D+4)y=x\cos 2x$$

and

$$u_2(x)=\left(\tfrac{1}{8}x+\tfrac{1}{16}\right)\cos 2x-\tfrac{1}{16}\sin 2x$$

is a particular solution of

$$(D^2-4D+4)y=x\sin 2x$$

The auxiliary equation for equation (2.50) has 2 as a root of multiplicity two. Hence a general solution of equation (2.50) is

$$y(x)=c_1 e^{2x}+c_2 xe^{2x}-\tfrac{1}{16}\cos 2x-\left(\tfrac{1}{8}x+\tfrac{1}{16}\right)\sin 2x \qquad \blacktriangleleft$$

▶ **Example 2**   Consider the equation

$$(D^2+3)y=e^x\sin 5x+x \tag{2.53}$$

By the principle of superposition (Theorem 2.5) we can obtain a particular solution of equation (2.53) from the sum of particular solutions of

$$(D^2+3)y=e^x\sin 5x \tag{2.54}$$

and

$$(D^2+3)y=x \tag{2.55}$$

respectively. Equation (2.55) has a particular solution of the form $z_1(x)=ax+b$. Inserting z_1 into equation (2.55) yields

$$x=(D^2+3)z_1$$

$$=(D^2+3)(ax+b)$$

$$=3ax+3b$$

Thus $a=\tfrac{1}{3}$ and $b=0$, so that $z_1(x)=x/3$.

To find a particular solution of equation (2.54), we begin by finding a particular solution of

$$(D^2+3)y=e^{(1+5i)x} \tag{2.56}$$

The change of variable $y = v e^{(1+5i)x}$ transforms equation (2.56) into

$$\left(D^2 + (2+10i)D + (-21+10i) \right)v = 1$$

This equation has a particular solution of the form $v_1(x) = d_0$. A short calculation shows that

$$d_0 = \frac{1}{-21+10i} = \frac{-21-10i}{541}$$

so that

$$u(x) = v(x)e^{(1+5i)x}$$

$$= \left(-\tfrac{21}{541} - \tfrac{10}{541}i \right)e^x(\cos 5x + i\sin 5x)$$

$$= e^x\left(-\tfrac{21}{541}\cos 5x + \tfrac{10}{541}\sin 5x \right) + ie^x\left(-\tfrac{21}{541}\sin 5x - \tfrac{10}{541}\cos 5x \right)$$

is a particular solution of equation (2.56). By Theorem 2.6

$$z_2(x) = e^x\left(-\tfrac{21}{541}\sin 5x - \tfrac{10}{541}\cos 5x \right)$$

is a particular solution of equation (2.54). By Theorem 2.5, $z_1 + z_2$ is a particular solution of equation (2.53). The auxiliary equation $r^2 + 3 = 0$ for equation (2.53) has $\sqrt{3}\,i$ and $-\sqrt{3}\,i$ as its roots. Hence a general solution of equation (2.53) is

$$y(x) = c_1\cos\sqrt{3}\,x + c_2\sin\sqrt{3}\,x + e^x\left(-\tfrac{21}{541}\sin 5x - \tfrac{10}{541}\cos 5x \right) + \tfrac{1}{3}x \quad \blacktriangleleft$$

The method of undetermined coefficients may be summarized as follows. If

$$g(x) = e^{ax}\left(b_0 + b_1 x + \cdots + b_n x^n \right)\begin{cases} \sin bx \\ \cos bx \end{cases}$$

and $a+bi$ is a root of multiplicity k of the auxiliary equation

$$r^2 + a_1 r + a_0 = 0$$

then the differential equation

$$\left(D^2 + a_1 D + a_0 \right)y = g(x) \qquad\qquad \textbf{(2.57)}$$

has a particular solution of the form

$$z(x)=x^k\left(c_0+c_1x+\cdots+c_nx^n\right)e^{ax}\cos bx$$

$$+x^k\left(d_0+d_1x+\cdots+d_nx^n\right)e^{ax}\sin bx \qquad \textbf{(2.58)}$$

In fact, the method of undetermined coefficients is often presented without the change of variables $y=ve^{a+bi}$. In such a case one merely assumes that equation (2.57) has a solution of the form in (2.58), inserts such a solution into the differential equation, and solves for the coefficients $c_0, c_1\ldots, c_n, d_0, d_1,\ldots,$ d_n. The author believes that the method presented in the last two sections is computationally simpler than this alternative method.

Exercises   SECTION 2.8

In Exercises 1–20 find a general solution of the given differential equation.

1. $(D^2-1)y=\cos 2x$

2. $(D^2-1)y=e^x\sin x$

3. $(D^2+1)y=\sin x+x^2$

4. $(D^2+4)y=e^{2x}\cos x$

5. $(D^2+3D-10)y=e^{-x}\sin 3x$

6. $(D+5D+4)y=x$

7. $(D^2-4D+4)y=(2x+4)\cos x$

8. $(D^2-4D+4)y=e^x$

9. $(D^2-5D+6)y=\cos 2x+e^{4x}$

10. $(D^2-6D+7)y=\sin x$

11. $(D^2-D+2)y=xe^{2x}\sin 3x$

12. $(D^2-6D+25)y=e^{3x}\cos 4x$

13. $(D^2-16)y=\cos x+\sin 4x$

14. $(D^2+4)y=\cos x+\sin 4x$

15. $(D^2+D+1)y=x+\sin x$

16. $(D^2-2D+2)y=5\cos 3x$

17. $(D^2+2D+2)y=e^{-x}(2\cos x+3\sin x)$

18. $(D^2-2D+2)y=e^{-x}(2\cos x+3\sin x)$

19. $(D^2+1)y=\cos x+e^{2x}+x^2+\sin 2x$

20. $(D^2+4D+5)y=e^{-2x}\sin x+5$

21. Find a solution to the equation in Exercise 1 such that $y(0)=0$, $y'(0)=1$.
22. Find a solution to the equation in Exercise 2 such that $y(0)=1$, $y'(0)=0$.
23. Find a solution to the equation in Exercise 3 such that $y(0)=1$, $y'(0)=1$.

2.9 Resonance and Other Applications

Consider the mechanical system consisting of a mass m connected to a spring that is connected to a crankshaft rotating with angular velocity w radians/minute (Figure 2.8). We will assume that there is a frictional force

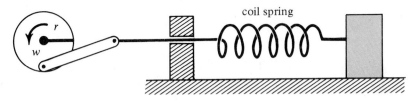

FIGURE 2.8

proportional to the velocity of the mass but oppositely directed. Suppose at time $t=0$ the system is in the position shown in Figure 2.8, and that the mass is at its equilibrium position. If the crankshaft now begins to rotate, the total

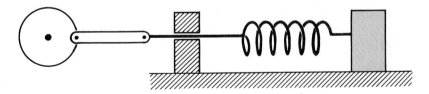

FIGURE 2.9

force acting on the mass is

$$F=-k(x-y)-a\frac{dx}{dt}$$

where k is the spring constant, a is a constant of proportionality, x is the distance the mass is displaced from its equilibrium position, and y is the displacement of the point p on the spring. If the radius, r, of the crankshaft is considerably shorter than the length of the bar, then the displacement, y, is approximately $r\sin wt$, so that the total force on the spring may be taken to be

$$F=-kx-a\frac{dx}{dt}+kr\sin wt$$

Using Newton's second law of motion, $F = m\,d^2x/dt^2$, we find that

$$m\frac{d^2x}{dt^2} + a\frac{dx}{dt} + kx = kr\sin wt \qquad (2.59)$$

For simplicity of notation we will set $2b = a/m$, $p^2 = k/m$, and $A = kr/m$. With this notation and setting $D = d/dt$, equation (2.59) becomes

$$(D^2 + 2bD + p^2)x = A\sin wt \qquad (2.60)$$

From the discussion at the end of Section 2.8, we know that there is a particular solution of equation (2.60) having the form

$$x_1(t) = c_1\cos wt + c_2\sin wt$$

whenever iw is not a solution of the auxiliary equation $r^2 + 2br + p^2 = 0$; i.e., whenever $b \neq 0$ and $w^2 \neq p^2$. Inserting x_1 into the differential equation yields

$$A\sin wt = (D^2 + 2bD + p^2)(c_1\cos wt + c_2\sin wt)$$

$$= \left[(p^2 - w^2)c_1 + 2bwc_2\right]\cos wt + \left[(p^2 - w^2)c_2 - 2bwc_1\right]\sin wt$$

Thus we must have

$$(p^2 - w^2)c_1 + 2bwc_2 = 0$$

$$-2bwc_1 + (p^2 - w^2)c_2 = A$$

This system of equations has

$$c_1 = \frac{-2bw}{(p^2 - w^2)^2 + (2bw)^2}A, \quad c_2 = \frac{p^2 - w^2}{(p^2 - w^2)^2 + (2bw)^2}A$$

as its solution whenever $b \neq 0$ and $w^2 \neq p^2$. Let θ be an angle such that

$$\sin\theta = \frac{-2bw}{\left[(p^2 - w^2)^2 + (2bw)^2\right]^{1/2}}, \quad \cos\theta = \frac{p^2 - w^2}{\left[(p^2 - w^2) + (2bw)^2\right]^{1/2}}$$

Then

$$x_1(t) = \frac{-2bwA}{\left(p^2 - w^2\right)^2 + (2bw)^2} \cos wt + \frac{\left(p^2 - w^2\right)A}{\left(p^2 - w^2\right)^2 + (2bw)^2} \sin wt$$

$$= \frac{A}{\left[\left(p^2 - w^2\right)^2 + (2bw)^2\right]^{1/2}} (\sin\theta\cos wt + \cos\theta\sin wt)$$

$$= \frac{A}{\left[\left(p^2 - w^2\right)^2 + (2bw)^2\right]^{1/2}} \sin(wt + \theta) \qquad\qquad \textbf{(2.61)}$$

We will consider the behavior of the amplitude

$$A(w) = \frac{A}{\left[\left(p^2 - w^2\right)^2 + (2bw)^2\right]^{1/2}}$$

of this solution as w varies in the interval $[0, \infty)$. Clearly $A(0) = A/p^2$ and $A(w) \to 0$ as $w \to \infty$. A straightforward calculation shows that

$$A'(w) = -2wA \frac{w^2 - p^2 + 2b^2}{\left[\left(p^2 - w^2\right)^2 + (2bw)^2\right]^{3/2}}$$

Therefore

(i) $A'(w) > 0$ whenever $w^2 < p^2 - 2b^2$

(ii) $A'(w) = 0$ whenever $w = 0$ or $w^2 = p^2 - 2b^2$

(iii) $A'(w) < 0$ whenever $w^2 > p^2 - 2b^2$

If $p^2 - 2b^2 \leq 0$, then $A(w)$ is a strictly decreasing function beginning at $A(0) = A/p^2$ and tending toward zero as $w \to \infty$. The case which is of interest is $p^2 - 2b^2 > 0$. In this case $p^2 > b^2$, so (Case 1 of Section 2.6) the solutions of the associated homogeneous equation represent damped harmonic motion. The maximum value of the amplitude $A(w)$ occurs when $w = (p^2 - 2b^2)^{1/2}$, and the maximum value is

$$A\left(\sqrt{p^2 - 2b^2}\right) = \frac{A}{2b\left(p^2 - b^2\right)^{1/2}}$$

When the frequency w of the forcing function $A \sin wt$ is such that $w = (p^2 - 2b^2)^{1/2}$, then the forcing function is said to be in **resonance** with the system. That is, the forcing function is in resonance with the system if the frequency w in the forcing term maximizes the amplitude of the solution in equation (2.61). The value $(p^2 - 2b)^{1/2}/2\pi$ is called the **resonance frequency** of the system. In Section 2.6 we found that the solutions of the associated homogeneous equation have frequency $(p^2 - b^2)^{1/2}/2\pi$. Thus the resonance frequency is less than that of the unforced, damped oscillation. Notice that as the coefficient b of the damping term in the differential equation tends to zero, the maximum of $A(w)$ tends to infinity, and the value of w at which this maximum is attained tends to p. For example, Figure 2.10 shows the graphs of $A(w)$ when $A = 1$ and $p = 1$ with various choices for b.

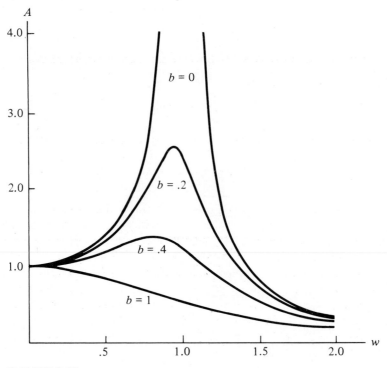

FIGURE 2.10

In the limiting case, called **pure resonance**, $w = p$, $b = 0$, and equation (2.60) becomes

$$(D^2 + w^2)y = A \sin wt \tag{2.62}$$

Since $r = iw$ is a root of the auxiliary equation $r^2 + w^2 = 0$, there is a particular solution of equation (2.62) having the form

$$x_2(t) = d_1 t \cos wt + d_2 t \sin wt$$

Inserting x_2 into equation (2.62) yields

$$A \sin wt = (D^2 + w^2)(d_1 t \cos wt + d_2 t \sin wt)$$

$$= -2d_1 w \sin wt + 2d_2 w \cos wt$$

Therefore $d_1 = -A/2w$ and $d_2 = 0$, so

$$x_2(t) = -\frac{A}{2w} t \cos wt$$

Notice that the solution x_2 is unbounded as $t \to \infty$.

The preceding discussion shows that the bounds on solutions of the forced equation

$$(D^2 + 2bD + p^2)x = A \sin wt$$

may greatly exceed the bounds on the solutions of the undamped equation

$$(D^2 + 2bD + p^2)x = 0$$

even when A is relatively small. It is because of this phenomenon that a seemingly strong structure will collapse under what appears to be a relatively small force. For example, in 1831 near Manchester, England, a column of soldiers marched in cadence across a suspension bridge. The frequency of the resulting periodic force must have equaled or nearly equaled the natural frequency of the bridge because the bridge underwent large oscillations and finally collapsed. This is why soldiers do not march in cadence across a bridge. A similar disaster in 1940 involved a bridge near Tacoma, Washington. This time, however, the periodic force was caused by wind blowing against the superstructure of the bridge.

▶ **Example 1**   Suppose that in the previous discussion $m = .5$ kilogram, $k = 4$ newtons/meter, the frictional force is twice the velocity, $r = .25$ meter, and $w = 2\pi$ radians/minute. In such a case equation (2.59) becomes

$$.5\frac{d^2x}{dt^2} + 2\frac{dx}{dt} + 4x = \sin 2\pi t$$

or, equivalently,

$$\frac{d^2x}{dt^2} + 4\frac{dx}{dt} + 8x = 2\sin 2\pi t \qquad (2.63)$$

This is equation (2.60) with $b=2$, $p^2=8$, $A=2$, and $w=2\pi$. Using equation (2.61), we find that a particular solution of equation (2.63) is

$$x_1(t) = \frac{1}{2(4+\pi^4)^{1/2}}\sin(2\pi t + \theta) \qquad (2.64)$$

where θ is such that

$$\sin\theta = \frac{-2\pi}{(4+\pi^4)^{1/2}} \quad \text{and} \quad \cos\theta = \frac{2-\pi^2}{(4+\pi^4)^{1/2}} \qquad (2.65)$$

The roots of the auxiliary equation $r^2+4r+8=0$ for equation (2.63) are $-2\pm2i$. Hence a general solution of equation (2.63) is

$$x(t) = c_1 e^{-2t}\cos 2t + c_2 e^{-2t}\sin 2t + \frac{1}{2(4+\pi^4)^{1/2}}\sin(2\pi t + \theta)$$

Suppose that at $t=0$ the system is at rest and in the position shown in Figure 2.9. Then $x(0)=0$ and $x'(0)=0$. We now determine the coefficients c_1 and c_2. We have

$$0 = x(0) = c_1 + \frac{1}{2(4+\pi^4)^{1/2}}\sin\theta$$

$$= c_1 + \frac{1}{2(4+\pi^4)^{1/2}}\frac{-2\pi}{(4+\pi^4)^{1/2}}$$

and

$$0 = x'(0) = -2c_1 + 2c_2 + \frac{2\pi}{2(4+\pi^4)^{1/2}}\cos\theta$$

$$= -2c_1 + 2c_2 + \frac{\pi}{(4+\pi^4)^{1/2}}\frac{2-\pi^2}{(4+\pi^4)^{1/2}}$$

Solving these equations for c_1 and c_2 yields

$$c_1 = \frac{\pi}{4+\pi^4} \quad \text{and} \quad c_2 = \frac{\pi^3}{8+2\pi^4}$$

Thus a solution of the initial-value problem is

$$x(t) = \frac{\pi}{4+\pi^4}e^{-2t}\cos 2t + \frac{\pi^3}{8+2\pi^4}e^{-2t}\sin 2t + \frac{1}{2(4+\pi^4)^{1/2}}\sin(2\pi t + \theta)$$

In this case

$$A(w) = \frac{2}{\left[(8-w^2)^2 + (4w)^2\right]^{1/2}}$$

$$= \frac{2}{[64+w^4]^{1/2}}$$

Evidently $A(w)$ is strictly decreasing and the forcing term $2\sin 2\pi t$ is not in resonance with the system since the maximum of $A(w)$ occurs when $w=0$. Note that there is no nonzero forcing term of the form $A\sin wt$ in resonance with the system, even though the system is not critically damped. ◀

Exercises   SECTION 2.9

1. Consider the electrical circuit shown in Figure 2.11, where R is a resistor measured in ohms, C a capacitor measured in farads, L an inductor measured in henries, and E an electromotive force measured in volts. It can be shown by Kirchhoff's laws that the charge q on the capacitor satisfies

$$L\frac{d^2q}{dt^2} + R\frac{dq}{dt} + \frac{1}{C}q = E(t)$$

(a) The portion of the solution that approaches zero as $t \to \infty$ is called the **transient** portion of the solution. The remaining portion is called the **steady-state** portion of the solution. For the case $L=1$ henry, $R=2$ ohms, $C=.1$ farad, and $E(t)=1$ volt, determine the transient and steady-state portions of the solution if $q(0)=0$ and $q'(0)=0$.

(b) For what value of R is the forcing term in resonance with the resulting system if $E(t)=\sin t$?

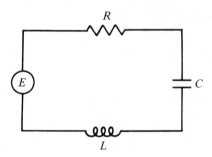

R

E

C

L

FIGURE 2.11

2. Suppose a scientific instrument is placed on blocks of rubber to absorb vibrations. If the spring force of the rubber is taken to be $-100y$, the damping force to be $-10\,dy/dt$, and we suppose that the vibrations are sinusoidal of the form $k+.3\sin 100\pi t$, then it can be shown that the deflection y of a 64 pound instrument satisfies

$$\frac{d^2y}{dt^2}+60\frac{dy}{dt}+600y=600k+180(10\pi\cos 100\pi t+\sin 100\pi t)$$

[R. Oldenburger, *Mathematical Engineering Analysis*, Macmillan, New York, 1950, page 38.] Find a general solution of this equation.

3. If P denotes the aggregate production in units per year, measured from an initial equilibrium value of a company, and if there are two consecutive time lags of $\frac{1}{8}$ year each in the response of production to demand, then it can be shown that [A. W. Phillips, Stabilization policy in a closed economy, Economic Journal, **64** (1954), 290–323]

$$(D+8)^2 P=-8^2$$

with initial conditions $P(0)=P'(0)=0$. Find P.

4. Under appropriate assumptions on the demand policy, it can be shown that when there are two consecutive time lags of $\frac{1}{4}$ year each, the equation for aggregate production in Exercise 3 becomes modified to

$$(D^2+3D+6)P=-8$$

with initial conditions $P(0)=0$, $P'(0)=-4$. Find P.

5. One model for investment growth is

$$\frac{d^2k}{dt^2}+c^2k=c^2\bar{k}$$

where k is the capital stock, $\bar{k}$ the desired level of capital stock, and c a positive constant. Show that the capital stock oscillates about the desired level $\bar{k}$. Hint: Show that a general solution of the equation is $k(t)=A\cos(ct+a)+\bar{k}$ for arbitrary constants A, a. [J. P. Lewis, *Introduction to Mathematics for Students of Economics*, Macmillan, New York, 1969, page 388.]

6. A simple pendulum of mass m and length L is hinged at a point P as shown in Figure 2.12. If the hinged point P undergoes a harmonic displacement $x(t)=A\sin wt$, then the equation of motion is

$$L\frac{d^2\theta}{dt^2}+g\sin\theta=w^2A\sin wt$$

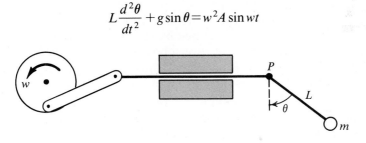

FIGURE 2.12

For small values of θ, $\sin\theta\approx\theta$, so the equation of motion may be approximated by

$$L\frac{d^2\theta}{dt^2}+g\theta=w^2A\sin wt$$

Find the solution that satisfies the initial condition $\theta(0)=\theta'(0)=0$ for the special case $L=1$ meter, $g=9.8$ meter/second^2, $A=1$, and $w=1$ radian/second.

7. Suppose that a straight, light, elastic rod has one end clamped in the chuck of a lathe. To the other end is attached an imperfectly balanced disk of mass m. The distance from the center of mass of the disk to the axis of the rod will be denoted by h. (Since the rod is imperfectly balanced, $h\neq0$.) We wish to determine the motion of the disk when the rod is rotated. This system has two degrees of freedom, x and y, in the plane in which the rotating disk vibrates. We will assume that a force kz is required to deflect the rod a distance z in any direction. If the rod is rotated with a constant angular velocity w, then the horizontal, x, and vertical, y, deflections of the disk can be shown to satisfy the equations

$$m\frac{d^2x}{dt^2} + kx = hw^2\cos wt$$

$$m\frac{d^2y}{dt^2} + ky = hw^2\sin wt$$

[H. McCallion, *Vibration of Linear Mechanical Systems*, Halsted Press, New York, 1973, pages 207–210.] Assuming that the disk begins at rest (that is, $x(0)=x'(0)=0$ and $y(0)=y'(0)=0$), find the deflections $x(t)$ and $y(t)$ at time t.

8. A flagellum is a taillike part of an organism such as a bacterium or protozoan. The organism propels itself by wavelike motions of the flagellum. These waves can be generated by the propagation along the flagellum of rather abrupt transitions between bent and straight states. One study of this wave motion [C. J. Brokow, Bend propagation along flagella, Nature, **209** (1966), 161–163] gives the following differential equation for the curvature $C(s)$ of the flagellum at a distance s along the flagellum:

$$\frac{d^2C}{ds^2} - a^2C = \begin{cases} -a^2C_0 & \text{if } s<0 \\ 0 & \text{if } 0<s \end{cases}$$

where C_0 is the curvature that will result if there are only contractive forces, and $0<a$ is a constant depending on physical properties of the flagellum. (The curvature C of a curve at a point P on it is the rate of change in the angle of inclination of the tangent line to the curve at P with respect to arc length.) Find the solution to this equation if we require that $\lim_{s\to-\infty}C(s)=C_0$, $\lim_{s\to+\infty}C(s)=0$, and both C and dC/ds are continuous. Hint: Consider the cases $s<0$ and $0<s$ separately. Then choose the constants so that the two functions can be pieced together in a continuously differentiable fashion at $s=0$.

2.10 Inverse Differential Operators
(OPTIONAL)

The method of undetermined coefficients requires tedious calculations even in relatively simple problems. In this section we present a method that eliminates most of these calculations.

Suppose we wish to find a particular solution of the differential equation

$$f(D)=g(x) \tag{2.66}$$

where $f(z)=z^2+a_1z+a_0$. It is natural to write

$$y(x)=\frac{1}{f(D)}g(x) \tag{2.67}$$

and try to define $1/f(D)$ so that the function given in (2.67) is a solution of equation (2.66). That is, we want to define $1/f(D)$ so that

$$f(D)\left[\frac{1}{f(D)}g(x)\right]=g(x)$$

The quantity $1/f(D)$ is called an **inverse differential operator**. In Exercises 15 and 16 the reader is asked to verify the following fundamental properties of inverse differential operators:

(i) $\dfrac{1}{f(D)}[cg(x)]=c\dfrac{1}{f(D)}g(x)$ for every constant c

(ii) $\dfrac{1}{f(D)}[g(x)+h(x)]=\dfrac{1}{f(D)}g(x)+\dfrac{1}{f(D)}h(x)$

Rather than give a general theory of inverse differential operators, we will proceed by merely manipulating symbols and considering only a few specific choices for the function $g(x)$. This will lead to a tentative value of $[1/f(D)]g(x)$. We then verify that this tentative choice is a solution of equation (2.66).

Before beginning this discussion, we list three identities that can be easily verified by direct computation:

$$(D-b)(e^{ax})=(a-b)e^{ax} \tag{2.68}$$

$$(D-a)(xe^{ax})=e^{ax} \tag{2.69}$$

$$(D-a)^2(\tfrac{1}{2}x^2e^{ax})=e^{ax} \tag{2.70}$$

We now determine $[1/f(D)]g(x)$ in the special case $g(x)=e^{ax}$. Notice that

$$f(D)e^{ax}=(D^2+a_1D+a_0)e^{ax}$$

$$=D^2e^{ax}+a_1De^{ax}+a_0e^{ax}$$

$$=a^2e^{ax}+a_1ae^{ax}+a_0e^{ax}$$

$$=f(a)e^{ax} \tag{2.71}$$

This suggests that

$$\frac{1}{f(D)}e^{ax}=\frac{1}{f(a)}e^{ax} \tag{2.72}$$

whenever $f(a)\neq 0$. Using the identities in (2.71) and (2.72), we have

$$f(D)\left(\frac{1}{f(D)}e^{ax}\right)=f(D)\left(\frac{1}{f(a)}e^{ax}\right)$$

$$=\frac{1}{f(a)}\left(f(D)e^{ax}\right)$$

$$=e^{ax}$$

so that $[1/f(D)]e^{ax}$ as given in (2.72) is a particular solution of

$$f(D)y=e^{ax}$$

whenever $f(a)\neq 0$.

If $f(a)=0$, then there are two cases to consider: $f(z)=(z-b)(z-a)$ with $a\neq b$ and $f(z)=(z-a)^2$. In the first case we can use the identities in (2.68) and (2.69) to obtain

$$f(D)(xe^{ax})=(D-b)(D-a)(xe^{ax})$$

$$=(D-b)e^{ax}$$

$$=(a-b)e^{ax}$$

Thus

$$\frac{1}{f(D)}e^{ax}=\frac{1}{a-b}xe^{ax} \tag{2.73}$$

whenever $f(z)=(z-b)(z-a)$ with $a\neq b$. Similarly, if $f(z)=(z-a)^2$ we can use the identity in (2.70) to obtain

$$f(D)\left(\tfrac{1}{2}xe^{ax}\right)=(D-a)^2\left(\tfrac{1}{2}xe^{ax}\right)$$

$$=e^{ax}$$

Thus

$$\frac{1}{f(D)}e^{ax}=\tfrac{1}{2}xe^{ax}$$

whenever $f(z)=(z-a)^2$.

▶ **Example 1** Consider the differential equation

$$(D^2+3D+1)y=4e^{-2x} \qquad\qquad \textbf{(2.74)}$$

Here $f(z)=z^2+3z+1$ and $f(-2)=-1\neq0$. Using equation (2.72) we have

$$y(x)=\frac{1}{D^2+3D+1}(4e^{-2x})$$

$$=4\frac{1}{f(-2)}e^{-2x}$$

$$=-4e^{-2x}$$

as a particular solution of equation (2.74). ◀

▶ **Example 2** Consider the differential equation

$$(D^2-D-6)y=e^{3x} \qquad\qquad \textbf{(2.75)}$$

Here $f(z)=z^2-z-6=(z+2)(z-3)$. Using equation (2.73) we have

$$y(x)=\frac{1}{D^2-D-6}e^{3x}$$

$$=\frac{1}{(D+2)(D-3)}e^{3x}$$

$$=\tfrac{1}{5}xe^{3x}$$

as a particular solution of equation (2.75). ◀

The identity in 2.72 holds even if a is a complex number. In particular, when $a=ic$ and $a=-ic$, where c is a real number, we have that the functions

$$y_1(x)=\frac{1}{f(ic)}e^{icx}=\frac{1}{f(ic)}(\cos cx+i\sin cx)$$

and

$$y_2(x)=\frac{1}{f(-ic)}e^{-icx}=\frac{1}{f(-ic)}(\cos cx-i\sin cx)$$

are particular solutions of

$$f(D)y=e^{icx}=\cos cx+i\sin cx$$

and

$$f(D)y=e^{-icx}=\cos cx-i\sin cx$$

respectively. By the principle of superposition (Theorem 2.5), the function

$$y(x)=y_1(x)+y_2(x)$$

$$=\frac{\cos cx+i\sin cx}{f(ic)}+\frac{\cos cx-i\sin cx}{f(-ic)}$$

$$=\frac{\cos cx+i\sin cx}{(a_0-c^2)+a_1ci}\frac{(a_0-c^2)-a_1ci}{(a_0-c^2)-a_1ci}$$

$$+\frac{\cos cx-i\sin cx}{(a_0-c^2)-a_1ci}\frac{(a_0-c^2)+a_1ci}{(a_0-c^2)+a_1ci}$$

$$=2\frac{(a_0-c^2)\cos cx+a_1c\sin cx}{(a_0-c^2)^2+(a_1c)^2}$$

is a particular solution of

$$f(D)y=e^{icx}+e^{-icx}$$

$$=2\cos cx$$

whenever $f(\pm ic)\neq0$. Therefore

$$\frac{1}{f(D)}\cos cx=\frac{(a_0-c^2)\cos cx+a_1c\sin cx}{(a_0-c^2)^2+(a_1c)^2}\qquad(2.76)$$

whenever $f(\pm ic)\neq0$. By considering the difference between $y_1(x)$ and $y_2(x)$, one can show that

$$\frac{1}{f(D)}\sin cx=\frac{(a_0-c^2)\sin cx-a_1c\cos cx}{(a_0-c^2)^2+(a_1c)^2}$$

whenever $f(\pm ic)\neq0$.

▶ **Example 3** Consider the differential equation

$$(D^2 + 2D + 3)y = \cos 4x \qquad (2.77)$$

Here $f(z) = z^2 + 2z + 3$, $a_1 = 2$, $a_0 = 3$, and $c = 4$. Evidently $f(\pm 4i) \neq 0$, so we may use the identity in (2.76) to find that a particular solution of equation (2.77) is given by

$$y(x) = \frac{1}{D^2 + 2D + 3} \cos 4x$$

$$= \frac{-13\cos 4x + 8\sin 4x}{233} \qquad ◀$$

Exercises SECTION 2.10

In Exercises 1–14 use the techniques of this section to find a ~~particular~~ *general* solution of the given differential equation.

1. $(D^2 + 2D + 1)y = e^{3x}$
2. $(D^2 + 3D - 1)y = 5e^{-2x}$
3. $(D^2 + 3D + 2)y = 7e^{-x}$
4. $(D^2 - 5D + 6)y = 3e^{3x}$
5. $(D^2 + 1)y = 2e^x$
6. $(D^2 - 4)y = e^{3x}$
7. $(D^2 + 6D + 9)y = 9e^{-3x}$
8. $(D^2 - 10D + 25)y = e^{5x}$
9. $(D^2 + D + 1)y = \cos 3x$
10. $(D^2 - 4)y = 7\sin 2x$
11. $(D^2 + 2D - 5)y = \sin x$
12. $(D^2 + 5D + 7)y = 5\cos 4x$
13. $(D^2 - D - 1)y = 2\cos x$
14. $(D^2 + 1)y = 4\sin 5x$

15. Show that $\dfrac{1}{f(D)}[cg(x)] = c\dfrac{1}{f(D)}g(x)$ for every constant c.

16. Use the principle of superposition (Theorem 2.5) to show that

$$\frac{1}{f(D)}[g(x) + h(x)] = \frac{1}{f(D)}g(x) + \frac{1}{f(D)}h(x)$$

In Exercises 17–30 use the techniques of this section and Exercise 16 to find a particular solution of the given differential equation.

17. $(D^2-3D+2)y=3e^{2x}+2e^{-x}$

18. $(D^2+3D-2)y=2e^{3x}-7e^{5x}$

19. $(D^2-3D+2)y=4e^x-5e^{2x}$

20. $(D^2+5D+4)y=e^{-x}+3e^{-4x}$

21. $(D^2+1)y=2e^{6x}+\sin 2x$

22. $(D^2+4D+1)y=e^{-3x}+\cos 3x$

23. $(D^2-4)y=e^{2x}-\sin x$

24. $(D^2-4)y=e^{-2x}+\cos x$

25. $(D^2+2D+1)y=\cos x+3\sin x$

26. $(D^2+11D-7)y=2\cos 3x-\sin 4x$

27. $(D^2-D-6)y=\cos x-\cos 2x$

28. $(D^2+D+2)y=\sin x-3\cos 2x$

29. $(D^2+7D+5)y=2e^x-4\cos 2x+\sin x$

30. $(D^2-5D+6)y=3e^x+\sin 2x-4e^{-2x}$

31. Use inverse differential operators to solve the problem in Example 1 of Section 2.9.

32. Use inverse differential operators to solve the problem in Exercise 6 of Section 2.9.

33. Use inverse differential operators to solve the problem in Exercise 7 of Section 2.9.

2.11 Reduction of Order

Thus far we have learned how to find a general solution of a nonhomogeneous differential equation

$$\left(D^2+a_1(x)D+a_0(x)\right)y=g(x) \tag{2.78}$$

whenever a_1 and a_0 are constant functions and g has one of the special forms that permit the use of the method of undetermined coefficients. In general, there is no method for finding a particular solution of equation (2.78), let alone a general solution. However, if one nonzero solution y_1 (obtained by some unspecified method or divination) of the homogeneous equation

$$\left(D^2+a_1(x)D+a_0(x)\right)y=0 \tag{2.79}$$

is known, it is possible to construct a second solution that is linearly independent of y_1. Thus a general solution of equation (2.79) can be constructed from any nonzero solution of the equation. A generalization of this method can be used to construct a particular solution of equation (2.78) from a general solution of equation (2.79). We will first consider a method of obtaining a general solution of (2.79) from a nonzero solution, and then in the next section go on to find a particular solution of (2.78) from a general solution of (2.79).

Suppose we know one nonzero solution y_1 of equation (2.79) on an open interval I, and we wish to find a second solution y_2 that is linearly independent of y_1. We begin by assuming that y_2 can be written in the form

$$y_2(x)=v(x)y_1(x) \tag{2.80}$$

where v is a function to be determined subsequently. If this choice for y is inserted into equation (2.79), we obtain

$$0=\left(D^2+a_1(x)D+a_0(x)\right)vy_1$$

$$=\left(D^2y_1+a_1(x)Dy_1+a_0(x)y_1\right)v+\left(y_1D^2+\left(2y_1'+y_1a_1(x)\right)D\right)v \tag{2.81}$$

Since y_1 is a solution of equation (2.79), (2.81) becomes

$$\left(y_1D^2+\left(2y_1'+a_1(x)y_1\right)D\right)v=0 \tag{2.82}$$

Thus y_2 is a solution of (2.79) if and only if v is a solution of (2.82). Equation (2.82) is a first-order linear differential equation for Dv. Setting $w=Dv$, equation (2.82) becomes

$$Dw+\left[2\frac{y_1'}{y_1}+a_1(x)\right]w=0$$

which is a first-order linear differential equation for w. If $P(x)$ is any antiderivative of

$$p(x)=2\frac{y_1'}{y_1}+a_1(x)$$

on the interval I, then

$$w(x)=e^{-P(x)}+c \tag{2.83}$$

The antiderivative $P(x)$ may be determined as follows:

$$P(x)=\int\left[2\frac{y_1'(x)}{y_1(x)}+a_1(x)\right]dx$$

$$=\ln\left(y_1(x)\right)^2+\int a_1(x)\,dx \tag{2.84}$$

If $A(x)$ is any antiderivative of $a_1(x)$ on the interval I, then from equations (2.83) and (2.84) we have

$$Dv=w(x)=\frac{1}{\left(y_1(x)\right)^2}e^{-A(x)}+c \tag{2.85}$$

Thus we are able to determine Dv and, by integration, v. When this choice for the function v is used in equation (2.80), we obtain a solution y_2 of equation (2.79) that is linearly independent of y_1, provided the constants of integration are chosen so that v is not a constant.

In the following example we will illustrate the steps leading to equation (2.85) using specific choices for the functions a_1 and y_1.

▶ **Example 1**　Consider the differential equation

$$(D^2+x^{-1}D-x^{-2})y \tag{2.86}$$

By inspection, $y_1(x)=x$ is a solution of equation (2.86). If we assume that another solution has the form $y_2(x)=vx$, then (2.86) becomes

$$0=(D^2+x^{-1}D-x^{-2})vx$$

$$=(xD^2+3D)v$$

or, equivalently,

$$\left(D^2+\frac{3}{x}D\right)v=0 \tag{2.87}$$

which is a first-order linear equation for Dv. Setting $w=Dv$, we can rewrite equation (2.87) as

$$\frac{dw}{dx}+\frac{3}{x}w=0$$

The solution of this equation is easily shown to be $w(x) = cx^{-3}$. Then

$$\frac{dv}{dx} = Dv = w = cx^{-3}$$

and

$$v(x) = -\frac{c}{2}x^{-2} + d$$

Since we are looking for a nonconstant function v, any choice of d and any nonzero choice of c will suffice. For simplicity we will set $c = -2$ and $d = 0$ so that

$$v(x) = x^{-2}$$

Then $y_2(x) = xv(x) = x^{-1}$ is a solution of equation (2.86) that is linearly independent of y_1. A general solution of equation (2.86) is

$$y(x) = c_1 x + c_2 x^{-1} \qquad \blacktriangleleft$$

The method described above and illustrated in Example 1 is often called **reduction of order**.

▶ **Example 2** It is not always possible to evaluate the indefinite integral in order to obtain the function v in equation (2.85). For example, it can be verified directly that $y_1(x) = e^{x^2}$ is a solution on $(0, \infty)$ of the homogeneous differential equation

$$(D^2 - 2xD - 2)y = 0 \qquad \textbf{(2.88)}$$

We will use the method of reduction of order to find a solution that has the form $y_2 = vy_1$ and that is linearly independent of y_1. From equation (2.85) we have

$$Dv = \frac{1}{e^{2x^2}}e^{-\int 2x\,dx + c}$$

$$= \frac{e^{-x^2 + c}}{e^{2x^2}}$$

$$= e^{-3x^2}e^{c}$$

Since we are looking for a nonconstant function v, any choice of c will suffice. For simplicity we set $c=0$. Then

$$v(x)=\int e^{-3x^2}\,dx$$

This integral cannot be evaluated in terms of finite sums, products, or quotients of polynomials, exponentials, logarithms, or trigonometric functions. A general solution of equation (2.88) is

$$y(x)=c_1 e^{x^2}+c_2 e^{x^2}\int e^{-3x^2}\,dx \qquad\qquad \blacktriangleleft$$

Exercises   SECTION 2.11

In Exercises 1–10 use the method of reduction of order to find a general solution of the given differential equation.

1. $(D^2+x^{-1}D-4x^{-2})y=0,\quad y_1(x)=x^2$
2. $(D^2-x^{-1}D+x^{-2})y=0,\quad y_1(x)=x$
3. $(D^2+3x^{-1}D+x^{-2})y=0,\quad y_1(x)=x^{-1}$
4. $(D^2-2x^{-2})y=0,\quad y_1(x)=x^{-1}$
5. $(D^2-6x^{-2})y=0,\quad y_1(x)=x^3$
6. $(D^2+x^{-1}D+(1-\tfrac{1}{4}x^{-2}))y=0,\quad y_1(x)=x^{-1/2}\cos x$
7. $(D^2-2x^{-1}D+\tfrac{9}{4}x^{-2})y=0,\quad y_1(x)=x^{3/2}$
8. $(D^2+3x^{-1}D+x^{-2})y=0,\quad y_1(x)=x^{-1}$
9. $(D^2-3x^{-1}D+4x^{-2})y=0,\quad y_1(x)=x^2$
10. $(D^2+5x^{-1}D+4x^{-2})y=0,\quad y_1(x)=x^{-2}$

2.12 Variation of Parameters

We will now show that the method of reduction of order can be modified so that a solution on an interval I of the nonhomogeneous differential equation

$$\left(D^2+a_1(x)D+a_0(x)\right)y=f(x) \qquad\qquad \textbf{(2.89)}$$

can be obtained whenever we know a general solution on I of the homogeneous differential equation

$$\left(D^2 + a_1(x)D + a_0(x)\right)y = 0 \qquad \textbf{(2.90)}$$

Suppose that y_1 and y_2 are linearly independent solutions of equation (2.90). We begin by assuming that a solution of equation (2.89) has the form

$$\psi(x) = v_1(x)y_1(x) + v_2(x)y_2(x) \qquad \textbf{(2.91)}$$

where v_1 and v_2 are functions to be determined subsequently. If y is replaced by ψ in equation (2.89), we obtain (where $'$ and $''$ denote the first and second derivatives with respect to x, respectively)

$$f(x) = \left(D^2 + a_1(x)D + a_0(x)\right)\left(v_1 y_1 + v_2 y_2\right)$$

$$= v_1\left(y_1'' + a_1(x)y_1' + a_0(x)y_1\right) + v_2\left(y_2'' + a_1(x)y_2' + a_0(x)y_2\right)$$

$$+ \left(v_1'' y_1 + v_1' y_1' + v_2'' y_2 + v_2' y_2'\right) + v_1' y_1' + v_2' y_2' + a_1(x)\left(v_1' y_1 + v_2' y_2\right)$$

$$\textbf{(2.92)}$$

Since y_1 and y_2 are solutions of equation (2.90) and since $v_1'' y_1 + v_1' y_1' + v_2'' y_2 + v_2' y_2' = (v_1' y_1 + v_2' y_2)'$, equation (2.92) can be written as

$$f(x) = \left(v_1' y_1 + v_2' y_2\right)' + a_1(x)\left(v_1' y_1 + v_2' y_2\right) + v_1' y_1' + v_2' y_2' \qquad \textbf{(2.93)}$$

We wish to determine two functions, v_1 and v_2. Equation (2.93) gives us only one equation with which to determine these two functions. We would like to find a second equation that v_1 and v_2 satisfy so that we have two equations and two unknowns. In equation (2.93) we notice that two of the three summands contain the term $v_1' y_1 + v_2' y_2$ or its derivative. If we make the additional assumption that $v_1' y_1 + v_2' y_2 \equiv 0$, then equation (2.93) becomes $f(x) = v_1' y_1' + v_2' y_2'$ and we have the following pair of equations for v_1' and v_2':

$$v_1' y_1 + v_2' y_2 = 0$$

$$v_1' y_1' + v_2' y_2' = f(x) \qquad \textbf{(2.94)}$$

This is a system of two equations in two unknowns, v_1' and v_2', that can be solved for v_1' and v_2'. Doing this, we have

$$v_1'(x) = -\frac{f(x)y_2(x)}{y_1(x)y_2'(x) - y_1'(x)y_2(x)}$$

$$v_2'(x) = \frac{f(x)y_1(x)}{y_1(x)y_2'(x) - y_1'(x)y_2(x)} \qquad \textbf{(2.95)}$$

Note that the denominators of v_1' and v_2' are identical. This expression, denoted by $W[y_1, y_2](x)$, is called the **Wronskian** of y_1 and y_2:

$$W[y_1, y_2](x) = y_1(x)y_2'(x) - y_1'(x)y_2(x)$$

In Exercise 8 it is shown that $W[y_1, y_2](x) \neq 0$ for every x in I whenever y_1 and y_2 are linearly independent. Using equation (2.91) along with the integrals of the functions in (2.95), we find that

$$y(x) = -y_1(x)\int \frac{f(x)y_2(x)}{W[y_1, y_2](x)} dx + y_2(x)\int \frac{f(x)y_1(x)}{W[y_1, y_2](x)} dx \quad \text{(2.96)}$$

is a particular solution of (2.89).

The method described above and illustrated in the following examples is called **variation of parameters**. In the following example we will illustrate the steps leading from the equations in (2.94) to equation (2.96) with specific choices for the functions f, y_1, and y_2.

▶ **Example 1**   Consider the equation

$$(D^2 + 1)y = \tan x \qquad \left(-\frac{\pi}{2} < x < \frac{\pi}{2}\right) \qquad \text{(2.97)}$$

Evidently $y_1(x) = \cos x$ and $y_2(x) = \sin x$ are solutions of the homogeneous equation $(D^2 + 1)y = 0$. With $y_1(x) = \cos x$, $y_2(x) = \sin x$, and $f(x) = \tan x$, the equations in (2.94) become

$$v_1' \cos x + v_2' \sin x = 0 \qquad \text{(2.98)}$$

$$-v_1' \sin x + v_2' \cos x = \tan x \qquad \text{(2.99)}$$

From equation (2.98) we obtain

$$v_2' = -v_1' \frac{\cos x}{\sin x} \qquad \text{(2.100)}$$

Inserting this value of v_2' into equation (2.99) yields

$$-v_1' \sin x + \left(-v_1' \frac{\cos x}{\sin x}\right)\cos x = \tan x$$

If we multiply each side of this equation by $-\sin x$ and use the trigonometric identity $\sin^2 x + \cos^2 x = 1$, we find that

$$v_1' = -\tan x \sin x$$

$$= \cos x - \sec x$$

Then, using equation (2.100),

$$v_2' = -(\tan x \sin x)\frac{\cos x}{\sin x}$$

$$= \sin x$$

Hence

$$v_1(x) = \sin x - \ln|\sec x + \tan x| + b_1$$

$$v_2(x) = -\cos x + b_2$$

where b_1 and b_2 are constants of integration. Since we need only one particular choice for v_1 and v_2, we set $b_1 = b_2 = 0$ for simplicity. Consequently

$$\psi(x) = v_1(x)\cos x + v_2(x)\sin x$$

$$= (\sin x - \ln|\sec x + \tan x|)\cos x + (-\cos x)\sin x$$

$$= -(\cos x)\ln|\sec x + \tan x|$$

is a particular solution of equation (2.97). A general solution is

$$y(x) = c_1\cos x + c_2\sin x - (\cos x)\ln|\sec x + \tan x| \qquad \blacktriangleleft$$

▶ **Example 2**   Consider the equation

$$(D^2 + 2D + 1)y = \frac{e^{-x}}{x} \qquad (0 < x) \qquad \text{(2.101)}$$

Evidently $y_1(x) = e^{-x}$ and $y_2(x) = xe^{-x}$ are solutions of the homogeneous equation $(D^2 + 2D + 1)y = 0$. The Wronskian of these two solutions is

$$W[y_1, y_2](x) = y_1(x)y_2'(x) - y_1'(x)y_2(x)$$

$$= e^{-x}(e^{-x} - xe^{-x}) - (-e^{-x})(xe^{-x})$$

$$= e^{-2x}$$

With $f(x) = x^{-1}e^{-x}$, equation (2.96) becomes

$$y(x) = -e^{-x}\int dx + xe^{-x}\int x^{-1}dx$$

$$= -e^{-x}b_1 + xe^{-x}(\ln x + b_2)$$

where b_1 and b_2 are constants of integration. Since we need only one particular solution, we will set $b_1 = b_2 = 0$. Consequently a general solution of equation (2.101) is

$$y(x) = c_1 e^{-x} + c_2 xe^{-x} + xe^{-x}\ln x \qquad \blacktriangleleft$$

▶ **Example 3**   Consider the differential equation

$$(D^2 + x^{-1}D - x^{-2})y = x^{-2}e^x \qquad \text{(2.102)}$$

In Example 1 of Section 2.11 we found that $y_1(x) = x$ and $y_2(x) = x^{-1}$ are linearly independent solutions of the homogeneous equation. We will use the method of variation of parameters to find a particular solution of equation (2.102). Before using equation (2.96) to obtain a particular solution, we will compute the Wronskian of y_1 and y_2:

$$W[y_1, y_2](x) = y_1(x)y_2'(x) - y_1'(x)y_2(x)$$

$$= x(-x^{-2}) - 1(x^{-1}) = -2x^{-1}$$

Now equation (2.96) becomes

$$y(x) = -x \int \frac{x^{-2}e^x x^{-1}}{-2x^{-1}} dx + x^{-1} \int \frac{x^{-2}e^x x}{-2x^{-1}} dx$$

$$= \frac{x}{2} \int x^{-2}e^x dx - \frac{x^{-1}}{2} \int e^x dx$$

$$= \frac{x}{2} \int x^{-2}e^x dx - \frac{x^{-1}}{2} e^x + cx^{-1}$$

where c is a constant of integration. Since we need only one particular solution, we set $c = 0$. Consequently, a general solution of equation (2.102) is

$$y(x) = c_1 x + c_2 x^{-1} + \frac{x}{2} \int x^{-2}e^x dx - \frac{x^{-1}}{2} e^x \qquad \blacktriangleleft$$

Exercises   SECTION 2.12

In Exercises 1–5 find a general solution of the given equation.

1. $(D^2 + 1)y = \sec x$, $-\pi/2 < x < \pi/2$

2. $(D^2 - 4D + 4)y = \dfrac{e^{2x}}{x}$, $0 < x$

3. $(D^2 - 6D + 9)y = (\ln x)e^{3x}, \quad 0 < x$

4. $(D^2 + 9)y = \tan^2(3x), \quad -\pi/6 < x < \pi/6$

5. $(D^2 - 2x^{-1}D + 2x^{-2})y = x^{-3}\ln x, \quad 0 < x$

Hint: Find solutions of the associated homogeneous equation having the form $y(x) = x^r$.

6. Given that $y_1(t) = t^{1/2}$ is a solution of

$$\left(D^2 + \frac{1}{4t^2}\right)y = 0, \quad 0 < t$$

Find a general solution of

$$\left(D^2 + \frac{1}{4t^2}\right)y = \frac{1}{t^{1/2}}$$

7. Show that a general solution of

$$(D^2 + 1)y = f(x)$$

is

$$y(x) = c_1\cos x + c_2\sin x + \int_0^x f(t)\sin(x - t)\, dt$$

Hint: Use variation of parameters and the trigonometric identity $\cos t \sin x - \sin t \cos x = \sin(x - t)$.

8. Let y_1 and y_2 be solutions of equation (2.90) on an open interval I. Show that either $W[y_1, y_2](x) = 0$ for every x in I or $W[y_1, y_2](x) \neq 0$ for every x in I. Hint: Show that $W[y_1, y_2](x)$ is a solution of $y' = -a_1 y$.

9. Let y_1 and y_2 be linearly independent solutions on an open interval I of $[D^2 + a_1(x)D + a_0(x)]y = 0$, where a_1 and a_0 are continuous functions on I. Show that $W[y_1, y_2](x) \neq 0$ for every x in I. Hint: Suppose that $W[y_1, y_2](x_0) = 0$ for some x_0 in I. Use Exercise 2 of Appendix 2 to show that there are constants c_1 and c_2, not both zero, such that $c_1 y_1(x_0) + c_2 y_2(x_0) = 0$ and $c_1 y_1'(x_0) + c_2 y_2'(x_0) = 0$. Consider the initial-value problem $[D^2 + a_1(x)D + a_0(x)]y = 0, \ y(x_0) = 0, \ y'(x_0) = 0$.

10. Let y_1 and y_2 be solutions on an open interval I of $(D^2 + a_1(x)D + a_0(x))y = 0$, where a_0 and a_1 are continuous functions on I. Suppose that $W[y_1, y_2](x_0) \neq 0$ for some x_0 in I. Show that y_1 and y_2 are linearly independent. Hint: If $c_1 y_1(x) + c_2 y_2(x) = 0$ for every x in I, then $c_1 y_1'(x) +$

$c_2 y_2'(x) = 0$ for every x in I. Solve the system of equations

$$c_1 y_1(x_0) + c_2 y_2(x_0) = 0$$

$$c_1 y_1'(x_0) + c_2 y_2'(x_0) = 0$$

for c_1 and c_2.

2.13 The Delta Function

In physics and engineering one encounters forces that act for a very short period of time. For example, if a mass m is set in motion by a sudden blow, it attains a momentum mv equal to the impulse of the blow. That is,

$$mv = \int_0^{t_0} F(t)\, dt$$

where F is the force exerted on the object by the blow and t_0 is the length of time the force acts on the object. The term "blow" implies that t_0 is so small that the change in momentum occurs almost instantaneously. Since the change in momentum is finite and positive, the force F must be very large during the time interval in which it acts. The graph of such a function is shown in Figure 2.13.

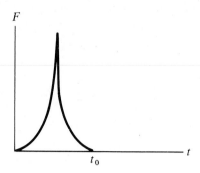

FIGURE 2.13

It turns out to be easier to deal with an idealized situation in which F is a "function" such that

(i) $F(t) = 0$ for $t \neq 0$

(ii) $\int_{-\infty}^{\infty} F(t)\, dt$ is finite and nonzero

Clearly no such function exists. But physicists have shown that meaningful results can be obtained by ignoring the logical difficulties and proceeding as if such a function existed. At the urging of these physicists, a theory was constructed by L. Schwartz and others to make rigorous sense of the calculations. We will proceed informally and ignore the rather complicated rigorous theory. The best known such function is the (Dirac) **delta function**, $\delta(t)$, which the physicist P. A. M. Dirac introduced to ease calculations in mathematical physics and quantum mechanics. The delta function is a "function" such that

$$\delta(t)=0 \quad \text{for } t\neq0$$

and

$$\int_{-\infty}^{\infty} \delta(t)\,dt=1 \tag{2.103}$$

The basic property of the delta function is that

$$\int_{-\varepsilon}^{\varepsilon} f(t)\delta(t)\,dt=f(0) \tag{2.104}$$

for every $\varepsilon>0$ and every continuous function f on $(-\infty,\infty)$. For our purposes we may treat the delta function as though it were a function in the usual sense (even though it is not one) and still obtain valid results.

Consider the simple mechanical system (Figure 2.14) consisting of a mass m of .5 kilogram attached to a spring with constant $k=4.5$ newtons/meter.

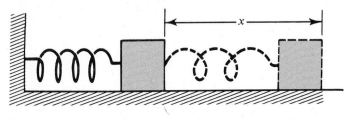

FIGURE 2.14

Suppose that the mass is initially at rest at its equilibrium position and that it is then set into motion by being struck with a force of 2 newtons. Then the equation of motion is

$$\left(D^2+\frac{4.5}{.5}\right)x=\frac{2}{.5}\delta(t) \tag{2.105}$$

(where $D=d/dt$). The functions $x_1(t)=\cos 3t$ and $x_2(t)=\sin 3t$ are linearly independent solutions of the differential equation $(D^2+9)x=0$, whose

Wronskian is

$$W[y_1, y_2](x) = x_1(t)x_2'(t) - x_1'(t)x_2(t)$$

$$= \cos 3t(3\cos 3t) - (-3\sin 3t)\sin 3t$$

$$= 3$$

We now use the method of variation of parameters to find a particular solution of equation (2.105). With $f(x) = 4\delta(t)$, the equations in (2.95) become

$$v_1'(t) = -\tfrac{4}{3}(\sin 3t)\delta(t)$$

$$v_2'(t) = \tfrac{4}{3}(\cos 3t)\delta(t)$$

When $t<0$ the mass is at rest. Hence $x(t)\equiv 0$ is the solution of equation (2.105) whenever $t<0$. This means that $v_1(t)\equiv 0$ and $v_2(t)\equiv 0$ for $t<0$. Then for $t>0$

$$v_1(t) - v_1(-t) = \int_{-t}^{t} v_1'(s)\, ds$$

$$= \int_{-t}^{t} -\tfrac{4}{3}(\sin 3s)\delta(s)\, ds$$

$$= -\tfrac{4}{3}\sin(0)$$

$$= 0$$

and

$$v_2(t) - v_2(-t) = \int_{-t}^{t} v_2'(s)\, ds$$

$$= \int_{-t}^{t} \tfrac{4}{3}(\cos 3s)\delta(s)\, ds$$

$$= \tfrac{4}{3}\cos(0) = \tfrac{4}{3}$$

Consequently

$$v_1(t) = 0 \quad \text{for all } t$$

and

$$v_2(t) = \frac{4}{3}\begin{cases} 0 & \text{for } t<0 \\ 1 & \text{for } 0<t \end{cases}$$

Therefore

$$x(t)=c_1\cos 3t+c_2\sin 3t+v_2(t)\sin 3t$$

is a general solution of equation (2.105) on $(0,\infty)$. We note that $v_2(t)$ is multiplied by $\sin 3t$ so that the value assigned to $v_2(0)$ is immaterial so far as $x(t)$ is concerned. We will set $v_2(0)=0$ so that $v_2(t)=\frac{4}{3}H(t)$ where

$$H(t)=\begin{cases} 0 & \text{for } t\leqslant 0 \\ 1 & \text{for } 0<t \end{cases}$$

is the **Heaviside function.** Proceeding formally, we have

$$\int_{-x}^{x} H'(t)\,dt=H(t)\Big|_{-x}^{x}$$

$$=H(x)-H(-x)=1$$

for every positive x. Letting $x\to\infty$, we find that

$$\int_{-\infty}^{\infty} H'(t)\,dt=1$$

Thus $H'(t)$ satisfies the properties of the delta function $\delta(t)$ in (2.103). In fact, it can be shown that in an appropriate sense $H'(t)=\delta(t)$.

We now need to calculate the constants c_1 and c_2 in the solution:

$$x(t)=c_1\cos 3t+c_2\sin 3t+\tfrac{4}{3}H(t)\sin 3t \qquad \textbf{(2.106)}$$

Since the mass was initially at its equilibrium position, we have $x(0)=0$. For very small positive values of t the forces on the mass due to the spring are negligible. Hence for very small values of t,

$$mx'(t)-mx'(-t)=\int_{-t}^{t} mx''(s)\,ds$$

$$=\int_{-t}^{t} (\text{force})\,ds$$

$$\approx\int_{-t}^{t} 2\delta(s)\,ds=2$$

Since the mass, $m=.5$, was initially at rest, we have $x'(s)=0$ for $s<0$, so $x'(t)\approx 4$ for very small values of t. We conclude that $x'(0)=4$. In addition we will require that x' is continuous at $t=0$ in the sense that $\lim_{t\to 0_+}x'(t)=4$.

From (2.106) we have

$$x'(t) = -3c_1 \sin 3t + 3c_2 \cos 3t + 4H'(t)\sin 3t + 4H(t)\cos 3t$$

so

$$4 = \lim_{t \to 0_+} x'(t) = 3c_2 + 4$$

Hence $c_2 = 0$. Moreover, $0 = x(0) = c_1$, so the solution of the problem is

$$x(t) = \tfrac{4}{3} H(t)\sin 3t$$

The graph of the function x is shown in Figure 2.15.

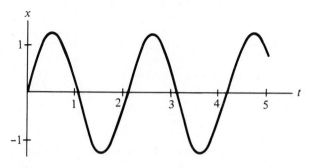

FIGURE 2.15

Exercises   SECTION 2.13

1. Let f be a continuous function on $(-\infty, \infty)$. Show that $\int_{-\varepsilon}^{\varepsilon} f(t)\delta(t-a)\,dt = f(a)$. Hint: Use the change of variable $t \to t+a$ and equation (2.104).

2. Let $u(t)$ be the solution of

$$\left(D^2 + a_1 D + a_0\right)x = \delta(t)$$

Assuming the differentiation and integration can be interchanged, show that

$$U(t) = \int_{-\infty}^{\infty} f(s)u(t-s)\,ds$$

is a solution of

$$\left(D^2 + a_1 D + a_0\right)x = f(t)$$

3. Let g be a continuous function on $(-\infty, \infty)$ such that $g(0)=1$. Show that $g(t)\delta(t)=\delta(t)$ in the sense that $g(t)\delta(t)$ satisfies the same properties (the equations in (2.103) and (2.104)) as $\delta(t)$.

4. Suppose that the mass in the mechanical system described in this section is displaced a distance $x=.2$ meter, held stationary, and then released. At time $t=\pi$ the mass is struck with a force of 2 newtons; i.e., $\delta(t)$ is replaced by $\delta(t-\pi)$ in equation (2.105). Determine the motion of the mass.

3 Higher-Order Linear Differential Equations

3.1 Introduction

In this chapter we extend the results of the previous chapter for second-order linear differential equations to linear differential equations of orders greater than two. Let $a_0, a_1, \ldots, a_{n-1}$ and r be functions defined on an open interval I. A differential equation of the form

$$\left(D^n + a_{n-1}(x)D^{n-1} + \cdots + a_1(x)D + a_0(x)\right)y = r(x) \tag{3.1}$$

where $D^k = d^k/dx^k$, is called an nth-order linear differential equation on I. If $r(x) \equiv 0$ on I, the equation is called **homogeneous**. Otherwise the equation is said to be **nonhomogeneous**.

The simplest nth-order linear differential equation is

$$D^n y = 0 \tag{3.2}$$

If we integrate each side of this equation, we obtain

$$D^{n-1} y = c_{n-1}$$

where c_{n-1} is an arbitrary constant. Repeated integrations show that

$$y(x) = c_0 + c_1 x + \cdots + c_{n-1} x^{n-1} \tag{3.3}$$

where $c_0, c_1, \ldots$, and c_{n-1} are arbitrary constants. In order to obtain a unique solution to equation (3.2), we need to impose constraints on the solution that will enable us to determine specific values for the constants $c_0, c_1, \ldots$, and c_{n-1}. In Chapter 2 we saw that prescribed initial values of the solution and its first derivative determine a unique solution of a second-order linear differential equation. We now investigate the use of initial values of y and some of its derivatives to determine a unique solution of equation (3.2). For algebraic simplicity we will assume that the initial values are given at $x=0$. Clearly

$$y(0)=c_0$$

Since

$$y'(x)=c_1+2c_2x+ \cdots +(n-1)c_{n-1}x^{n-2}$$

we have

$$y'(0)=c_1$$

If we continue taking derivatives of y and evaluating them at $x=0$, we find that

$$y^{(k)}(0)=k!c_k$$

for $k=1,2,\ldots,n-1$, so

$$c_k=y^{(k)}(0)/k!$$

Thus prescribed values of $y(0)$, $y'(0),\ldots$, and $y^{(n-1)}(0)$ uniquely determine the coefficients $c_0, c_1, \ldots$, and c_{n-1}. Since the coefficients in equation (3.3) are uniquely determined, the prescribed values of $y(0)$, $y'(0),\ldots$ and $y^{(n-1)}(0)$ determine a unique solution of equation (3.2). We have just proved a special case of the following theorem.

THEOREM 3.1 Let $a_0, a_1,\ldots, a_{n-1}$ and r be continuous functions on an open interval I. Let x_0 be any number in I and let $b_0, b_1,\ldots, b_{n-1}$ be arbitrary real numbers. Then there is a unique function y such that

$$\left(D^n+a_{n-1}(x)D^{n-1}+ \cdots +a_1(x)D+a_0(x)\right)y=r(x) \qquad \text{(3.4)}$$

on the interval I, and

$$y(x_0)=b_0, \, y'(x_0)=b_1,\ldots, \, y^{(n-1)}(x_0)=b_{n-1} \qquad \text{(3.5)}$$

The problem of finding a solution of equation (3.4) satisfying the conditions in (3.5) is called the **initial-value problem** for equation (3.4).

Notice that if $n=2$, then Theorem 3.1 is precisely Theorem 2.1. In the following section we will show that all of the theorems for second-order linear differential equations given in Section 2.2 are special cases of theorems for nth-order linear differential equations.

Exercises   SECTION 3.1

1. Let $a_0, a_1, \ldots, a_{n-1}$ be functions on an interval I and let x_0 be any number in I. Show that $u(x) \equiv 0$ on I is the solution of the initial-value problem

$$\left(D^n + a_{n-1}(x)D^{n-1} + \cdots + a_1(x)D + a_0(x)\right)y = 0$$

$$y(x_0) = 0, \ y'(x_0) = 0, \ldots, \ y^{(n-1)}(x_0) = 0$$

2. Find all possible values of r such that $u(x) = e^{rx}$ is a solution of $(D^3 + 3D^2 - 4D - 12)y = 0$.

3. Let y_1 and y_2 be solutions on an interval I of

$$\left(D^n + a_{n-1}(x)D^{n-1} + \cdots + a_1(x)D + a_0(x)\right)y = 0$$

Show that $c_1 y_1 + c_2 y_2$ is also a solution on I for every choice of the constants c_1 and c_2.

4. Let u be a solution on an interval I of

$$\left(D^3 + a_2(x)D^2 + a_1(x)D + a_0(x)\right)y = 0$$

and let z be a solution on I of

$$\left(D^3 + a_2(x)D^2 + a_1(x)D + a_0(x)\right)y = f(x)$$

Show that $u+z$ is a solution on I of

$$\left(D^3 + a_2(x)D^2 + a_1(x)D + a_0(x)\right)y = f(x)$$

5. Show that $u(x) = x$ is the only solution of $(D^3 + x^{-2}D - x^{-3})y = 0$ having the form $y(x) = x^r$ where r is a number.

3.2 General Solutions

In Theorem 2.2 we proved that if u_1 and u_2 are solutions on an interval J of the differential equation

$$(D^2 + a_1(x)D + a_0(x))y = 0$$

then $c_1 u_1 + c_2 u_2$ is also a solution on J for every choice of the constants c_1 and c_2. A similar proof shows that if $y_1, y_2, \ldots, y_k$ are solutions on an interval I of the differential equation (where k is not necessarily equal to n)

$$(D^n + a_{n-1}(x)D^{n-1} + \cdots + a_1(x)D + a_0(x))y = 0 \qquad \textbf{(3.6)}$$

then $c_1 y_1 + c_2 y_2 + \cdots + c_k y_k$ is also a solution on I for every choice of the constants $c_1, c_2, \ldots, c_k$.

In order to discuss further the nature of solutions of equation (3.6), we need to extend our concept of linear independence to include more than two functions.

DEFINITION 3.1 Let $f_1, f_2, \ldots, f_k$ be functions defined on an interval I. These functions are said to be **linearly dependent** if there are constants $c_1, c_2, \ldots, c_k$, not all zero, such that

$$c_1 f_1(x) + c_2 f_2(x) + \cdots + c_k f_k(x) = 0$$

for every x in I. If the functions are not linearly dependent, they are said to be **linearly independent**.

By Exercise 6 of Section 2.2, Definition 3.1 is equivalent to Definition 2.1 for the special case $k = 2$.

▶ **Example 1** The functions $f_1(x) = x$, $f_2(x) = 3x - 2$, $f_3(x) = x + 1$ on $(-\infty, \infty)$ are linearly dependent since

$$-5f_1(x) + f_2(x) + 2f_3(x) = 0$$

for every x. ◀

▶ **Example 2** Consider the functions $f_1(x) = \cos x$, $f_2(x) = \sin x$, $f_3(x) = \sin(x+1)$ on $(-\infty, \infty)$. Using the trigonometric identity for the sine of the sum of two angles, we have

$$\sin(x+1) = \sin x \cos 1 + \cos x \sin 1$$

so

$$(\cos 1)f_1(x)+(\sin 1)f_2(x)-f_3(x)=0$$

for all x. Hence the functions f_1, f_2, f_3 are linearly dependent.   ◀

▶ **Example 3**   Consider the functions $f_1(x)=x$, $f_2(x)=\cos x$, $f_3(x)=\sin x$ on $(-\infty,\infty)$. We will show that these functions are linearly independent by showing that $c_1=0, c_2=0, c_3=0$ are the only constants for which

$$c_1 f_1(x)+c_2 f_2(x)+c_3 f_3(x)=0 \tag{3.7}$$

for every x. Let c_1, c_2, c_3 be any constants such that equation (3.7) holds for every x. Then

$$c_1 x+c_2\cos x+c_3\sin x=0$$

for every x. In particular, for the three cases $x=0$, $\pi/2$, and π, we obtain

$$c_2=0 \tag{3.8}$$

$$c_1\frac{\pi}{2}+c_3=0 \tag{3.9}$$

$$c_1\pi-c_2=0 \tag{3.10}$$

respectively. From equations (3.8) and (3.10) we have that $c_1=0$. This along with equation (3.9) yields $c_3=0$. Thus $c_1=0, c_2=0, c_3=0$ are the only constants for which equation (3.7) holds for all x. Therefore the functions f_1, f_2, and f_3 are linearly independent.   ◀

▶ **Example 4**   Consider the functions $f_1(x)=e^{ax}$, $f_2(x)=e^{bx}$, $f_3(x)=e^{cx}$ on $(-\infty,\infty)$, where $a<b<c$ are distinct numbers. We will show that these functions are linearly independent by showing that $c_1=0, c_2=0$, $c_3=0$ are the only constants for which

$$c_1 f_1(x)+c_2 f_2(x)+c_3 f_3(x)=0 \tag{3.11}$$

for every x. Let c_1, c_2, c_3 be any constants for which equation (3.11) holds for every x. Then

$$c_1 e^{ax}+c_2 e^{bx}+c_3 e^{cx}=0 \tag{3.12}$$

or, equivalently,

$$c_1 e^{(a-c)x}+c_2 e^{(b-c)x}+c_3=0 \tag{3.13}$$

for every x. Notice that $a-c<0$ and $b-c<0$. If we now let $x\to\infty$ in equation (3.13), we have $c_3=0$. Equation (3.12) now becomes

$$c_1e^{ax}+c_2e^{bx}=0$$

for every x. Since the functions $f_1(x)=e^{ax}$ and $f_2(x)=e^{bx}$ are linearly independent (Example 3 of Section 2.2), we must have $c_1=0$ and $c_2=0$. Thus $c_1=0, c_2=0, c_3=0$ are the only constants that satisfy equation (3.12). The functions $f_1(x)=e^{ax}$, $f_2(x)=e^{bx}$, $f_3(x)=e^{cx}$ are linearly independent whenever a, b, c are distinct. In fact, it can be shown that the functions

$$g_1(x)=e^{b_1x}, g_2(x)=e^{b_2x},\ldots, g_k(x)=e^{b_kx}$$

are linearly independent on any interval whenever $b_1, b_2,\ldots, b_k$ are distinct numbers. ◀

It is now possible to state a theorem that will serve as a basis for determining all of the solutions of equation (3.6).

THEOREM 3.2 Let $a_0, a_1,\ldots, a_n$ be continuous functions on an open interval I. Then there are n linearly independent solutions on I of

$$\left(D^n+a_{n-1}(x)D^{n-1}+\cdots+a_1(x)D+a_0(x)\right)y=0 \qquad \textbf{(3.14)}$$

Moreover, if $y_1, y_2,\ldots, y_n$ are any n linearly independent solutions on I and if u is any solution on I, then there are constants $c_1, c_2,\ldots, c_n$ such that

$$u(x)=c_1y_1(x)+c_2y_2(x)+\cdots+c_ny_n(x)$$

for every x in I.

Proof We will do the proof for the special case $n=2$. Let x_0 be any number in the interval I. By Theorem 3.1 there are solutions u and v of equation (3.14) such that

$$u(x_0)=1 \qquad u'(x_0)=0 \qquad \textbf{(3.15)}$$

$$v(x_0)=0 \qquad v'(x_0)=1 \qquad \textbf{(3.16)}$$

We will show that u and v are linearly independent by showing that $c_1=0, c_2=0$ are the only constants for which

$$c_1u(x)+c_2v(x)=0 \qquad \textbf{(3.17)}$$

for every x in I. Since equation (3.17) holds for every x in I, we can take the derivative of each side of equation (3.17) to obtain

$$c_1 u'(x) + c_2 v'(x) = 0 \qquad (3.18)$$

for every x in I. In particular, equations (3.17) and (3.18) hold when $x = x_0$:

$$c_1 u(x_0) + c_2 v(x_0) = 0$$

$$c_1 u'(x_0) + c_2 v'(x_0) = 0$$

Using the values from (3.15) and (3.16), we get $c_1 = 0$, $c_2 = 0$. Thus $c_1 = 0, c_2 = 0$ are the only constants for which equation (3.17) holds for every x in I. The solutions u and v are two linearly independent solutions of equation (3.14) for the special case $n = 2$.

Now let y_1 and y_2 be linearly independent solutions and u any solution on the interval I of equation (3.14) for the special case $n = 2$. Let x_0 be any number in I and consider the system of linear equations for the unknowns d_1 and d_2:

$$d_1 y_1(x_0) + d_2 y_2(x_0) = u(x_0)$$

$$d_1 y_1'(x_0) + d_2 y_2'(x_0) = u'(x_0) \qquad (3.19)$$

By Exercise 9 of Section 2.12, the determinant of the coefficient matrix

$$\begin{bmatrix} y_1(x_0) & y_2(x_0) \\ y_1'(x_0) & y_2'(x_0) \end{bmatrix}$$

is nonzero. Therefore (see Appendix 2) there is a unique solution of the system of equations in (3.19). If d_1 and d_2 satisfy the system of equations in (3.19), then u and $d_1 y_1 + d_2 y_2$ are solutions of the initial-value problem

$$(D^2 + a_1(x)D + a_0(x))y = 0$$

$$y(x_0) = u(x_0)$$

$$y'(x_0) = u'(x_0)$$

Since this initial-value problem has a unique solution on the interval I (Theorem 3.1), we must have

$$u(x) = d_1 y_1(x) + d_2 y_2(x)$$

for every x in I. This completes the proof for the case $n = 2$. ■

The preceding theorem tells us that if we can find n linearly independent solutions of equation (3.14), then we can find all of the solutions by considering sums of the linearly independent solutions. This leads us to the following definition.

DEFINITION 3.2   Let $a_0, a_1, \ldots, a_{n-1}$ be continuous functions on an open interval I and let $y_1, y_2, \ldots, y_n$ be linearly independent solutions on I of

$$\left(D^n + a_{n-1}(x)D^{n-1} + \cdots + a_1(x)D + a_0(x)\right)y = 0$$

The expression

$$y(x) = c_1 y_1(x) + c_2 y_2(x) + \cdots + c_n y_n(x)$$

where $c_1, c_2, \ldots, c_n$ are arbitrary constants, is called a **general solution** on I.

We conclude this section with theorems concerning the nature of solutions of the nonhomogeneous nth-order linear differential equation.

THEOREM 3.3   Let $a_0, a_1, \ldots, a_{n-1}$ and r be functions on an open interval I with $a_0, a_1, \ldots, a_{n-1}$ continuous, let $y_1, y_2, \ldots, y_n$ be linearly independent solutions on I of

$$\left(D^n + a_{n-1}(x)D^{n-1} + \cdots + a_1(x)D + a_0(x)\right)y = 0$$

and let z be any solution on I of

$$\left(D^n + a_{n-1}(x)D^{n-1} + \cdots + a_1(x)D + a_0(x)\right)y = r(x) \qquad \textbf{(3.20)}$$

If w is any solution of equation (3.20), then there are constants $c_1, c_2, \ldots, c_n$ such that

$$w(x) = c_1 y_1(x) + c_2 y_2(x) + \cdots + c_n y_n(x) + z(x)$$

This theorem prompts the following definition.

DEFINITION 3.3   Let $a_0, a_1, \ldots, a_{n-1}$ and r be defined on an open interval I with $a_0, a_1, \ldots, a_{n-1}$ continuous on I. Let y_h be a general solution of the homogeneous differential equation

$$\left(D^n + a_{n-1}(x)D^{n-1} + \cdots + a_1(x)D + a_0(x)\right)y = 0$$

and let z be any solution on I (referred to as a **particular solution**) of the nonhomogeneous differential equation

$$\left(D^n + a_{n-1}(x)D^{n-1} + \cdots + a_1(x)D + a_0(x)\right)y = f(x)$$

The expression

$$y(x) = y_h(x) + z(x)$$

is called a **general solution** on I of the nonhomogeneous differential equation.

THEOREM 3.4   (Principle of Superposition)   Let $a_0, a_1, \ldots, a_{n-1}, f$, and g be functions on an open interval I with $a_0, a_1, \ldots, a_{n-1}$ continuous on I. If y_1 and y_2 are solutions on I of

$$\left(D^n + a_{n-1}(x)D^{n-1} + \cdots + a_1(x)D + a_0(x)\right)y = f(x)$$

and

$$\left(D^n + a_{n-1}(x)D^{n-1} + \cdots + a_1(x)D + a_0(x)\right)y = g(x)$$

then $y_1 + y_2$ is a solution of

$$\left(D^n + a_{n-1}(x)D^{n-1} + \cdots + a_1(x)D + a_0(x)\right)y = f(x) + g(x)$$

Theorems 3.3 and 3.4 are generalizations of Theorems 2.4 and 2.5 in which $n = 2$. The proofs of the theorems given above are only slight modifications of the proofs of Theorems 2.4 and 2.5. In Exercises 13 and 14 the reader is asked to supply these proofs.

Exercises   SECTION 3.2

1. (a) Find all values of r such that e^{rx} is a solution of $(D^3 - 7D + 6)y = 0$.
   (b) Find a general solution.
   (c) Find the solution that satisfies the initial conditions $y(0) = 3$, $y'(0) = 0$, $y''(0) = 14$.

2. (a) Show that $(D^4 - 5D^2 + 4)y = 0$ has solutions of the form $y(x) = e^{rx}$.
   (b) Find a general solution.
   (c) Find the solution that satisfies the initial conditions $y(0) = 0$, $y'(0) = 6$, $y''(0) = 0$, $y'''(0) = 18$.

3. (a) Show that $(D^3 + 8x^{-1}D^2 + 5x^{-2}D - 5x^{-3})y = 0$ has solutions of the form $y(x) = x^r$.

(b) Find a general solution.

(c) Find the solution that satisfies the initial conditions $y(1) = 1$, $y'(1) = 3$, $y''(1) = -26$.

4. (a) Show that $y_1(x) = e^x$, $y_2(x) = e^{-x}$, $y_3(x) = \cos x$, and $y_4(x) = \sin x$ are linearly independent solutions of $(D^4 - 1)y = 0$.

(b) Find a general solution of $(D^4 - 1)y = 0$.

(c) Show that $z(x) = -x$ is a solution of $(D^4 - 1)y = x$.

(d) Find a general solution of $(D^4 - 1)y = x$.

5. (a) Show that $y_1(x) = e^x$, $y_2(x) = e^{2x}$, and $y_3(x) = e^{3x}$ are linearly independent solutions of $(D^3 - 6D^2 + 11D - 6)y = 0$.

(b) Find a general solution of $(D^3 - 6D^2 + 11D - 6)y = 0$.

(c) Find the solution of $(D^3 - 6D^2 + 11D - 6)y = 0$ that satisfies the initial conditions $y(0) = 1$, $y'(0) = 0$, $y''(0) = 0$.

(d) Show that $z(x) = -\frac{1}{6}x - \frac{11}{36}$ is a solution of $(D^3 - 6D^2 + 11D - 6)y = x$.

(e) Find a general solution of $(D^3 - 6D^2 + 11D - 6)y = x$.

(f) Find the solution of $(D^3 - 6D^2 + 11D - 6)y = x$ that satisfies the initial conditions $y(0) = 1$, $y'(0) = 0$, $y''(0) = 0$.

6. (a) Show that $y_1(x) = e^{-x}$, $y_2(x) = e^x$, and $y_3(x) = xe^x$ are linearly independent solutions of $(D^3 - D^2 - D + 1)y = 0$.

(b) Find a general solution of $(D^3 - D^2 - D + 1)y = 0$.

(c) Find the solution of $(D^3 - D^2 - D + 1)y = 0$ that satisfies the initial conditions $y(0) = 0$, $y'(0) = -2$, $y''(0) = 0$.

(d) Show that $z(x) = \frac{1}{3}e^{2x}$ is a solution of $(D^3 - D^2 - D + 1)y = e^{2x}$.

(e) Find a general solution of $(D^3 - D^2 - D + 1)y = e^{2x}$.

(f) Find the solution of $(D^3 - D^2 - D + 1)y = e^{2x}$ that satisfies the initial conditions $y(0) = \frac{1}{3}$, $y'(0) = -\frac{4}{3}$, $y''(0) = \frac{4}{3}$.

7. (a) Show that $y_1(x) = 1$, $y_2(x) = x$, $y_3(x) = e^{2x}$, and $y_4(x) = e^{-2x}$ are linearly independent solutions of $(D^4 - 4D^2)y = 0$.

(b) Find a general solution of $(D^4 - 4D^2)y = 0$.

(c) Find the solution of $(D^4 - 4D^2)y = 0$ that satisfies the initial conditions $y(0) = 3$, $y'(0) = 1$, $y''(0) = 8$, $y'''(0) = 0$.

(d) Show that $z(x) = -x^3$ is a solution of $(D^4 - 4D^2)y = 24x$.

(e) Find a general solution of $(D^4 - 4D^2)y = 24x$.

(f) Find the solution of $(D^4 - 4D^2)y = 24x$ that satisfies the initial conditions $y(0) = 1$, $y'(0) = 1$, $y''(0) = 0$, $y'''(0) = 6$.

8. An equation of the form

$$\left(x^n D^n + a_{n-1}x^{n-1}D^{n-1} + \cdots + xa_1 D + a_0\right)y = 0$$

is called a Cauchy-Euler equation. Show that such an equation has solutions of the form $y(x)=x^r$.

In Exercises 9–12 use Exercise 8 to find a general solution of the given Cauchy-Euler equation. Hint: In Exercises 11 and 12 use Exercise 14(a) of Section 2.2.

9. $(D^3-3x^{-1}D^2+6x^{-2}D-6x^{-3})y=0$
10. $(D^3+5x^{-1}D^2+2x^{-2}D-2x^{-3})y=0$
11. $(D^3+2x^{-1}D^2+x^{-2}D-x^{-3})y=0$
12. $(D^3+x^{-1}D^2+x^{-2}D-x^{-3})y=0$

13. Prove Theorem 3.3.
14. Prove Theorem 3.4.

3.3 The Wronskian

Let $y_1, y_2,\ldots, y_n$ be linearly independent solutions on an interval I of the differential equation

$$\left(D^n+a_{n-1}(x)D^{n-1}+a_1(x)D+a_0(x)\right)y=0 \qquad \textbf{(3.21)}$$

Then any solution of equation (3.21) can be written as $c_1y_1+c_2y_2+\cdots+c_ny_n$ for an appropriate choice of the constants $c_1, c_2,\ldots, c_n$. In particular, if x_0 is any number in I, then the solution to the initial-value problem involving equation (3.21) and the initial conditions

$$y(x_0)=b_0,\ y'(x_0)=b_1,\ldots,\ y^{(n-1)}(x_0)=b_{n-1}$$

can be written in this form. Since the solution to the initial-value problem is unique (Theorem 3.1), there is a unique choice of the constants $c_1, c_2,\ldots, c_n$ such that

$$c_1y_1(x_0)+c_2y_2(x_0)+\cdots+c_ny_n(x_0)=b_0$$
$$c_1y_1'(x_0)+c_2y_2'(x_0)+\cdots+c_ny_n'(x_0)=b_1$$
$$\cdots\cdots\cdots\cdots\cdots\cdots\cdots\cdots\cdots\cdots$$
$$c_1y_1^{(n-1)}(x_0)+c_2y_2^{(n-1)}(x_0)+\cdots+c_ny_n^{(n-1)}(x_0)=b_{n-1}$$

It is known that a system of equations, such as that given above, has a unique solution if and only if the determinant of the coefficients of the unknowns (in this case $c_1, c_2, \ldots, c_n)$ is nonzero. Thus we must have

$$\det \begin{bmatrix} y_1(x_0) & y_2(x_0) & \cdots & y_n(x_0) \\ y_1'(x_0) & y_2'(x_0) & \cdots & y_n'(x_0) \\ \cdots\cdots\cdots\cdots\cdots\cdots\cdots\cdots\cdots\cdots \\ y_1^{(n-1)}(x_0) & y_2^{(n-1)}(x_0) & \cdots & y_n^{(n-1)}(x_0) \end{bmatrix} \neq 0 \qquad \textbf{(3.22)}$$

The determinant in (3.22) is called the **Wronskian** of $y_1, y_2, \ldots, y_n$ at x_0 and is denoted by $W[y_1, y_2, \ldots, y_n](x_0)$. We have shown that if $y_1, y_2, \ldots, y_n$ are linearly independent solutions, then $W[y_1, y_2, \ldots, y_n](x_0) \neq 0$ for every x_0 in I. It can be proved that the converse is also true. (The special case $n=2$ is considered in Exercises 9 and 10 of Section 2.12.) These results are of such importance that we state them as a theorem.

THEOREM 3.5 Let $a_0, a_1, \ldots, a_{n-1}$ be continuous functions on an open interval I, and let $y_1, y_2, \ldots, y_n$ be solutions on I of

$$\left(D^n + a_{n-1}(x)D^{n-1} + \cdots + a_1(x)D + a_0(x)\right)y = 0$$

Then $y_1, y_2, \ldots, y_n$ are linearly independent if and only if $W[y_1, y_2, \ldots, y_n](x) \neq 0$ for at least one x in I. In fact, the Wronskian is either zero for every x in I or nonzero for every x in I.

It should be noted that the functions $y_1, y_2, \ldots, y_n$ in Theorem 3.5 must be solutions on an interval I of an nth-order linear differential equation. If they are not, the Wronskian may not have the stated properties. (See Exercise 1.)

▶ **Example 1** Consider the differential equation $D^3y = 0$ and its solutions $y_1(x) = 1$, $y_2(x) = x$, and $y_3(x) = x^2$ on $(-\infty, \infty)$. Then

$$W[y_1, y_2, y_3](0) = \det \begin{bmatrix} 1 & 0 & 0 \\ 0 & 1 & 0 \\ 0 & 0 & 2 \end{bmatrix} = 2$$

so that the solutions y_1, y_2, and y_3 are linearly independent. ◀

▶ **Example 2** Consider the differential equation $D^3y = 0$ and its solutions $y_1(x) = x$, $y_2(x) = x^2 + x$, and $y_3(x) = x^2$ on $(-\infty, \infty)$. Then

$$W[y_1, y_2, y_3](0) = \det \begin{bmatrix} 0 & 0 & 0 \\ 1 & 1 & 0 \\ 0 & 2 & 2 \end{bmatrix} = 0$$

so that the solutions y_1, y_2, and y_3 are linearly dependent. ◀

▶ **Example 3**   Consider the third-order differential equation

$$(D-a)(D-b)(D-c)y=0 \tag{3.23}$$

where a, b, and c are constants. The auxiliary equation for equation (3.23) is $(r-a)(r-b)(r-c)=0$. Therefore $y_1(x)=e^{ax}$, $y_2(x)=e^{bx}$, and $y_3(x)=e^{cx}$ are solutions of equation (3.23). The Wronskian of y_1, y_2, y_3 is

$$W[y_1, y_2, y_3](x)=\det\begin{bmatrix} y_1(x) & y_2(x) & y_3(x) \\ y_1'(x) & y_2'(x) & y_3'(x) \\ y_1''(x) & y_2''(x) & y_3''(x) \end{bmatrix}$$

$$=\det\begin{bmatrix} e^{ax} & e^{bx} & e^{cx} \\ ae^{ax} & be^{bx} & ce^{cx} \\ a^2e^{ax} & b^2e^{bx} & c^2e^{cx} \end{bmatrix}$$

When $x=0$ we have

$$W[y_1, y_2, y_3](0)=\det\begin{bmatrix} 1 & 1 & 1 \\ a & b & c \\ a^2 & b^2 & c^2 \end{bmatrix}$$

$$=bc^2+ca^2+ab^2-cb^2-ac^2-ba^2$$

$$=(a-b)(b-c)(c-a)$$

By Theorem 3.5, the functions y_1, y_2, y_3 are linearly independent if and only if a, b, and c are distinct.   ◀

Due to the computational difficulties in evaluating the determinant of an $n \times n$ matrix, computing the Wronskian is usually not a practical test of linear independence if $n \geq 4$. Nonetheless, the Wronskian is a valuable theoretical tool in the study of differential equations.

Exercises   SECTION 3.3

1. Consider the functions $y_1(x)\equiv 1$ and $y_2(x)=x^2$ on the interval $(-\infty, \infty)$. Show that the Wronskian of y_1 and y_2 assumes both zero and nonzero values on $(-\infty, \infty)$ even though y_1 and y_2 are linearly independent on $(-\infty, \infty)$. Can y_1 and y_2 be solutions of a second-order linear differential equation? Why or why not?

2. Is there a second-order linear differential equation that has the following pairs of functions as solutions on $(-\infty, \infty)$?
   (a) $y_1(x) \equiv 1$, $y_2(x) = \cos x$
   (b) $y_1(x) = x$, $y_2(x) = \cos x$
   Explain your answer.

3. By Exercise 5 of Section 3.2, the functions $u_1(x) = e^x$, $u_2(x) = e^{2x}$, $u_3(x) = e^{3x}$ are solutions on $(-\infty, \infty)$ of $(D^3 - 6D^2 + 11D - 6)y = 0$. Hence the functions $y_1(x) = e^x + e^{2x}$, $y_2(x) = e^{2x} + e^{3x}$, $y_3(x) = e^x + e^{3x}$ are also solutions on $(-\infty, \infty)$. Use the Wronskian to determine whether y_1, y_2, y_3 are linearly independent.

4. Suppose that one solution of

$$\left(D^2 + a_1(x)D + a_0(x)\right)y = 0$$

is $y_1(x) = 1 + x$, and that the Wronskian of any two solutions is constant. Show that $y_2(x) \equiv 1$ is also a solution. Find a general solution.

3.4 Higher-Order Linear Differential Equations with Constant Coefficients

In Section 2.3 we found that if a_0 and a_1 are constants, then the differential equation

$$\left(D^2 + a_1 D + a_0\right)y = 0$$

has solutions of the form $y = e^{rx}$, where r satisfies the equation

$$r^2 + a_1 r + a_0 = 0$$

Now consider the differential equation

$$\left(D^n + a_{n-1}D^{n-1} + \cdots + a_1 D + a_0\right)y = 0 \qquad \text{(3.24)}$$

where $a_0, a_1, \ldots, a_{n-1}$ are constants. Setting

$$f(r) = r^n + a_{n-1}r^{n-1} + \cdots + a_1 r + a_0 \qquad \text{(3.25)}$$

we are able to rewrite equation (3.24) as

$$f(D)y = 0 \qquad \text{(3.26)}$$

As in the second-order case, we will find solutions of the form $y(x) = e^{rx}$. Inserting this choice for y into equation (3.26), we find that

$$f(D)e^{rx} = \left(D^n + a_{n-1}D^{n-1} + \cdots + a_1 D + a_0 \right)e^{rx}$$

$$= D^n e^{rx} + a_{n-1}D^{n-1}e^{rx} + \cdots + a_1 De^{rx} + a_0 e^{rx}$$

$$= r^n e^{rx} + a_{n-1}r^{n-1}e^{rx} + \cdots + a_1 re^{rx} + a_0 e^{rx}$$

$$= f(r)e^{rx}$$

Thus $y = e^{rx}$ is a solution of $f(D)y = 0$ whenever $f(r) = 0$. The equation

$$f(r) = 0$$

is called the **auxiliary equation** for $f(D)y = 0$.

It can be shown that if $r_1, r_2, \ldots, r_k$ are distinct real roots of the auxiliary equation, then

$$y_1(x) = e^{r_1 x},\ y_2(x) = e^{r_2 x}, \ldots,\ y_k(x) = e^{r_k x}$$

are linearly independent solutions of $f(D)y = 0$. Accepting this result without proof, we state the following theorem.

THEOREM 3.6 If $r_1, r_2, \ldots, r_n$ are distinct real roots of the polynomial

$$f(r) = r^n + a_{n-1}r^{n-1} + \cdots + a_1 r + a_0$$

then

$$y(x) = c_1 e^{r_1 x} + c_2 e^{r_2 x} + \cdots + c_n e^{r_n x}$$

is a general solution of the differential equation

$$\left(D^n + a_{n-1}D^{n-1} + \cdots + a_1 D + a_0 \right)y = 0$$

▶ **Example 1** A general solution of the differential equation

$$(D^3 + 4D^2 - D - 4)y = 0$$

is

$$y(x) = c_1 e^x + c_2 e^{-x} + c_3 e^{-4x}$$

since the roots of the auxiliary equation, $r^3+4r^2-r-4=0$, are $1, -1$, and -4. ◀

If the auxiliary equation (3.25) has a complex root, then we find corresponding solutions of equation (3.24) precisely as we did in the second-order case in Section 2.5. In the second-order case we argued that if $a+bi$, $b\neq0$ is a root of the auxiliary equation $r^2+a_1r+a_0=0$, then

$$y_1(x)=e^{ax}\cos bx \quad \text{and} \quad z_1(x)=e^{ax}\sin bx \qquad (3.27)$$

are linearly independent solutions of $(D^2+a_1D+a_0)y=0$. Exactly the same argument shows that if $a+bi$, $b\neq0$ is a root of the auxiliary equation (3.25), then the functions in (3.27) are solutions of the differential equation in (3.24).

▶ **Example 2** A general solution of the differential equation

$$(D^4+11D^3+61D^2+161D+150)y=0$$

is

$$y(x)=c_1e^{-2x}+c_2e^{-3x}+c_3e^{-3x}\cos 4x+c_4e^{-3x}\sin 4x$$

since the roots of the auxiliary equation, $r^4+11r^3+61r^2+161r+150=0$, are $-2, -3$, and $-3\pm4i$. ◀

In Section 2.3 we found that if the auxiliary equation is $(r-p)^2=0$, then a general solution of $(D-p)^2y=0$ is

$$y(x)=c_1e^{px}+c_2xe^{px}$$

A generalization of this argument may be used to show that if p is a real root of multiplicity k of the auxiliary equation $f(r)=0$, then

$$y_1(x)=e^{px}, \ y_2(x)=xe^{px}, \ldots, \ y_k(x)=x^{k-1}e^{px} \qquad (3.28)$$

are linearly independent solutions of $f(D)y=0$. Similarly if $a\pm bi$, $b\neq0$ are roots of multiplicity k of the auxiliary equation $f(r)=0$, then

$$y_1(x)=e^{ax}\cos bx, \ y_2(x)=xe^{ax}\cos bx, \ldots, \ y_k(x)=x^{k-1}e^{ax}\cos bx$$

$$z_1(x)=e^{ax}\sin bx, \ z_2(x)=xe^{ax}\sin bx, \ldots, \ z_k(x)=x^{k-1}e^{ax}\sin bx$$

(3.29)

are linearly independent solutions of $f(D)y=0$. Proceeding in the above manner, using each root of the auxiliary, we are able to find the same number of solutions as the order of the differential equation. The solutions found in this manner are linearly independent and therefore can be used to construct a general solution of the differential equation.

▶ **Example 3**   Consider the differential equation

$$(D-5)^3(D+2)(D^2+1)^2 y=0 \tag{3.30}$$

The roots of the auxiliary equation, $(r-5)^3(r+2)(r^2+1)^2=0$, are 5, 5, 5, -2, $\pm i$, and $\pm i$. Since -2 and 5 are roots of multiplicities one and three, respectively, the functions $y_1(x)=e^{-2x}$, $y_2(x)=e^{5x}$, $y_3(x)=xe^{5x}$, $y_4(x)=x^2e^{5x}$ are solutions of equation (3.30). Since $\pm i$ are roots of multiplicity two, the functions $y_5(x)=\cos x$, $y_6(x)=x\cos x$, $y_7(x)=\sin x$, $y_8(x)=x\sin x$ are also solutions of equation (3.30). Note that (3.30) is a differential equation of order eight and that we have found eight solutions. A general solution may now be formed by taking a linear combination of these functions:

$$y(x)=c_1e^{-2x}+c_2e^{5x}+c_3xe^{5x}+c_4x^2e^{5x}+c_5\cos x$$

$$+c_6x\cos x+c_7\sin x+c_8x\sin x \qquad ◀$$

Exercises   SECTION 3.4

In Exercises 1–22 find a general solution of the given differential equation.

1. $(D-2)(D+5)(D+3)y=0$
2. $(D-3)(D-4)(D+2)y=0$
3. $(D^2+4)(D+5)^2y=0$
4. $(D-4)^2(D+1)^2y=0$
5. $(D^2+2)(D^2-2)y=0$
6. $(D^2+D+1)^2D^3y=0$
7. $(D^2+3D+4)(D^2-D+1)^3y=0$
8. $(D^2+3D+1)(D^2-1)^3y=0$
9. $(D^3+3D^2+2D)y=0$
10. $(D^4-1)y=0$
11. $(D^3-1)y=0$
12. $(D^3+3D^2+3D+1)y=0$
13. $(D^3+2D^2-5D-6)(D^2-1)y=0$

14. $(D^2 + \frac{1}{4})(D - \frac{1}{8})(D + \frac{2}{3})y = 0$

15. $(D^2 + 5)^4 y = 0$

16. $(D^2 - 7)^6 y = 0$

17. $D^6(D^2 + 3D + 1)^3 y = 0$

18. $(D^2 + 3D + 6)^4 y = 0$

19. $(D^2 + 5D + 6)^3(D^2 + 5D - 6)^3 y = 0$

20. $(D^2 + 1)^6(D^2 - 4)^3(D + 3)(D - 6)y = 0$

21. $(D^2 + (a + b)D + ab)^3 y = 0$

22. $(D + a)^3(D^2 + bD + 2b^2)y = 0$

23. One model for the deflection y of a suspension bridge under a load p distributed along the bridge is

$$\left(EID^4 - (H + h)D^2 \right)y = p - q\frac{h}{H}$$

where E is Young's modulus (a constant), I is the moment of inertia of the cross section of the bridge (assumed to be constant), H is the horizontal tension in the cables due to the load p, h is the tension in the cables when there is no load on the bridge, and q is the weight per unit length of the bridge. Find a general solution of this equation in terms of E, I, H, and h for the special case $p - q(h/H) = 0$.

24. One end $(x = 0)$ of a uniform cylindrical rod is clamped in the chuck of a lathe. The other end $(x = L)$ is supported in a bearing and is free to rotate. There is a load mass m at the midpoint. It can be shown that the deflection y of the rotating rod satisfies

$$EI\frac{d^3 y}{dx^3} + \frac{1}{2}mw^2 y = 0$$

where E is Young's modulus, I is the moment of inertia of the rod, and w is the angular velocity. Find a general solution of this equation.

25. The differential equation for the deflection y of an elastically supported uniform beam with a constant axial force p is

$$\frac{d^4 y}{dx^4} + \frac{p}{EI}\frac{d^2 y}{dx^2} + \frac{k}{EI}y = 0$$

where $k > 0$ is a constant of proportionality, E is Young's modulus, and I is the moment of inertia of the beam. For which values of k do the solutions involve only trigonometric functions?

26. Consider the following system of first-order chemical reactions:

$$A \overset{k_1}{\to} B \overset{k_2}{\to} C \overset{k_3}{\to} A$$

Then

$$\frac{dA}{dt} = k_3 C - k_1 A$$

$$\frac{dB}{dt} = k_1 A - k_2 B$$

$$\frac{dC}{dt} = k_2 B - k_3 C$$

(a) Use the first equation, then the third, and then the second to show that

$$C = \frac{1}{k_3} \left(\frac{dA}{dt} + k_1 A \right)$$

$$B = \frac{1}{k_2 k_3} \frac{d^2 A}{dt^2} + \left(\frac{k_1}{k_2 k_3} + \frac{1}{k_2} \right) \frac{dA}{dt} + \frac{k_1}{k_2} A$$

$$\frac{d^3 A}{dt^3} + (k_1 + k_2 + k_3) \frac{d^2 A}{dt^2} + (k_1 k_3 + k_1 k_2 + k_2 k_3) \frac{dA}{dt} = 0$$

(b) Suppose that $A(0) = A_0$, $B(0) = 0$, and $C(0) = 0$. For the special case $k_1 = k_3 = \frac{1}{4} k_2$, find $A(t)$, $B(t)$, and $C(t)$.
(c) Find $\lim_{t \to \infty} A(t)$, $\lim_{t \to \infty} B(t)$, and $\lim_{t \to \infty} C(t)$.

3.5 An Example from Economics

In this section we introduce a simple model relating production and demand in the economy. We will suppose there is a sudden change in demand such as that caused by a strike or by a government edict such as that prohibiting the sale of U.S. wheat to Russia in 1980. Barring some other change in the economy, it turns out that production will not return to its previous level. We will then invent a simple model for government intervention intended to return the economy to its equilibrium. The following discussion is based on a portion of "Stabilization policy in a closed economy" [Economics Journal, **64** (1954), 290–323] by A. W. Phillips.

The basic quantities in this model are the production $P(t)$ and the demand $E(t)$ at time t, both of which are given in units per year and measured from initial equilibrium values. ($E(t)$ is used to represent the demand at time t instead of the more suggestive $D(t)$ because operator notation simplifies the calculations and the symbol D is needed in that context.) Thus P and E measure the deviation from the equilibrium values of the production and demand. The initial assumption is that the rate of change of production with respect to time, DP, is proportional to the difference between production and demand; that is,

$$DP = -\alpha(P-E) \tag{3.31}$$

where α is a positive constant. Intuitively, equation (3.31) tells us that production increases when demand exceeds production and decreases when production exceeds demand. Next it is assumed that demand and production are related by

$$E = (1-L)P - H(t) \tag{3.32}$$

where

$$H(t) = \begin{cases} 1 & \text{if } t > 0 \\ 0 & \text{if } t \leqslant 0 \end{cases}$$

is the Heaviside function and L is a positive constant representing a leakage of income from the system. The Heaviside function in equation (3.32) represents a spontaneous change in demand occurring at time $t=0$, where the units have been chosen so that this change in demand is one unit. Insertion of the value of E in (3.32) into equation (3.31) gives the first-order linear differential equation

$$DP + \alpha LP = -\alpha H(t) \tag{3.33}$$

which may be solved using the technique described in Section 1.2 to yield

$$P(t) = \left(P(0) + \frac{1}{L} \right) e^{-\alpha L t} - \frac{1}{L} \qquad (t>0)$$

Since $P(t)$ approaches $-1/L$ as $t \to \infty$, the spontaneous drop in demand, represented by the term $-H(t)$ in equation (3.31), causes the production to approach a value less than its previous equilibrium value. To prevent this from occurring, a policy for stabilizing production must be adopted. The adoption of a policy for stabilizing production implies that there is some desirable level of production to maintain. The difference between the actual production and the desired production will be called the error in production. Stabilization policy involves taking steps to correct errors in production, i.e., taking steps that will cause $P(t)$ to return to zero. Such steps could include altering government expenditures or changing rates of taxation.

Stabilization policy is based (at least in part) upon the deviation of production from its equilibrium level. Our measure of this deviation at time t will include the deviation at time t, $P(t)$, along with the total deviation up to time t, $\int_0^t P(s)\,ds$. One stabilization policy assumes that the relationship between production and demand in equation (3.32) is altered by adding a policy demand term Q to the right side of this equation. It is assumed that the rate of change of this policy demand function is proportional to the sum of itself and the terms reflecting the current and total productions:

$$\frac{dQ}{dt} = -\beta\left(Q + f_1 P + f_2 \int_0^t P(s)\,ds\right)$$

or, equivalently,

$$(D+\beta)Q = -\beta\left(f_1 P + f_2 \int_0^t P(s)\,ds\right)$$

where β, f_1, and f_2 are positive constants. We now modify equation (3.32) so that demand and production are related by

$$E = (1-L)P - H(t) + Q$$

or, equivalently,

$$E - (1-L)P + H(t) = Q$$

If we now multiply each side of this equation by $(D+\beta)$, we obtain

$$(D+\beta)\big(E - (1-L)P + H(t)\big) = (D+\beta)Q$$

$$= -\beta\left(f_1 P + f_2 \int_0^t P(s)\,ds\right) \qquad \textbf{(3.34)}$$

From equation (3.31) we have $-\alpha^{-1}DP = P - E$, so equation (3.34) may be rewritten as

$$(D+\beta)\big(\alpha^{-1}DP + LP + H(t)\big) = -\beta f_1 P - \beta f_2 \int_0^t P(s)\,ds$$

Upon rearranging terms, this equation becomes (for $t>0$)

$$D^2 P + (\alpha L + \beta)DP + \alpha\beta(L + f_1)P + \alpha\beta f_2 \int_0^t P(s)\,ds = -\alpha\beta \qquad \textbf{(3.35)}$$

If we now take the derivative of each side, we have

$$\left(D^3+(\alpha L+\beta)D^2+\alpha\beta(L+f_1)D+\alpha\beta f_2\right)P=0 \tag{3.36}$$

whenever $t>0$.

We assume that we begin at the equilibrium position. Thus $P(0)=0$. If we let $t\to 0^+$ in equation (3.32), we obtain $E(0)=-1$. Using this value and letting $t\to 0^+$ in equation (3.31), we find that $P'(0)=-\alpha$. If we let $t\to 0^+$ in equation (3.35), we obtain $P''(0)+(\alpha L+\beta)P'(0)+\alpha\beta(L+f_1)P(0)=-\alpha\beta$. It follows that $P''(0)=\alpha^2 L$. Thus we wish to solve the initial-value problem involving the differential equation in (3.36) and the initial conditions

$$P(0)=0 \qquad P'(0)=-\alpha \qquad P''(0)=\alpha^2 L$$

In the article referred to at the beginning of this section, the constants $f_2=f_1=.5$, $\alpha=4$, $\beta=8$, and $L=.25$ are used. With these choices for the constants, the initial-value problem becomes

$$(D^3+9D^2+24D+16)P=0 \tag{3.37}$$

$$P(0)=0, \quad P'(0)=-4, \quad P''(0)=4$$

The auxiliary equation $r^3+9r^2+24r+16=0$ has -1, -4, and -4 as its roots, so a general solution of equation (3.37) is

$$P(t)=c_1e^{-t}+(c_2+c_3t)e^{-4t}$$

We will now choose the constants c_1, c_2, and c_3 so that the initial conditions are satisfied. That is, we want to choose c_1, c_2, and c_3 so that

$$0=P(0)=c_1+c_2$$

$$-4=P'(0)=-c_1-4c_2+c_3$$

$$4=P''(0)=c_1+16c_2-8c_3$$

A short calculation shows that $c_1=-c_2=-\frac{28}{9}$ and $c_3=\frac{16}{3}$. Hence

$$P(t)=\frac{28}{9}e^{-t}+\left(\frac{28}{9}+\frac{16}{3}t\right)e^{-4t}$$

is a solution of the initial-value problem. Notice that $P(t)\to 0$ as $t\to\infty$. Thus the stabilization policy causes production to return to its equilibrium. The graph of $P(t)$ is shown in Figure 3.1.

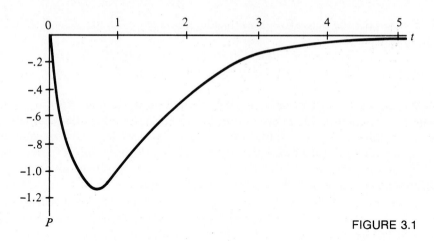

FIGURE 3.1

3.6 The Method of Undetermined Coefficients

In practice, it often happens that the right-hand side of a linear differential equation with constant coefficients has a simple form, such as x, $\sin 3x$, or $e^{-2x}\cos 5x$. In such cases there is an easy way to find a solution of the equation. We studied this method in Sections 2.7 and 2.8 for second-order equations, and luckily there is no change in the nth-order case. It is convenient to lump all special cases together in one general theorem.

THEOREM 3.7 Let

$$g(x)=e^{ax}\left(b_0+b_1x+\cdots+b_mx^m\right)\begin{cases}\sin bx\\\cos bx\end{cases}\qquad\text{(3.38)}$$

and suppose that $a+bi$ is a root of multiplicity k ($k=0$ if $a+bi$ is not a root) of the auxiliary equation

$$r^n+a_{n-1}r^{n-1}+\cdots+a_1r+a_0=0$$

where $a_0, a_1,\ldots, a_{n-1}$ are constants. Then the differential equation

$$\left(D^n+a_{n-1}D^{n-1}+\cdots+a_1D+a_0\right)y=g(x)$$

has a solution of the form

$$z(x)=x^k\left(c_0+c_1x+\cdots+c_mx^m\right)e^{ax}\cos bx$$
$$+x^k\left(d_0+d_1x+\cdots+d_mx^m\right)e^{ax}\sin bx\qquad\text{(3.39)}$$

Frequently $a+bi$ is not a root of the auxiliary equation, in which case $k=0$ and a solution of the equation has the form

$$z(x)=(c_0+c_1x+\cdots+c_mx^m)e^{ax}\cos bx+(d_0+d_1x+\cdots+d_m)e^{ax}\sin bx$$

Another special case of Theorem 3.7 is when $b=0$, i.e., when the root under consideration is real. Then the differential equation has a solution of the form

$$z(x)=x^k(c_0+c_1x+\cdots+c_mx^m)e^{ax}$$

As above, the multiplicity k of the root may be zero.

Conceptually, it is easy to find a general solution of a nonhomogeneous linear differential equation with constant coefficients when the right-hand side $g(x)$ has the form in (3.38). To do this, we proceed as follows:

1. Find the roots of the auxiliary polynomial.

2. Use the roots of the auxiliary equation to construct a general solution y_h of the homogeneous differential equation.

3. Use Theorem 3.7 to find the form of a solution z, called a **particular solution**, of the nonhomogeneous differential equation.

4. Insert z into the nonhomogeneous differential equation and evaluate the coefficients of z.

5. Add y_h and z to obtain a general solution.

In practice, difficulties can arise in steps 1 and 4. Difficulties associated with finding the roots of the auxiliary equation are discussed in Section 3.9. In order to evaluate the coefficients of the solution z, a system of m linear equations in m unknowns must be solved. Usually m is a small integer and there is no trouble in solving this system of equations. If m is large, there may be computational difficulties in evaluating these coefficients. However, there are very efficient numerical procedures for solving such systems that eliminate almost all of these difficulties.

We now give three examples to illustrate the above procedure which is usually called the **method of undetermined coefficients**.

▶ **Example 1** Consider the differential equation

$$(D^4-D^2)y=x \qquad\qquad \textbf{(3.40)}$$

Equation (3.38) has the form $g(x)=x$ when $a=0$, $b=0$, $m=1$, $b_0=0$, and $b_1=1$. Since 0 is a root of multiplicity two of the auxiliary equation

$$r^4-r^2=0$$

equation (3.40) has a particular solution of the form

$$z(x)=x^2(c_0+c_1x)$$

Inserting z into equation (3.40) yields

$$x=(D^4-D^2)z$$

$$=(D^4-D^2)(c_0x^2+c_1x^3)$$

$$=-6c_1x-2c_0$$

Equating the coefficients of like powers of x, we find that $c_1=-\frac{1}{6}$ and $c_0=0$. Therefore

$$z(x)=-\tfrac{1}{6}x^3$$

is a particular solution of equation (3.40). Since the roots of the auxiliary equation are 0, 0, 1 and -1, a general solution of (3.40) is

$$y(x)=C_1+C_2x+C_3e^x+C_4e^{-x}-\tfrac{1}{6}x^3 \qquad \blacktriangleleft$$

▶ **Example 2**   Consider the differential equation

$$(D-1)(D-2)(D-3)y=e^{-x}+e^x \tag{3.41}$$

We may use the method of undetermined coefficients to solve this equation. We begin by considering the two equations

$$(D-1)(D-2)(D-3)y=e^{-x} \tag{3.42}$$

$$(D-1)(D-2)(D-3)y=e^x \tag{3.43}$$

By the principle of superposition (Theorem 3.4) the sum of particular solutions of equations (3.42) and (3.43) is a particular solution. Equation (3.38) has the forms $g(x)=e^{-x}$ and $g(x)=e^x$ when $m=0$, $a=-1$, $b=0$, $b_0=1$ and $m=0$, $a=1$, $b=0$, $b_0=1$, respectively. Since the roots of the auxiliary equation

$$(r-1)(r-2)(r-3)=0$$

are 1, 2, and 3, equations (3.42) and (3.43) have particular solutions of the form

$$z_1(x)=c_0e^{-x}$$

$$z_2(x)=c_0'xe^x$$

respectively. Inserting these functions into equations (3.42) and (3.43) yields

$$e^{-x}=(D-1)(D-2)(D-3)z_1$$

$$=(D-1)(D-2)(D-3)\left(c_0e^{-x}\right)$$

$$=-24c_0e^{-x}$$

and

$$e^x=(D-1)(D-2)(D-3)z_2$$

$$=(D-1)(D-2)(D-3)\left(c_0'xe^x\right)$$

$$=2c_0'e^x$$

Therefore $c_0=-\frac{1}{24}$ and $c_0'=\frac{1}{2}$, so

$$z_1(x)=-\frac{1}{24}e^{-x}$$

and

$$z_2(x)=\frac{1}{2}xe^x$$

are particular solutions of equations (3.42) and (3.43), respectively. Hence

$$z(x)=\frac{1}{2}xe^x-\frac{1}{24}e^{-x}$$

is a particular solution of equation (3.41), and

$$y(x)=C_1e^x+C_2e^{2x}+C_3e^{3x}+\frac{1}{2}xe^x-\frac{1}{24}e^{-x}$$

is a general solution of equation (3.41). ◀

▶ **Example 3**   Consider the differential equation

$$(D^2+1)Dy=\cos x \tag{3.44}$$

Equation (3.38) has the form $g(x)=\cos x$ when $m=0$, $a=0$, $b=1$, and $b_0=1$. Since i is a root of multiplicity one of the auxiliary equation

$$(r^2+1)r=0$$

equation (3.44) has a particular solution of the form

$$z(x)=c_0 x\cos x+d_0 x\sin x$$

Inserting z into equation (3.44) yields

$$\cos x=(D^2+1)Dz$$

$$=(D^2+1)D(c_0 x\cos x+d_0 x\sin x)$$

$$=-2c_0\cos x-2d_0\sin x$$

Equating the coefficients of $\cos x$ and $\sin x$, we find that $c_0=-\frac{1}{2}$ and $d_0=0$. Thus

$$z(x)=-\tfrac{1}{2}x\cos x$$

is a particular solution of equation (3.44), and

$$y(x)=C_1+C_2\cos x+C_3\sin x-\tfrac{1}{2}x\cos x$$

is a general solution of equation (3.44).   ◀

Exercises   SECTION 3.6

In Exercises 1–14 find a general solution of the given differential equation.

1. $(D-1)^2(D+1)y=x^2+1$
2. $(D-1)^3(D+1)y=e^{2x}$
3. $(D-2)^2(D+1)y=e^{2x}$
4. $(D-2)^2(D+1)y=e^x$
5. $(D^2+D+1)^2y=e^x+x$
6. $(D^2+1)^3y=e^x-x$

7. $(D^2-4D+5)(D+1)y=e^x\cos x$

8. $(D^2+1)^2y=\cos 2x$

9. $(D^2+4)(D-2)y=\cos x$

10. $(D^2-4)(D-2)y=\sin x+e^x$

11. $(D-1)^3y=e^x$

12. $(D-1)^2(D-2)^2y=e^{2x}+e^{-x}$

13. $D^3(D-3)^2y=x$

14. $(D^4-1)y=\sin x+e^{-x}$

15. If in Section 3.5 there were three consecutive lags of $\frac{1}{12}$ year each in the response of production to demand, then under certain assumptions the equation for P becomes

$$(D+12)^3P=(-12)^3$$

with initial conditions $P(0)=P'(0)=P''(0)=0$. Find P.

16. Consider the equation for the deflection of a suspension bridge given in Exercise 23 of Section 3.4. Find a general solution to this equation in terms of E, I, H, and h for the special case $p-q(h/H)=x(L-x)$, where L is the length of the bridge.

17. Consider a pipe of infinite length laid along the x-axis. The pipe has wall thickness T, and let R denote the average of the internal and external radii of the pipe. Suppose that a liquid being pumped through the pipe exerts a constant pressure p on the wall of the pipe. This pressure causes a slight radial deflection of the wall of the pipe. In order to strengthen the walls of the pipe, a reinforcing ring is placed around the pipe at $x=0$. A partial cross-sectional view of the pipe at a distance x from the reinforcing ring is shown in Figure 3.2. The original outer and inner radii of the pipe are r_1 and r_2, respectively, and the deflected pipe wall is shown by the dashed

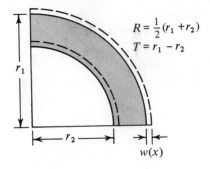

$$R=\frac{1}{2}(r_1+r_2)$$
$$T=r_1-r_2$$

FIGURE 3.2

lines. The radial deflection of the pipe wall at a distance x from the reinforcing ring, $w(x)$, is given by

$$\frac{d^4 w}{dx^4} + \frac{12}{R^2 T^2} w = \frac{12}{ET^3} P$$

where E is the modulus of elasticity for the pipe. At the ring we assume that there is no deflection of the pipe. Find the deflection w for any choice of the constants E, R, T, and P. Hint: The roots of the equation $z^4 + A^4 = 0$ are $A(1 + i)/\sqrt{2}$, $A(1 - i)/\sqrt{2}$, $A(-1 + i)/\sqrt{2}$, and $A(-1-i)/\sqrt{2}$.

18. The deflection y of a beam with uniform cross section is

$$EI \frac{d^4 y}{dx^4} = p(x)$$

where I is the moment of inertia of the cross section of the beam, E is Young's modulus, and $p(x)$ is the load per unit length. There are three natural end conditions:

(i) hinged support: $y = 0$ and $dy^2/dx^2 = 0$
(ii) clamped support: $y = 0$ and $dy/dx = 0$
(iii) free end: $d^2 y/dx^2 = 0$ and $d^3 y/dx^3 = 0$

In the case $p(x) = p_0$, a constant, find the deflection of a beam of length 1 which is

(a) clamped at both ends
(b) clamped at $x = 0$ and free at $x = 1$
(c) hinged at $x = 0$ and clamped at $x = 1$

19. A particular vibration absorber consists of a gyrostat suspended about an axis perpendicular to that of the torsional vibration it is supposed to suppress. The gyrostat is attached to springs and a viscous damping mechanism that limits the vibration about its axis. This device is a more efficient version of one that was used in 1904 in an attempt to stabilize the rolling of ships at sea. If a disturbing torque $T = \sin pt$ is applied to this vibration absorber, rotating it through an angle θ, then the oscillation about the equilibrium satisfies

$$\frac{d^4 \theta}{dt^4} + 43{,}517.65 \frac{d^2 \theta}{dt^2} + 156{,}960\theta = \left(1.57 - .04 p^2\right)\sin pt$$

[R. N. Arnold, The turning and damped gyrostatic vibration absorber, Proc. Inst. Mech. Engrs., **157** (1947), 1–19; repeated in H. McCallion,

Vibration of Linear Mechanical Systems, Halsted Press, New York, 1973, page 252.] Find a general solution of this equation for the special case $p=1$.

3.7 Inverse Differential Operators

(OPTIONAL)

In Section 2.10 we used inverse differential operators to find particular solutions of certain second-order linear differential equations. Inverse differential operators can also be used to find particular solutions of higher-order linear differential equations. In fact, the identities we used in Section 2.10 can also be used to find particular solutions for higher-order equations. We repeat these identities here:

$$(D-b)e^{ax}=(a-b)e^{ax}$$

$$(D-a)(xe^{ax})=e^{ax}$$

$$(D-a)^2(\tfrac{1}{2}x^2e^{ax})=e^{ax}$$

$$\frac{1}{D^2+a_1D+a_0}e^{ax}=\frac{1}{a^2+a_1a+a_0}e^{ax} \quad \text{if } a^2+a_1a+a_0\neq0 \qquad \textbf{(3.45)}$$

$$\frac{1}{D^2+a_1D+a_0}\cos cx=\frac{(a_0-c^2)\cos cx+a_1c\sin cx}{(a_0-c^2)^2+(a_1c)^2} \qquad \textbf{(3.46)}$$

$$\text{if } (a_0-c^2)^2+(a_1c)^2\neq0$$

$$\frac{1}{D^2+a_1D+a_0}\sin cx=\frac{(a_0-c^2)\sin cx-a_1c\cos cx}{(a_0-c^2)^2+(a_1c)^2} \qquad \textbf{(3.47)}$$

$$\text{if } (a_0-c^2)^2+(a_1c)^2\neq0$$

It will be convenient to rewrite the first three of these identities as

$$\frac{1}{D-b}e^{ax}=\frac{1}{a-b}e^{ax} \quad \text{if } a\neq b \qquad \textbf{(3.48)}$$

$$\frac{1}{D-a}e^{ax}=xe^{ax} \qquad \textbf{(3.49)}$$

$$\frac{1}{(D-a)^2}e^{ax}=\tfrac{1}{2}x^2e^{ax} \qquad \textbf{(3.50)}$$

We will also need the identities

$$\frac{1}{D-a}\cos cx = \frac{-a}{a^2+c^2}\cos cx + \frac{c}{a^2+c^2}\sin cx \qquad \text{(3.51)}$$

$$\frac{1}{D-a}\sin cx = \frac{-c}{a^2+c^2}\cos cx - \frac{a}{a^2+c^2}\sin cx \qquad \text{(3.52)}$$

which can be easily established by showing that

$$(D-a)\left(\frac{-a}{a^2+c^2}\cos cx + \frac{c}{a^2+c^2}\sin cx\right) = \cos cx$$

$$(D-a)\left(\frac{-c}{a^2+c^2}\cos cx - \frac{a}{a^2+c^2}\sin cx\right) = \sin cx$$

In Exercises 15 and 16 of Section 2.10 the reader was asked to show that

$$\frac{1}{f(D)}[cg(x)] = c\frac{1}{f(D)}g(x) \quad \text{for every constant } c$$

and

$$\frac{1}{f(D)}[g(x)+h(x)] = \frac{1}{f(D)}g(x) + \frac{1}{f(D)}h(x)$$

whenever $f(z)$ is a second-degree polynomial. In fact, these identities hold whenever $f(z)$ is any nonzero polynomial.

▶ **Example 1**   Consider the differential equation

$$(D+2)(D-3)(D+4)y = e^{5x} \qquad \text{(3.53)}$$

Then

$$y(x) = \frac{1}{(D+2)(D-3)(D+4)}e^{5x}$$

is a particular solution of equation (3.53). Repeated use of the identity in

(3.48) yields

$$y(x) = \left(\frac{1}{D+2}\right)\left(\frac{1}{D-3}\right)\left(\frac{1}{D+4}\right)e^{5x}$$

$$= \frac{1}{9}\left(\frac{1}{D+2}\right)\left(\frac{1}{D-3}\right)e^{5x}$$

$$= \left(\frac{1}{2}\right)\left(\frac{1}{9}\right)\left(\frac{1}{D+2}\right)e^{5x}$$

$$= \left(\frac{1}{7}\right)\left(\frac{1}{2}\right)\left(\frac{1}{9}\right)e^{5x} = \frac{1}{126}e^{5x} \qquad \blacktriangleleft$$

▶ **Example 2** Consider the differential equation

$$(D-2)(D+1)(D-1) = e^x + \cos x \qquad \textbf{(3.54)}$$

Then

$$y(x) = \frac{1}{(D-2)(D+1)(D-1)}(e^x + \cos x)$$

$$= \frac{1}{(D-2)(D+1)(D-1)}e^x + \frac{1}{(D-2)(D+1)(D-1)}\cos x$$

is a particular solution of equation (3.54). Using the above identities, we have

$$\frac{1}{(D-2)(D+1)(D-1)}e^x = \left(\frac{1}{D-1}\right)\left(\frac{1}{D-2}\right)\left(\frac{1}{D+1}\right)e^x$$

$$= \frac{1}{2}\left(\frac{1}{D-1}\right)\left(\frac{1}{D-2}\right)e^x$$

$$= (-1)\left(\frac{1}{2}\right)\left(\frac{1}{D-1}\right)e^x$$

$$= -\tfrac{1}{2}xe^x$$

and

$$\frac{1}{(D-2)(D+1)(D-1)}\cos x = \frac{1}{D-2}\left(\frac{1}{D^2-1}\cos x\right)$$

$$= -\frac{1}{2}\frac{1}{D-2}\cos x$$

$$= -\frac{1}{2}\left(-\frac{2}{5}\cos x + \frac{1}{5}\sin x\right)$$

$$= \frac{1}{5}\cos x - \frac{1}{10}\sin x$$

Thus

$$y(x) = -\frac{1}{2}xe^x + \frac{1}{5}\cos x - \frac{1}{10}\sin x \qquad \blacktriangleleft$$

Exercises   SECTION 3.7

In Exercises 1–16 use inverse differential operators to find a general solution of the given differential equation.

1. $(D-1)(D-2)(D-3)y = e^{4x}$
2. $(D+1)(D+2)(D-2)y = 3e^{3x}$
3. $(D+1)(D-1)(D-3)y = -5e^x$
4. $(D-6)(D-3)(D+4)y = 7e^{6x}$
5. $(D^2+D+1)(D-1)y = e^{2x}$
6. $(D^2+3D+2)(D+2)y = 2e^{-x}$
7. $(D^2-D+2)(D-7)y = \cos 3x$
8. $(D^2+2D+7)(D+5)y = \sin 2x$
9. $(D-1)^2(D+4)y = e^{3x} + 3\sin x$
10. $(D+1)^2(D+7)y = e^{-5x} + \cos 2x$
11. $(D^2+4D+1)(D+5)y = e^{4x} - e^x$
12. $(D^2+3D+5)(D-1)y = 2e^x$
13. $(D^2-1)(D^2+1)y = e^x + e^{-x}$
14. $(D^2-4)(D^2+2D+3)y = e^{2x} + 2$
15. $(D^2+1)(D^2+2D-1)y = e^x + 5e^{2x} + 3\cos x$
16. $(D^2+3D+5)(D^2+6D+1)y = e^{2x} + 5x$

17. Use inverse differential operators to solve the problem in Exercise 19 of Section 3.6.

3.8 Variation of Parameters

The method of variation of parameters discussed in Section 2.12 for second-order linear differential equations can be generalized to determine a particular solution of an nth-order linear differential equation

$$\left(D^n+a_{n-1}(x)D^{n-1}+\cdots+a_1(x)D+a_0(x)\right)y=f(x) \qquad \text{(3.55)}$$

Suppose that we know n linearly independent solutions $y_1, y_2, \ldots, y_n$ on I of the homogeneous differential equation

$$\left(D^n+a_{n-1}(x)D^n+\cdots+a_1(x)D+a_0(x)\right)y=0 \qquad \text{(3.56)}$$

The method of variation of parameters consists of finding functions $v_1, v_2, \ldots, v_n$ on I such that

$$z(x)=v_1(x)y_1(x)+v_2(x)y_2(x)+\cdots+v_n(x)y_n(x) \qquad \text{(3.57)}$$

is a particular solution of equation (3.55). In order to determine these n functions, we will need to specify n conditions that they must satisfy. One condition must be that the function z is a solution of equation (3.55). The remaining $n-1$ conditions will be chosen so that derivatives of the functions v_i of orders higher than the first do not appear in our calculations. From equation (3.57) we have

$$z'=(v_1y_1'+v_2y_2'+\cdots+v_ny'_n)+(v_1'y_1+v_2'y_2+\cdots+v_n'y_n) \qquad \text{(3.58)}$$

In order to prevent the second derivatives of the v_i from appearing when we compute z'', we will require that

$$v_1'y_1+v_2'y_2+\cdots+v_n'y_n=0$$

From equation (3.58) we can now compute z'':

$$z''=(v_1y_1''+v_2y_2''+\cdots+v_ny_n'')+(v_1'y_1'+v_2'y_2'+\cdots+v_n'y_n')$$

In order to prevent the second derivatives of the v_i from appearing when we compute z''', we will require that

$$v_1'y_1'+v_2'y_2'+\cdots+v_n'y_n'=0$$

Continuing in this manner, we obtain

$$z^{(k)}=v_1y_1^{(k)}+v_2y_2^{(k)}+\cdots+v_ny_n^{(k)} \qquad \text{(3.59)}$$

along with the $n-1$ equations

$$v_1' y_1^{(k)} + v_2' y_2^{(k)} + \cdots + v_n' y_n^{(k)} = 0 \tag{3.60}$$

for $k = 0, 1, \ldots, n-2$.

Taking the derivative of each side of equation (3.59) with $k = n-1$ yields

$$z^{(n)} = \left(v_1 y_1^{(n)} + v_2 y_2^{(n)} + \cdots + v_n y_n^{(n)} \right)$$

$$+ \left(v_1' y_1^{(n-1)} + v_2' y_2^{(n-1)} + \cdots + v_n' y_n^{(n-1)} \right)$$

This time we do not set $v_1' y_1^{(n-1)} + v_2' y_2^{(n-1)} + \cdots + v_n' y_n^{(n-1)}$ equal to zero. Instead, we choose a value so that the function z is a solution of the differential equation. Inserting the above values for $z^{(k)}$ into equation (3.55) and grouping terms with like functions v_i, equation (3.55) becomes

$$v_1' y_1^{(n-1)} + v_2' y_2^{(n-1)} + \cdots + v_n' y_n^{(n-1)} = f(x) \tag{3.61}$$

because $y_1, y_2, \ldots, y_n$ are solutions of the homogeneous differential equation (3.56). The equations in (3.59) and (3.61) are n linear equations for the n unknown functions $v_1', v_2', \ldots, v_n'$:

$$
\begin{aligned}
& y_1 v_1' + y_2 v_2' + \cdots + y_n v_n' = 0 \\
& y_1' v_1' + y_2' v_2' + \cdots + y_n' v_n' = 0 \\
& \cdots\cdots\cdots\cdots\cdots\cdots\cdots\cdots\cdots\cdots\cdots \\
& y_1^{(n-2)} v_1' + y_2^{(n-2)} v_2' + \cdots + y_n^{(n-2)} v_n' = 0 \\
& y_1^{(n-1)} v_1' + y_2^{(n-1)} v_2' + \cdots + y_n^{(n-1)} v_n' = f(x)
\end{aligned}
\tag{3.62}
$$

Notice that the determinant of the matrix formed by the coefficients of the system of equations in (3.62) is the Wronskian $W[y_1, y_2, \ldots, y_n](x)$. From Theorem 3.5 we know that this Wronskian is nonzero on the interval I. Recall that a system of equations, such as that in (3.62), has a unique solution whenever the determinant of the matrix formed by the coefficients is nonzero. Moreover, the solution of the system of equations in (3.62) may be written as

$$v_k'(x) = \frac{f(x) W_k(x)}{W[y_1, y_2, \ldots, y_n](x)}$$

where

$$
W_k(x) = \det
\begin{bmatrix}
y_1(x) & \cdots & y_{k-1}(x) & 0 & y_{k-1}(x) & \cdots & y_n(x) \\
y_1'(x) & \cdots & y_{k-1}'(x) & 0 & y_{k+1}'(x) & \cdots & y_n'(x) \\
\vdots & & \vdots & \vdots & \vdots & & \vdots \\
y_1^{(n-1)}(x) & \cdots & y_{k-1}^{(n-1)}(x) & 1 & y_{k+1}^{(n-1)}(x) & \cdots & y_n^{(n-1)}(x)
\end{bmatrix}
$$

for $k=0,1,\ldots,n$. With this notation

$$
v_k(x) = \int \frac{f(x)W_k(x)}{W[y_1, y_2, \ldots, y_n](x)}\, dx \tag{3.63}
$$

so a particular solution of equation (3.55) is

$$
z(x) = \sum_{k=1}^{n} v_k(x) y_k(x)
$$

$$
= \sum_{k=1}^{n} y_k(x) \int \frac{f(x)W_k(x)}{W[y_1, y_2, \ldots, y_n](x)}\, dx
$$

The integrals in (3.63) can be simplified somewhat by using the identity

$$
W[y_1, y_2, \ldots, y_n](x) = W[y_1, y_2, \ldots, y_n](x_0) e^{\int_{x_0}^{x} a_{n-1}(t)\, dt}
$$

where x_0 is any number in the interval I. The special case of this identity in which $n=2$ is considered in Exercise 8 of Section 2.12. The verification of this identity requires a knowledge of how to compute the derivative of a determinant and will be omitted.

▶ **Example 1**   Consider the differential equation

$$
(D^3 - 3x^{-1}D^2 + 6x^{-2}D - 6x^{-3})y = x^{-1} \tag{3.64}
$$

on $(0,\infty)$. The functions $y_1(x)=x$, $y_2(x)=x^2$, and $y_3(x)=x^3$ are linearly independent solutions on $(0,\infty)$ of the homogeneous equation

$$
(D^3 - 3x^{-1}D^2 + 6x^{-2}D - 6x^{-3})y = 0
$$

The Wronskian of these solutions is

$$W[y_1, y_2, \ldots, y_n](x) = \det \begin{bmatrix} x & x^2 & x^3 \\ 1 & 2x & 3x^2 \\ 0 & 2 & 6x \end{bmatrix} = 2x^3$$

while W_1, W_2, and W_3 are given by

$$W_1(x) = \det \begin{bmatrix} 0 & x^2 & x^3 \\ 0 & 2x & 3x^2 \\ 1 & 2 & 6x \end{bmatrix} = x^4$$

$$W_2(x) = \det \begin{bmatrix} x & 0 & x^3 \\ 1 & 0 & 3x^2 \\ 0 & 1 & 6x \end{bmatrix} = -2x^3$$

$$W_3(x) = \det \begin{bmatrix} x & x^2 & 0 \\ 1 & 2x & 0 \\ 0 & 2 & 1 \end{bmatrix} = x^2$$

so that

$$v_1(x) = \int \frac{x^{-1}x^4}{2x^3} \, dx = \tfrac{1}{2}x + c_1$$

$$v_2(x) = \int \frac{x^{-1}(-2x^3)}{2x^3} \, dx = -\ln x + c_2$$

$$v_3(x) = \int \frac{x^{-1}x^2}{2x^3} \, dx = -\tfrac{1}{2}x^{-1} + c_3$$

Since we need only one specific choice of v_1, v_2, v_3, we will choose the constants of integration to be zero. Then

$$z(x) = v_1(x)y_1(x) + v_2(x)y_2(x) + v_3(x)y_3(x)$$

$$= \tfrac{1}{2}x^2 - x^2\ln x - \tfrac{1}{2}x^2$$

$$= -x^2\ln x$$

is a particular solution of equation (3.64) and

$$y(x) = C_1 x + C_2 x^2 + C_3 x^3 - x^2\ln x$$

is a general solution.

◀

Exercises SECTION 3.8

In Exercises 1–3 use the method of variation of parameters to find a general solution of the given differential equation.

1. $(D^3 + D)y = \sec x, \quad -\pi/2 < x < \pi/2$
2. $(D^3 - 2D^2 + D)y = e^x$
3. $(D^3 + 5x^{-1}D^2 + 2x^{-2}D - 2x^{-3})y = \ln x, \quad 0 < x$
   See Exercise 9 of Section 3.2.

In Exercises 4–7 find a formula (involving integrals) for a particular solution of the given differential equation.

4. $(D^3 - D)y = \ln x, \quad 0 < x$
5. $(D+1)(D-1)^2 y = x^{-1}, \quad 0 < x$
6. $(D^4 - 1)y = f(x)$
7. $(D-1)(D+1)^2 y = f(x)$

3.9 Computational Difficulties

In Sections 2.3, 2.5, and 3.4 we showed that a general solution of a linear differential $f(D)y = 0$ with real constant coefficients can be obtained if we know the roots of the polynomial equation $f(r) = 0$. Finding the roots of a polynomial is not an elementary problem. If the polynomial has degree $n = 1$, 2, 3, or 4, then the roots may be found by rational operations and the extraction of roots. For the case $n = 1$ or $n = 2$, the roots are easily determined. If $n = 3$ or $n = 4$, then there are formulas for the roots similar to, but more complicated than, the quadratic formula [G. Birkhoff and S. MacLane, *A Survey of Modern Algebra*, Macmillan, New York, 1965, pages 105–108]. However, it is not possible to solve all polynomial equations of degree $n \geqslant 5$ in this manner.

Suppose that

$$f(r) = a_n r^n + a_{n-1} r^{n-1} + \cdots + a_1 r + a_0$$

where $a_0, a_1, \ldots, a_n$ are integers. Then it is well known that a rational root p/q has the property that q is a divisor of a_n and p is a divisor of a_0. This allows all of the rational roots to be found. If $n \geqslant 5$ and not all of the roots are rational, then, in general, some approximation method, such as Newton's method, must be used to approximate the irrational roots. In fact, there are numerical procedures that (in theory) determine all of the roots, both real and complex, including multiplicities. However, these procedures are very sensitive to round-off errors in the calculations.

Other difficulties may arise if the coefficients in the differential equation are determined experimentally. For example, suppose that the coefficient a_0 in

$$\left(D^2 - 4D + a_0\right)y = 0$$

is found experimentally to be 4. Then $r = 2$ is a root of multiplicity two of the auxiliary equation. However, suppose that there was an error of 10^{-7} in obtaining a_0, so that in fact $a_0 = 3.9999999$. The rounded roots of

$$x^2 - 4x + 3.9999999 = 0$$

are $r = 1.999683772$ and 2.000316228. Thus an error of 10^{-7} in the coefficient a_0 has caused an error in the roots of approximately $.000316228$. The error in the roots is approximately 3162 times as great as the error in the coefficient. In many applications, difficulties similar to that discussed here must be considered.

The above example and other difficulties in computation may be found in "Pitfalls in Computation, or Why a Math Book Isn't Enough" by G. Forsythe [American Mathematical Monthly, Vol. 7, No. 9, November 1970, pages 931–956].

Exercise   SECTION 3.9

Suppose that a physical phenomenon is modeled by a differential equation of the form

$$\left(D^2 + a_1 D + a_0\right)y = 0$$

where the coefficients a_1 and a_0 are determined experimentally. Suppose that the true values for a_1 and a_0 and 0 and 4, respectively, but due to experimental error the value for a_1 contains a small error. Discuss the difference between the solutions of the differential equation using the true values of the coefficients and the solutions of the equation using the following values for a_1:

(a) .01   (b) −.01

4 Laplace Transforms

4.1 Introduction

In this chapter we will reduce the problem of solving the initial-value problem

$$\left(D^n + a_{n-1}D^{n-1} + \cdots + a_0\right)y = g(t) \tag{4.1}$$

$$y(0) = y_0, \, y'(0) = y_1, \ldots, \, y^{(n-1)}(0) = y_{n-1} \tag{4.2}$$

to a problem which frequently can be solved using elementary algebra and data taken from tables. The basic procedure of this new technique for solving an initial-value problem is as follows. With each function f we associate a new function $L[f]$, called the Laplace transform of f. If f is an unknown solution of the initial-value problem in (4.1) and (4.2), it will turn out that $L[f]$ is an unknown function satisfying an algebraic equation of the form

$$AL[f] + B = C$$

where A, B, and C are constants. Hence $L[f]$ can be found easily. Once $L[f]$ is known, we can recover the function f by techniques to be explained subsequently, and thus we have solved the initial-value problem.

We begin a detailed development of the technique described above by giving the definition of the Laplace transform of a function.

DEFINITION 4.1   Let f be a real-valued function on the interval $(0, \infty)$. If for some real number s the integral

$$\int_0^\infty e^{-st}f(t)\,dt \tag{4.3}$$

exists and is finite, then it is called the **Laplace transform** of f. We will denote the Laplace transform of a function f by $L[f](s)$, though other notations are also in use.

Notice that the Laplace transform of a function f is a function whose domain is the set of all real numbers s for which the integral in (4.3) exists and is finite. For the purposes of this book the precise domain is not important and will not be discussed directly. To keep the notation as simple as possible, we will write $L[f]$ instead of $L[f](s)$.

▶ **Example 1**   We will compute the Laplace transforms of the functions 1, t, e^{at}, and $\cos at$. Using integration by parts to evaluate the Laplace transforms of t and $\cos at$, we have

(i) $L[1]=\displaystyle\int_0^\infty 1\cdot e^{-st}\,dt=-\frac{1}{s}e^{-st}\bigg|_0^\infty=\frac{1}{s}$ for every $s>0$

(ii) $L[t]=\displaystyle\int_0^\infty te^{-st}\,dt=t\left(-\frac{1}{s}e^{-st}\right)\bigg|_0^\infty-\int_0^\infty\left(-\frac{1}{s}\right)e^{-st}\,dt$

$$=-\frac{1}{s^2}e^{-st}\bigg|_0^\infty=\frac{1}{s^2}$$

for every $s>0$

(iii) $L[e^{at}]=\displaystyle\int_0^\infty e^{at}e^{-st}\,dt=\int_0^\infty e^{(a-s)t}\,dt=\frac{1}{a-s}e^{(a-s)t}\bigg|_0^\infty=\frac{1}{s-a}$

for every $s>a$

(iv) $L[\cos at] = \int_0^\infty e^{-st} \cos at \, dt = \left. \dfrac{e^{-st} \sin at}{a} \right|_0^\infty$

$$- \int_0^\infty (-s) e^{-st} \frac{\sin at}{a} \, dt$$

$$= \frac{s}{a} \int_0^\infty e^{-st} \sin at \, dt$$

$$= \left. \left(-\frac{s}{a} e^{-st} \right) \frac{\cos at}{a} \right|_0^\infty - \frac{s}{a} \int_0^\infty (-s) e^{-st} \frac{\cos at}{-a} \, dt$$

$$= \frac{s}{a^2} - \frac{s^2}{a^2} \int_0^\infty e^{-st} \cos at \, dt$$

$$= \frac{s}{a^2} - \frac{s^2}{a^2} L[\cos at] \quad \text{for every } s > 0$$

This last equality can be used to determine $L[\cos at]$:

$$L[\cos at] = \frac{s}{s^2 + a^2} \qquad \blacktriangleleft$$

Before discussing properties of the Laplace transform of a function f, we will present two properties that assure the existence of the Laplace transform $L[f]$.

Suppose that f is a function on $[0, \infty)$ and that there are constants k and M such that $|f(t)| \leq Me^{kt}$ for every $t \geq 0$. Such a function is said to be of **exponential order**. If f is also continuous on $[0, \infty)$, then

$$\left| \int_0^\infty e^{-st} f(t) \, dt \right| \leq \int_0^\infty e^{-st} |f(t)| \, dt$$

$$\leq \int_0^\infty e^{-st} Me^{kt} \, dt$$

$$\leq M \int_0^\infty e^{(k-s)t} \, dt$$

This last integral is finite whenever $s > k$. Therefore a continuous function of exponential order has a Laplace transform. In fact, continuity is not essential.

A function g is said to be **piecewise continuous** on an interval $[a, b]$ if

(a) g is continuous except at a finite number of points, and

(b) if g is not continuous at t_0, then each of the following limits exists:

$$\lim_{t \to t_0^-} g(t) \qquad (t_0 \neq a)$$

$$\lim_{t \to t_0^+} g(t) \qquad (t_0 \neq b)$$

A function g is said to be piecewise continuous on $[0, \infty)$ if it is piecewise continuous on $[0, c]$ for each $c > 0$.

▶ **Example 2**   The functions

$$h(t) = \begin{cases} 0 & \text{if } t \leq 1 \\ 1 & \text{if } 1 < t \end{cases}$$

and

$$f(t) = \begin{cases} 0 & \text{if } 1 < t < 2,\ 3 < t < 4,\ 5 < t < 6, \ldots \\ 1 & \text{if } 0 \leq t \leq 1,\ 2 \leq t \leq 3,\ 4 \leq t \leq 5, \ldots \end{cases}$$

are piecewise continuous on $[0, \infty)$, while

$$g(t) = \begin{cases} 0 & \text{if } t = 1, \dfrac{1}{2}, \dfrac{1}{3}, \ldots \\ 1 & \text{otherwise} \end{cases}$$

is not piecewise continuous since it is discontinuous at infinitely many points in $[0, 1]$.   ◀

THEOREM 4.1   Let f be a piecewise-continuous function on $[0, \infty)$ of exponential order. Then the Laplace transform of f exists.

The proof of this theorem is beyond the scope of this text.

Table 4.1 lists the Laplace transforms of a dozen functions that commonly arise in the solutions of elementary differential equations. We will often need to refer to this table when we know $L[f]$ and wish to recover f. (The table is reproduced inside the back cover for the convenience of the reader.)

Table 4.1

$f(t)$	$L[f]$
1	$\dfrac{1}{s}$
e^{at}	$\dfrac{1}{s-a}$
$\sin at$	$\dfrac{a}{s^2+a^2}$
$\cos at$	$\dfrac{s}{s^2+a^2}$
$t^n(n=1,2,\dots)$	$\dfrac{n!}{s^{n+1}}$
$t^n e^{at}$	$\dfrac{n!}{(s-a)^{n+1}}$
$t\sin at$	$\dfrac{2as}{(s^2+a^2)^2}$
$t\cos at$	$\dfrac{s^2-a^2}{(s^2+a^2)^2}$
$e^{-at}\sin bt$	$\dfrac{b}{(s+a)^2+b^2}$
$e^{-at}\cos bt$	$\dfrac{s+a}{(s+a)^2+b^2}$
$\begin{cases} 1 & \text{if } 0\leqslant t\leqslant a \\ 0 & \text{if } a<t \end{cases}$	$\dfrac{1}{s}(1-e^{-as})$
$\delta(t)$	1

Exercises SECTION 4.1

In Exercises 1–6 use the definition of the Laplace transform to compute the transform of each function.

1. t^2

2. $f(t)=\begin{cases} 0 & \text{if } 0\leqslant t\leqslant a \\ 1 & \text{if } a<t \end{cases}$

3. $\sin at$

4. $f(t) = \begin{cases} 0 & \text{if } t \le a \\ 1 & \text{if } a < t < b \\ 0 & \text{if } b \le t \end{cases}$

5. $6 + t$

6. $t^2 + \cos t$

In Exercises 7–11 determine the values of s for which the Laplace transform $L[f](s)$ exists for each function f.

7. 1

8. e^{-t}

9. e^{at}

10. te^{at}

11. a piecewise continuous function f on $[0, \infty)$ such that $|f(t)| \le Me^{kt}$ for every $t \ge 0$

In Exercises 12–19 use Table 4.1 to find a function whose Laplace transform is:

12. $\dfrac{1}{s-3}$

13. $\dfrac{2}{(s+1)^2 + 4}$

14. $\dfrac{2s}{(s^2+1)^2}$

15. $\dfrac{6}{s^4}$

16. $\dfrac{1}{s^2 + 2s + 1}$

17. $\dfrac{s^2 - 1}{(s^2 + 1)^2}$

18. $\dfrac{s}{s^2 + 3}$

19. $\dfrac{1}{s^2 + 2s + 2}$

20. Let f be the square-wave function defined by

$$f(t) = \begin{cases} 1 & \text{if } 0 \le t \le 1,\, 2 \le t \le 3,\, 4 \le t \le 5,\ldots \\ 0 & \text{if } 1 \le t \le 2,\, 3 \le t \le 4,\, 5 \le t \le 6,\ldots \end{cases}$$

Show that $L[f]=1/s(1+e^{-s})$. Hint: If $|x|<1$, then $1/(1+x)= \sum_{n=0}^{\infty}(-1)^n x^n$.

21. Compute the Laplace transform of the delta function δ. Interpret $\int_0^\infty e^{-st}\delta(t)\,dt$ as $\lim_{a\to 0^-}\int_a^\infty e^{-st}\delta(t)\,dt$.

4.2 Elementary Properties of the Laplace Transform

In this and the following two sections we will obtain needed properties of the Laplace transform. We will return to our investigation of differential equations in Section 4.5.

One of the basic properties of the Laplace transform is isolated in the following theorem. It not only facilitates the computation of the Laplace transform of a sum, but, as we will see in Section 4.5, makes the Laplace transform a useful tool in solving linear differential equations with constant coefficients.

THEOREM 4.2   (Linear Property)   Let f and g be functions whose Laplace transforms exist, and let c_1, c_2 be any two numbers. Then

$$L[c_1 f+c_2 g]=c_1 L[f]+c_2 L[g]$$

Proof

$$L[c_1 f+c_2 g]=\int_0^\infty e^{-st}\left(c_1 f(t)+c_2 g(t)\right)dt$$

$$=\int_0^\infty \left(c_1 e^{-st}f(t)+c_2 e^{-st}g(t)\right)dt$$

$$=c_1\int_0^\infty e^{-st}f(t)\,dt+c_2\int_0^\infty e^{-st}g(t)\,dt$$

$$=c_1 L[f]+c_2 L[g] \quad\blacksquare$$

▶ **Example 1**   We will use Theorem 4.2 to find $L[\cos^2 t]$. Since $\cos^2 t= \frac{1}{2}(1+\cos 2t)$, we have

$$L[\cos^2 t]=L[\tfrac{1}{2}(1+\cos 2t)]$$

$$=\tfrac{1}{2}L[1]+\tfrac{1}{2}L[\cos 2t]$$

From Example 1 of Section 4.1 we have $L[1]=1/s$ and $L[\cos 2t]=s/(s^2+4)$. Hence

$$L[\cos^2 t] = \frac{1}{2}\frac{1}{s} + \frac{1}{2}\frac{s}{s^2+4}$$

◀

Repeated application of Theorem 4.2 yields

COROLLARY 4.1   Let $f_1, f_2, \ldots, f_n$ be functions whose Laplace transforms exist, and let $c_1, c_2, \ldots, c_n$ be numbers. Then

$$L[c_1 f_1 + c_2 f_2 + \cdots + c_n f_n] = c_1 L[f_1] + c_2 L[f_2] + \cdots + c_n L[f_n]$$

▶ **Example 2**   The Laplace transform of $f(t)=4t+7e^{2t}+5\cos 3t$ can be found using Example 1 of Section 4.1 and Corollary 4.1:

$$L[4t+7e^{2t}+5\cos 3t] = 4L[t] + 7L[e^{2t}] + 5L[\cos 3t]$$

$$= \frac{4}{s^2} + \frac{7}{s-2} + \frac{5s}{s^2+9}$$

◀

A portion of Table 4.1 was calculated in Example 1 of Section 4.1. The following theorem gives relatively easy methods for computing many of the other Laplace transforms in the table.

THEOREM 4.3   Let f be a piecewise-continuous function on $[0,\infty)$ of exponential order. If $F(s)$ is the Laplace transform of f, then

   (i) $F(s+b)$ is the Laplace transform of $e^{-bt}f(t)$,

   (ii) $(-1)^n F^{(n)}(s)$ is the Laplace transform of $t^n f(t)$ for $n=1,2,\ldots$

Proof   Let f be as stated in the theorem. Then

$$L[e^{-bt}f(t)] = \int_0^\infty e^{-st}e^{-bt}f(t)\,dt$$

$$= \int_0^\infty e^{-(s+b)t}f(t)\,dt$$

$$= F(s+b)$$

Since $\dfrac{d^n}{ds^n}(e^{-st})=(-1)^n t^n e^{-st}$, we have

$$F^{(n)}(s)=\int_0^\infty \frac{d^n}{ds^n}\left(e^{-st}f(t)\right)dt$$

$$=(-1)^n\int_0^\infty e^{-st}t^n f(t)\,dt$$

$$=(-1)^n L[t^n f(t)]$$

Hence $L[t^n f(t)]=(-1)^n F^{(n)}(s)$ ∎

▶ **Example 3**   In part (iv) of Example 1, Section 4.1, we showed that the Laplace transform of $\cos at$ is $s/(s^2+a^2)$. By Theorem 4.3 (i) the Laplace transform of $e^{-bt}\cos at$ is

$$\frac{s+b}{(s+b)^2+a^2}$$

By Theorem 4.3 (ii) the Laplace transform of $t\cos at$ is

$$-\frac{d}{ds}\left(\frac{s}{s^2+a^2}\right)$$

Hence

$$L[t\cos at]=\frac{s^2-a^2}{(s^2+a^2)^2} \qquad ◀$$

Step functions (such as the function h in Example 2 of Section 4.1) arise in various applications of differential equations. The following theorem gives an alternate method of computing the Laplace transform of a class of functions that includes the step functions.

THEOREM 4.4   Let f be a piecewise-continuous function on $[0,\infty)$ of exponential order. Define a function g by

$$g(t)=\begin{cases}0 & \text{if }0\leqslant t\leqslant a\\ f(t-a) & \text{if }a<t\end{cases}$$

where a is a positive number. Then $L[g]=e^{-as}L[f]$.

Proof

$$L[g] = \int_0^\infty e^{-st} g(t)\, dt$$

$$= \int_0^a e^{-st} \cdot 0 \, dt + \int_a^\infty e^{-st} f(t-a)\, dt$$

$$= \int_a^\infty e^{-st} f(t-a)\, dt$$

If we now change variables in this last integral according to $u = t - a$, we have

$$L[g] = \int_0^\infty e^{-s(u+a)} f(u)\, du$$

$$= e^{-sa} \int_0^\infty e^{-su} f(u)\, du$$

$$= e^{-sa} L[f] \quad \blacksquare$$

▶ **Example 4** Let g be the function defined by

$$g(t) = \begin{cases} 0 & \text{if } 0 \leqslant t \leqslant a \\ 1 & \text{if } a < t \end{cases}$$

In the notation of Theorem 4.4, $f(t) \equiv 1$. In part (i) of Example 1, Section 4.1, we found that $L[1] = 1/s$. Therefore $L[g] = (1/s)e^{-as}$. ◀

▶ **Example 5** Let g be the function defined by

$$g(t) = \begin{cases} 0 & \text{if } 0 \leqslant t \leqslant \pi \\ \cos(t-\pi) & \text{if } \pi < t \end{cases}$$

In the notation of Theorem 4.4, $f(t) = \cos t$. In part (iv) of Example 1, Section 4.1, we found that $L[\cos t] = s/(s^2+1)$. Therefore $L[g] = e^{-s\pi}s/(s^2+1)$. ◀

▶ **Example 6** Let h be the function defined by

$$h(t) = \begin{cases} 1 & \text{if } 0 \leqslant t \leqslant a \\ 0 & \text{if } a < t \end{cases}$$

Then

$$h(x) = 1 - g(x)$$

where g is the function in Example 4, and we have

$$L[h]=L[1]-L[g]$$

$$=\frac{1}{s}-\frac{1}{s}e^{-as}$$

$$=\frac{1}{s}(1-e^{-as})$$ ◀

Exercises   SECTION 4.2

In Exercises 1–12 use Table 4.1 and Theorem 4.2 to compute the Laplace transform of the given function.

1. $3+7t$
2. $6t^4+3\sin t$
3. $5\cos t+2t\sin t$
4. $e^{at}+te^{bt}$
5. $\sin 3t$
6. $t^4e^{5t}-e^{-t}\cos 4t+5$
7. $e^{-at}(\sin at-\cos at)$
8. $e^{-3t}\cos 5t-e^{-5t}\cos 3t$

9. $\begin{cases} 3 & \text{if } 0\leqslant t\leqslant 2 \\ 0 & \text{if } 2<t \end{cases}$

10. $\begin{cases} 7 & \text{if } 0\leqslant t\leqslant 1 \\ 0 & \text{if } 1<t \end{cases}$

11. $\begin{cases} 5 & \text{if } 0\leqslant t\leqslant 3 \\ 4 & \text{if } 3<t \end{cases}$

12. $\displaystyle\sum_{k=1}^{n}\frac{t^k}{k!}$

13. Given that $L[\sin at]=a/(s^2+a^2)$, use Theorem 4.3 to compute the Laplace transform of $e^{-bt}\sin at$, $t\sin at$, and $t^2\sin at$.

14. Given that $L[1]=1/s$, use Theorem 4.3 to compute $L[t]$, $L[t^2]$, $L[te^{at}]$, and $L[t^2e^{at}]$.

In Exercises 15 and 16 use Theorem 4.4 to compute the Laplace transform.

15. $g(t) = \begin{cases} 0 & \text{if } 0 \leqslant t \leqslant 1 \\ e^{t-1} & \text{if } 1 < t \end{cases}$

16. $g(t) = \begin{cases} 0 & \text{if } t \leqslant 2 \\ (t-2)^2 & \text{if } 2 < t \end{cases}$

In Exercises 17–24 you are given $L[f]$. Find f.

17. $\dfrac{2}{s-3}$

18. $\dfrac{3}{(s+1)^2 + 4}$

19. $\dfrac{s}{(s^2+3)^2}$

20. $\dfrac{3}{s^5}$

21. $\dfrac{5}{s^2 + 6s + 9}$

22. $\dfrac{s^2 - 1}{(s^2+1)^2}$

23. $\dfrac{2s+4}{s^2+3}$

24. $\dfrac{1}{s^2 - 1}$

4.3 Inverse Laplace Transforms

As indicated by the discussion at the beginning of Section 4.1, the Laplace transform method reduces the problem of solving a linear differential equation with initial conditions to the problem of finding a function whose Laplace transform is a given function of s. That is, given a function F, we want to find a function f such that $L[f] = F(s)$. Three questions immediately present themselves:

1. Is there a function f such that $L[f] = F(s)$?

2. If there is a function f such that $L[f] = F(s)$, is it unique?

3. If there is a unique function f such that $L[f] = F(s)$, how can it be found?

Not every function F is the Laplace transform of some function. However, if $F(s)=p(s)/q(s)$, where p and q are polynomials with the degree of q greater than the degree of p, then there is a function f such that $L[f]=F(s)$. The functions considered in this text will usually have Laplace transforms of this form or will be functions derived from functions that have Laplace transforms of this form.

It is certainly possible to have two functions with the same Laplace transform. In part (i) of Example 1, Section 4.1, we found that $L[1]=1/s$. Consider the function

$$f(t)=\begin{cases} 1 & \text{if } t\neq 2 \\ 0 & \text{if } t=2 \end{cases}$$

then

$$L[f]=\int_0^\infty e^{-st}f(t)\,dt$$

$$=\int_0^2 e^{-st}\cdot 1\,dt+\int_2^\infty e^{-st}\cdot 1\,dt$$

$$=-\frac{1}{s}(e^{-2s}-e^0)+\left(-\frac{1}{s}\right)(-e^{-2s})$$

$$=\frac{1}{s}=L[1]$$

In fact, if g is any function having a Laplace transform and h is a function such that $h(t)=g(t)$ except for a finite number of values of t, then $L[g]=L[h]$. However, it can be shown that there is at most one continuous function with a given Laplace transform. We state this result as a theorem to emphasize its importance.

THEOREM 4.5   Let f and g be continuous functions on $[0,\infty)$. If $L[f]=L[g]$, then $f(t)=g(t)$ for every $t\geqslant 0$.

DEFINITION 4.2   If $F(s)$ is the Laplace transform of a continuous function, then that continuous function will be denoted by $L^{-1}[F]$ and called the **inverse Laplace transform** of $F(s)$.

▶ **Example 1**   From Example 1 of Section 4.2 we have $L[1]=1/s$, $L[t]=1/s^2$, $L[e^{at}]=1/(s-a)$, and $L[\cos at]=s/(s^2+a^2)$. Hence $L^{-1}[1/s]=1$, $L^{-1}[1/s^2]=t$, $L^{-1}[1/(s-a)]=e^{at}$, and $L^{-1}[s/s^2+a^2]=\cos at$.   ◀

The inverse Laplace transform has a linear property just as does the Laplace transform:

$$L^{-1}[c_1 F + c_2 G] = c_1 L^{-1}[F] + c_2 L^{-1}[G]$$

for any numbers c_1 and c_2 whenever $L^{-1}[F]$ and $L^{-1}[G]$ exist. For example, using Table 4.1 and Theorem 4.2, we have

$$L^{-1}\left[\frac{2}{s} - \frac{3}{s^2}\right] = 2L^{-1}\left[\frac{1}{s}\right] - 3L^{-1}\left[\frac{1}{s^2}\right] = 2 - 3t$$

and

$$L^{-1}\left[\frac{5}{s-3} + \frac{s}{s^2+4}\right] = 5L\left[\frac{1}{s-3}\right] + L\left[\frac{s}{s^2+4}\right] = 5e^{3t} + \cos 2t$$

The third question may now be restated. Given a function $F(s)$ that is the Laplace transform of a continuous function, how is $L^{-1}[F]$ found? The direct computation of $L^{-1}[F]$ requires a knowledge of the theory of integration with respect to a complex variable and will not be considered in this text. The most common way of determining $L^{-1}[F]$ is to use tables, such as Table 4.1. Extensive tables of Laplace transforms may be found in most books of mathematical tables, e.g., *Standard Mathematical Tables* by the Chemical Rubber Publishing Company.

▶ **Example 2**   In practice it is frequently necessary to determine

$$L^{-1}\left[\frac{as+b}{(s-\alpha)(s-\beta)}\right]$$

where a, b, α, and β are real numbers with $\alpha \neq \beta$. We begin by using the method of partial fractions to find numbers A and B such that

$$\frac{as+b}{(s-\alpha)(s-\beta)} = \frac{A}{s-\alpha} + \frac{B}{s-\beta}$$

If we multiply each side of this equation by $(s-\alpha)(s-\beta)$, we obtain

$$as+b = A(s-\beta) + B(s-\alpha)$$

Setting $s = \alpha$ and then setting $s = \beta$, we find that $a\alpha + b = A(\alpha - \beta)$ and $a\beta + b = B(\beta - \alpha)$, respectively. Thus

$$A = \frac{a\alpha+b}{\alpha-\beta} \qquad B = -\frac{a\beta+b}{\alpha-\beta}$$

so that

$$L^{-1}\left[\frac{as+b}{(s-\alpha)(s-\beta)}\right]=L^{-1}\left[\frac{a\alpha+b}{\alpha-\beta}\frac{1}{s-\alpha}-\frac{a\beta+b}{\alpha-\beta}\frac{1}{s-\beta}\right]$$

$$=\frac{a\alpha+b}{\alpha-\beta}L^{-1}\left[\frac{1}{s-\alpha}\right]-\frac{a\beta+b}{\alpha-\beta}L^{-1}\left[\frac{1}{s-\beta}\right]$$

Using Table 4.1, we find that

$$L^{-1}\left[\frac{as+b}{(s-\alpha)(s-\beta)}\right]=\frac{a\alpha+b}{\alpha-\beta}e^{\alpha t}-\frac{a\beta+b}{\alpha-\beta}e^{\beta t} \qquad \blacktriangleleft$$

▶ **Example 3**   The inverse Laplace transform $L^{-1}[(2s+3)/(s-5)(s+4)]$ is easily determined using Example 2. Setting $a=2$, $b=3$, $\alpha=5$, and $\beta=-4$ in Example 2, we find that

$$L^{-1}\left[\frac{2s+3}{(s-5)(s+4)}\right]=\frac{13}{9}e^{5t}+\frac{5}{9}e^{-4t} \qquad \blacktriangleleft$$

▶ **Example 4**   The inverse Laplace transform $L^{-1}[1/s(s+3)]$ can be easily evaluated using Example 2. Setting $a=0$, $b=1$, $\alpha=0$, and $\beta=-3$, we find that

$$L^{-1}\left[\frac{1}{s(s+3)}\right]=\frac{1}{3}L^{-1}\left[\frac{1}{s}\right]-\frac{1}{3}L^{-1}\left[\frac{1}{s+3}\right]$$

$$=\frac{1}{3}-\frac{1}{3}e^{-3t} \qquad \blacktriangleleft$$

Exercises   SECTION 4.3

In Exercises 1–6 you are given $F(s)$. Use the method of partial fractions and Table 4.1 (page 195) to find $L^{-1}[F]$.

1. $\dfrac{1}{s^2-1}$

2. $\dfrac{2s+3}{(s-2)(s+1)}$

3. $\dfrac{5}{s^2(s^2+4)}$

4. $\dfrac{s-4}{(s^2+1)(s^2+4)}$

5. $\dfrac{s+2}{(s-3)(s^2+9)}$

6. $\dfrac{s^2+3s+1}{(s+2)(s^2+16)}$

In Exercises 7–12 you are given $F(s)$. Use Table 4.1 to find $L^{-1}[F]$.

7. $\dfrac{e^{as}}{s}$

8. $\dfrac{1}{s}(e^{-2s}-e^{-s})$

9. $\dfrac{s+5}{s^2+25}$

10. $\dfrac{3}{s^2+4s+7}$

11. $\dfrac{s}{s^2+6s+10}$

12. $\dfrac{s+4}{s^2+2s+5}$

4.4 Convolution

In Example 4 of Section 4.3 we used the method of partial fractions to find the inverse Laplace transform of a function that is the product of two Laplace transforms, namely $1/s$ and $1/(s+3)$. The next theorem gives another method of computing the inverse Laplace transform of a function that is the product of two Laplace transforms. In order to state this theorem in a concise form, we give the following definition.

DEFINITION 4.3   Let f and g be piecewise-continuous functions on $[0, \infty)$ of exponential order. The function called the **convolution** of f and g, denoted by $f*g$, is defined by

$$f*g(t)=\int_0^t f(x)g(t-x)\,dx \qquad \textbf{(4.4)}$$

If we change variables in the integral in (4.4) according to $u=t-x$, then

$$f*g(t)=\int_0^t f(x)g(t-x)\,dx$$

$$=-\int_t^0 f(t-u)g(u)\,du$$

$$=\int_0^t g(u)f(t-u)\,du$$

$$=g*f(t)$$

Hence $f*g=g*f$.

We give the following theorem without proof.

THEOREM 4.6 Let f and g be piecewise continuous on $[0,\infty)$ of exponential order. Then

$$L[f*g]=L[f]L[g]$$

or, equivalently,

$$L^{-1}[L[f]L[g]]=f*g$$

▶ **Example 1** In Example 4 of the preceding section, we showed that $L^{-1}[1/s(s+3)]=\frac{1}{3}(1-e^{-3t})$. We will now show the same result using Theorem 4.6. From Table 4.1 we have $L[1/s]=1$ and $L[1/s+3]=e^{-3t}$. By Theorem 4.6, $L[1*e^{-3t}]=L[1]L[e^{-3t}]=1/s(s+3)$. Hence

$$L^{-1}\left[\frac{1}{s(s+3)}\right]=1*e^{-3t}$$

$$=\int_0^t 1\cdot e^{-3(t-x)}\,dx$$

$$=\frac{1}{3}e^{-3(t-x)}\Big|_0^t$$

$$=\frac{1}{3}(1-e^{-3t})\qquad\qquad◀$$

▶ **Example 2** By Theorem 4.6, $L[e^{\alpha t} * e^{\beta t}] = L[e^{\alpha t}]L[e^{\beta t}] = 1/(s-\alpha)(s-\beta)$ for every pair of distinct real numbers α and β. Then

$$L^{-1}\left[\frac{1}{(s-\alpha)(s-\beta)}\right] = e^{\alpha t} * e^{\beta t}$$

$$= \int_0^t e^{\alpha x} e^{\beta(t-x)}\, dx$$

$$= e^{\beta t} \int_0^t e^{(\alpha-\beta)x}\, dx$$

$$= \frac{e^{\beta t}}{\alpha-\beta}\left(e^{(\alpha-\beta)x}\right)\Big|_0^t$$

$$= \frac{e^{\alpha t} - e^{\beta t}}{\alpha-\beta}$$

For the particular case that $\beta = -\alpha$, we have

$$L^{-1}\left[\frac{1}{s^2-\alpha^2}\right] = \frac{e^{\alpha t} - e^{-\alpha t}}{2\alpha}$$

$$= \frac{1}{\alpha}\sinh\alpha t \qquad\blacktriangleleft$$

▶ **Example 3** One model for the relationship between rainfall and runoff in a watershed is

$$\left(a_2 D^2 + a_1 D + 1\right)Q = I(t) \tag{4.5}$$

$$Q(0)=0, \quad Q'(0)=0$$

where $I(t)$ is the rainfall intensity at time t, and $Q(t)$ is the rate of runoff at time t. The coefficients a_2 and a_1 are determined empirically by considering previous rainfall-runoff relations of the watershed. Of particular interest is the case in which I is the delta function. (See Section 2.13.) Before determining the solution of the initial-value problem in (4.5) when I is the delta function, we will explain the importance of this solution. (This solution is determined in Example 4 of the following section.) Suppose that we have found the solution U of the initial-value problem in (4.5) when I is the delta function. Let f be any continuous function on $[0, \infty)$ with $f(0)=0$. Note that $U(0)=U'(0)=0$ and consider the function

$$z(t)=f*U(t)=\int_0^t f(s)U(t-s)\,ds$$

We will shown that $z(t)$ is a solution of the initial-value problem

$$(a_2 D^2 + a_1 D + 1)Q = f(t)$$

$$Q(0)=0, \quad Q'(0)=0$$

In order to do this, we need to compute the derivatives of $z(t)$. This can be done with a generalization of Leibnitz's rule.

Leibnitz's Rule (generalized version)   Let g and h be continuous functions on the interval $[a, b]$ with $g(t) \leqslant h(t)$ for every t in $[a, b]$. Let A denote the set of all points (t, s) such that $a \leqslant t \leqslant b$ and $g(t) \leqslant s \leqslant h(t)$. If $F(t, s)$ and $(\partial F/\partial t)(t, s)$ are continuous functions on a region containing A, then the function

$$G(t) = \int_{g(t)}^{h(t)} F(t, s)\,ds$$

is continuously differentiable and

$$G'(t) = F(t, h(t))h'(t) - F(t, g(t))g'(t) + \int_{g(t)}^{h(t)} \frac{\partial F}{\partial t}(t, s)\,ds$$

A proof of this result may be found in most advanced calculus texts. Using this version of Leibnitz's rule, we have

$$z'(t) = f(t)U(t-t)\frac{d}{dt}(t) - f(0)U(t-0)\frac{d}{dt}(0) + \int_0^t f(s)U'(t-s)\,ds$$

$$= \int_0^t f(s)U'(t-s)\,ds$$

and

$$z''(t) = f(t)U'(t-t)\frac{d}{dt}(t) - f(0)U'(t-0)\frac{d}{dt}(0) + \int_0^t f(s)U''(t-s)\,ds$$

$$= \int_0^t f(s)U''(t-s)\,ds$$

Evidently $z(0)=z'(0)=0$ and

$$a_2 z'' + a_1 z' + z = a_2 \int_0^t f(s) U''(t-s) \, ds + a_1 \int_0^t f(s) U'(t-s) \, ds$$

$$+ \int_0^t f(s) U(t-s) \, ds$$

$$= \int_0^t \left(a_2 U''(t-s) + a_1 U'(t-s) + U(t-s) \right) f(s) \, ds$$

$$= \int_0^t \delta(t-s) f(s) \, ds \qquad \textbf{(4.6)}$$

We defined f only on $[0, \infty)$. Now if we make f a continuous function on $(-\infty, \infty)$ by defining $f(s)=0$ whenever $s<0$, equation (4.6) can be rewritten as

$$a_2 z'' + a_1 z' + z = \int_{-t}^t \delta(t-s) f(s) \, ds$$

$$= f(t)$$

for each $t>0$ since $\int_{-\varepsilon}^{\varepsilon} \delta(t-s) f(s) \, ds = f(t)$ for every $\varepsilon>0$. (See Exercise 1 of Section 2.13.) We have shown that if U is the solution of the initial-value problem

$$a_2 Q'' + a_1 Q' + Q = \delta(t) \qquad \textbf{(4.7)}$$

$$Q(0)=0, \quad Q'(0)=0$$

then the solution z of the initial-value problem

$$a_2 Q'' + a_1 Q' + Q = f(t) \qquad \textbf{(4.8)}$$

$$Q(0)=0, \quad Q'(0)=0$$

is given by

$$z(t) = f * U(t)$$

whenever f is a continuous function on $[0, \infty)$ with $f(0)=0$. Thus knowing the solution of the initial-value problem in (4.7) enables us to find the solution of the initial-value problem in (4.8).

It can be shown that if U is the solution of the initial-value problem

$$a_n y^{(n)} + a_{n-1} y^{(n-1)} + \cdots + a_0 y = \delta(t)$$

$$y(0)=0, \, y'(0)=0,\ldots, \, y^{(n-1)}(0)=0$$

then the solution of the initial-value problem

$$a_n y^{(n)} + a_{n-1} y^{(n-1)} + \cdots + a_0 y = f(t)$$

$$y(0)=0, \, y'(0)=0,\ldots, \, y^{(n-1)}(0)=0$$

is

$$z(t)=f*U(t)=\int_0^t f(s)U(t-s)\,ds$$

This property is of great importance in certain aspects of hydrology [see *Linear Theory of Hydrological Systems*, Department of Agriculture Technical Bulletin No. 1468, by J. C. I. Dooge] as well as in other branches of engineering. ◀

Exercises SECTION 4.4

In Exercises 1–6 compute $f*g$.

1. $f(t)=1, \quad g(t)=1$
2. $f(t)=1, \quad g(t)=t$
3. $f(t)=t, \quad g(t)= \begin{cases} 1 & \text{if } t\leqslant 3 \\ 0 & \text{if } 3<t \end{cases}$
4. $f(t)=t, \quad g(t)=e^t$
5. $f(t)= \begin{cases} 0 & \text{if } t<2 \\ 1 & \text{if } 2\leqslant t\leqslant 3, \quad g(t)=t \\ 0 & \text{if } 3<t \end{cases}$
6. $f(t)=e^{2t}, \quad g(t)=e^{3t}$

In Exercises 7–10 you are given $F(s)$. Use Theorem 4.6 to compute $L^{-1}[F]$.

7. $\dfrac{1}{(s-1)(s-2)}$

8. $\dfrac{1}{s(s^2+4)}$

9. $\dfrac{s+1}{s(s^2+4)}$

10. $\dfrac{1}{s^2(s+3)}$

In Exercises 11–14 you are given Laplace transforms. Use Table 4.1 and Theorem 4.6 to find the continuous functions whose Laplace transforms they are.

11. $\dfrac{1-e^{-s}}{s^2}$

12. $\dfrac{1-e^{-s}}{s(s^2+\pi^2)}$

13. $\dfrac{1-e^{-3s}}{s(s+2)}$

14. $\dfrac{(1-e^{-s})(1-e^{-2s})}{s^2}$

4.5   Solution of Linear Differential Equations

The use of Laplace transforms enables us to reduce the problem of finding a solution y to a linear differential equation with constant coefficients to the problem of solving a linear algebraic equation for $L[y]$ and then determining y from tables of Laplace transforms. The following theorem is the basis for transforming the differential equation into a linear algebraic equation for $L[y]$.

THEOREM 4.7   Let f be a continuous function on $[0, \infty)$ of exponential order whose derivative f' is also of exponential order and piecewise continuous on $[0, \infty)$. Then

$$L[f']=sL[f]-f(0)$$

Proof Using integration by parts, we have

$$L[f'] = \int_0^\infty e^{-st}f'(t)\, dt$$

$$= e^{-st}f(t)\Big|_0^\infty - \int_0^\infty (-s)e^{-st}f(t)\, dt$$

$$= \lim_{t\to\infty} e^{-st}f(t) - f(0) + s\int_0^\infty e^{-st}f(t)\, dt$$

$$= \lim_{t\to\infty} e^{-st}f(t) + sL[f] - f(0) \tag{4.9}$$

Since f is of exponential order, there are numbers M and b such that $|f(t)| \le Me^{bt}$ for all t. Whenever $s > b$

$$0 \le \lim_{t\to\infty} |e^{-st}f(t)| \le \lim_{t\to\infty} e^{-st}Me^{bt} = \lim_{t\to\infty} Me^{(b-s)t} = 0$$

Hence $\lim_{t\to\infty} |e^{-st}f(t)| = 0$ and $\lim_{t\to\infty} e^{-st}f(t) = 0$. From equation (4.9) we now have

$$L[f'] = sL[f] - f(0)$$

which is the desired result. ∎

Repeated application of this theorem yields

> **THEOREM 4.8** Let f be a function on $[0, \infty)$ having a continuous $(n-1)$st derivative $f^{(n-1)}$ and a piecewise continuous nth derivative $f^{(n)}$. Assume that $f, f', \ldots, f^{(n-1)}$, and $f^{(n)}$ are of exponential order. Then
> $$L[f^{(n)}] = s^n L[f] - s^{n-1}f(0) - s^{n-2}f'(0) - \cdots - f^{(n-1)}(0)$$

Proof We will do the proof for the case $n=2$. A proof for an arbitrary positive integer n can be done using induction. Set $g = f'$ so that $g' = f''$. Then g satisfies the hypotheses of Theorem 4.7 and we have

$$L[f''] = L[g'] = sL[g] - g(0)$$

$$= sL[f'] - f'(0) \tag{4.10}$$

Noticing that f also satisfies the hypotheses of Theorem 4.7, we have

$$L[f'] = sL[f] - f(0) \tag{4.11}$$

Combining equations (4.10) and (4.11) gives

$$L[f''] = s(sL[f] - f(0)) - f'(0)$$

$$= s^2 L[f] - sf(0) - f'(0)$$

which is the desired result. ■

The method of Laplace transforms to solve an initial-value problem

$$y^{(n)} + a_{n-1} y^{(n)} + \cdots + a_0 y = g(t) \qquad \textbf{(4.12)}$$

$$y(0) = y_0, \ y'(0) = y_1, \ldots, \ y^{(n-1)}(0) = y_{n-1}$$

can be summarized as follows:

1. Take the Laplace transform of each side of equation (4.12).

2. Use the linear property of the Laplace transform (Theorem 4.2), Theorem 4.8, and the initial conditions to obtain a linear algebraic equation for the Laplace transform $L[y]$ of the solution.

3. Solve the algebraic equation for $L[y]$.

4. Use a table of Laplace transforms to determine the solution y of the initial-value problem (4.12).

We will now consider several detailed examples which will illustrate this procedure.

▶ **Example 1** Consider the initial-value problem

$$y' + 3y = 1, \quad y(0) = 2 \qquad \textbf{(4.13)}$$

To solve this problem, we begin by taking the Laplace transform of each side of the differential equation. Using the linear property of the Laplace transform, we have

$$L[y'] + 3L[y] = L[1] \qquad \textbf{(4.14)}$$

Using Theorem 4.7, or Theorem 4.8 with $n = 1$,

$$L[y'] = sL[y] - y(0)$$

$$= sL[y] - 2 \qquad \textbf{(4.15)}$$

From Table 4.1 we have $L[1]=1/s$. Using this identity and the one in (4.15), we can rewrite equation (4.14) as

$$(sL[y]-2)+3L[y]=\frac{1}{s}$$

or, equivalently,

$$(s+3)L[y]=2+\frac{1}{s}$$

Hence

$$L[y]=\frac{2}{s+3}+\frac{1}{s(s+3)}$$

The solution y can now be written in terms of inverse Laplace transforms (using the linear property of L^{-1}) as

$$y=L^{-1}\left[\frac{2}{s+3}+\frac{1}{s(s+3)}\right]$$

$$=2L^{-1}\left[\frac{1}{s+3}\right]+L^{-1}\left[\frac{1}{s(s+3)}\right]$$

In Example 1 of Section 4.4 we found that $L^{-1}[1/s(s+3)]=\frac{1}{3}-\frac{1}{3}e^{-3t}$. From Table 4.1 we find that $L^{-1}[1/(s+3)]=e^{-3t}$. Hence

$$y(t)=2e^{-3t}+\frac{1}{3}-\frac{1}{3}e^{-3t}$$

$$=\frac{5}{3}e^{-3t}+\frac{1}{3}$$

is the solution of the initial-value problem (4.13).   ◀

▶ **Example 2**   Consider the initial-value problem

$$y''-3y'+2y=e^{-t} \tag{4.16}$$

$$y(0)=3,\quad y'(0)=4$$

If we take the Laplace transform of each side of the differential equation, we obtain

$$L[y'']-3L[y']+2L[y]=L[e^{-t}] \tag{4.17}$$

Using Theorem 4.8 with $n=2$ and $n=1$ yields

$$L[y''] = s^2 L[y] - sy(0) - y'(0)$$

$$= s^2 L[y] - 3s - 4 \tag{4.18}$$

and

$$L[y'] = sL[y] - y(0)$$

$$= sL[y] - 3 \tag{4.19}$$

respectively. From Table 4.1 we find

$$L[e^{-t}] = \frac{1}{s+1} \tag{4.20}$$

The identities in (4.18), (4.19), and (4.20) enable us to rewrite equation (4.17) as

$$\left(s^2 L[y] - 3s - 4\right) - 3\left(sL[y] - 3\right) + 2L[y] = \frac{1}{s+1}$$

or, equivalently,

$$(s^2 - 3s + 2)L[y] = 3s - 5 + \frac{1}{s+1}$$

Hence

$$L[y] = \frac{3s-5}{(s-2)(s-1)} + \frac{1}{(s-2)(s-1)(s+1)}$$

The solution y can now be written in terms of inverse Laplace transforms as

$$y = L^{-1}\left[\frac{3s-5}{(s-2)(s-1)} + \frac{1}{(s-2)(s-1)(s+1)}\right]$$

$$= L^{-1}\left[\frac{3s-5}{(s-2)(s-1)}\right] + L^{-1}\left[\frac{1}{(s-2)(s-1)(s+1)}\right] \tag{4.21}$$

The first summand can be evaluated using Example 2 of Section 4.3:

$$L^{-1}\left[\frac{3s-5}{(s-2)(s-1)}\right]=e^{2t}+2e^t$$

We will evaluate the second summand using the method of partial fractions and Table 4.1. We begin by finding numbers A, B, and C such that

$$\frac{1}{(s-2)(s-1)(s+1)}=\frac{A}{s-2}+\frac{B}{s-1}+\frac{C}{s+1}$$

Multiplying each side of this equation by $(s-2)(s-1)(s+1)$, we obtain

$$1=A(s-1)(s+1)+B(s-2)(s+1)+C(s-2)(s-1)$$

Setting $s=2$, then $s=1$, and then $s=-1$, we find that $1=3A$, $1=-2B$, and $1=6C$, respectively. Thus $A=\frac{1}{3}$, $B=-\frac{1}{2}$, and $C=\frac{1}{6}$ so that

$$L^{-1}\left[\frac{1}{(s-2)(s-1)(s+1)}\right]=\frac{1}{3}L^{-1}\left[\frac{1}{s-2}\right]-\frac{1}{2}L^{-1}\left[\frac{1}{s-1}\right]=\frac{1}{6}L^{-1}\left[\frac{1}{s+1}\right]$$

$$=\tfrac{1}{3}e^{2t}-\tfrac{1}{2}e^t+\tfrac{1}{6}e^{-t}$$

We are now able to use equation (4.21) to obtain the solution:

$$y=(e^{2t}+2e^t)+\left(\tfrac{1}{3}e^{2t}-\tfrac{1}{2}e^t+\tfrac{1}{6}e^{-t}\right)$$

$$=\tfrac{4}{3}e^{2t}+\tfrac{3}{2}e^t+\tfrac{1}{6}e^{-t} \qquad \blacktriangleleft$$

▶ **Example 3**   Consider the initial-value problem

$$y''+2y'+5y=5 \tag{4.22}$$

$$y(0)=0, \quad y'(0)=0$$

If we take the Laplace transform of each side of this differential equation, we obtain

$$L[y'']+2L[y']+5L[y]=5L[1] \tag{4.23}$$

From Theorem 4.8 with $n=1,2$ and Table 4.1, we find that

$$L[y''] = s^2 L[y] - sy(0) - y'(0)$$

$$= s^2 L[y]$$

$$L[y'] = sL[y] - y(0)$$

$$= sL[y]$$

$$L[1] = \frac{1}{s}$$

Using these identities, equation (4.23) can be rewritten as

$$(s^2 + 2s + 5)L[y] = \frac{5}{s}$$

or, equivalently,

$$L[y] = \frac{5}{s(s^2 + 2s + 5)}$$

The solution y can now be written in terms of the inverse Laplace transform as

$$y = L^{-1}\left[\frac{5}{s(s^2 + 2s + 5)} \right]$$

Since the equation $s^2 + 2s + 5 = 0$ has no real solution, there are constants A, B, and C such that

$$\frac{5}{s(s^2 + 2s + 5)} = \frac{A}{s} + \frac{Bs + C}{s^2 + 2s + 5}$$

Multiplying each side of this equation by $s(s^2 + 2s + 5)$, we obtain

$$5 = A(s^2 + 2s + 5) + (Bs + C)s \tag{4.24}$$

Setting $s=0$, then $s=1$, and then $s=-1$, we find that $5=5A$, $5=8A+B+C$, and $5=4A+B-C$, respectively. Thus $A=1$ and

$$B+C = 5 - 8A = -3$$

$$B-C = 5 - 4A = 1$$

Easy calculations show that $B=-1$ and $C=-2$ so that

$$\frac{5}{s(s^2+2s+5)}=\frac{1}{s}-\frac{s+2}{s^2+2s+5}$$

Noticing that $s^2+2s+5=(s+1)^2+4$, we have

$$y=L^{-1}\left[\frac{5}{s(s^2+2s+5)}\right]$$

$$=L^{-1}\left[\frac{1}{s}\right]-L^{-1}\left[\frac{s+2}{s^2+2s+5}\right]$$

$$=L^{-1}\left[\frac{1}{s}\right]-L^{-1}\left[\frac{(s+1)+1}{(s+1)^2+4}\right]$$

$$=L^{-1}\left[\frac{1}{s}\right]-L^{-1}\left[\frac{s+1}{(s+1)^2+4}\right]-\frac{1}{2}L^{-1}\left[\frac{1}{(s^2+1)^2+4}\right]$$

Using Table 4.1, we find that

$$y=1-e^{-t}\cos 2t-e^{-t}\sin 2t$$

is the solution of the initial-value problem (4.22)　◀

▶ **Example 4**　In Example 3 of Section 4.4 we showed that if U is the solution of the initial-value problem

$$a_2Q''+a_1Q'+Q=\delta(t) \tag{4.25}$$

$$Q(0)=0, \quad Q'(0)=0$$

then the solution of the initial-value problem

$$a_2Q''+a_1Q'+Q=f(t) \tag{4.26}$$

$$Q(0)=0, \quad Q'(0)=0$$

is given by

$$z(t)=f*U(t)$$

whenever f is a continuous function on $[0, \infty)$ with $f(0)=0$. We will now determine the solution of the initial-value problem in (4.25).

Taking the Laplace transform of each side of the differential equation in (4.25), we have

$$a_2 L[Q''] + a_1 L[Q'] + L[Q] = L[\delta]$$

so that, using Theorem 4.8,

$$a_2 \left(s^2 L[Q] - sQ(0) - Q'(0) \right) + a_1 \left(sL[Q] - Q(0) \right) + L[Q] = 1$$

Since $Q(0)=0$ and $Q'(0)=0$, this equation gives

$$L[Q] = \frac{1}{a_2 s^2 + a_1 s + 1}$$

In practice, the equation $a_2 s^2 + a_1 s + 1 = 0$ usually has distinct real roots α and β. In such a case

$$L[Q] = \frac{1}{a_2 (s-\alpha)(s-\beta)}$$

From Example 2 of Section 4.4 we have

$$Q(t) = L^{-1} \left[\frac{1}{a_2 (s-\alpha)(s-\beta)} \right]$$

$$= \frac{1}{a_2} L^{-1} \left[\frac{1}{(s-\alpha)(s-\beta)} \right]$$

$$= \frac{1}{a_2 (\alpha - \beta)} \left(e^{\alpha t} - e^{\beta t} \right)$$

Since $a_2 s^2 + a_1 s + 1 = a_2 (s-\alpha)(s-\beta) = a_2 s^2 - a_2 (\alpha+\beta) + a_2 \alpha \beta$, we must have $a_2 \alpha \beta = 1$. Hence $\alpha \beta = 1/a_2$ and we have

$$Q(t) = \frac{\alpha \beta}{\alpha - \beta} \left(e^{\alpha t} - e^{\beta t} \right) \tag{4.27}$$

is the solution of the initial-value problem in (4.25).

The solution to the initial-value problem in (4.26) can now be written as

$$z(t)=f*U(t)$$

$$=\int_0^t f(s)U(t-s)\,ds$$

$$=\frac{\alpha\beta}{\alpha-\beta}\int_0^t f(s)(e^{\alpha(t-s)}-e^{\beta(t-s)})\,ds \qquad \textbf{(4.28)}$$

To be explicit, let us suppose that $a_2=2$, $a_1=1$, and

$$f(t)=\begin{cases} c & \text{if } 0\leqslant t\leqslant 1 \\ 0 & \text{if } 1<t \end{cases}$$

If the initial-value problem in (4.26) models the rainfall-runoff relationship in a watershed (see Example 3 of Section 4.4), then this choice of f means that we are considering a rainfall of constant intensity which lasts for one unit of time. Since $2s^2+3s+1=(2s+1)(s+1)$, the solution U of

$$2Q''+3Q'+Q=\delta(t)$$

$$Q(0)=0, \quad Q'(0)=0$$

can be obtained directly from equation (4.27):

$$Q(t)=e^{-t/2}-e^{-t}$$

The solution, which represents the runoff, of

$$2Q''+3Q'+Q=\begin{cases} c & \text{if } 0\leqslant t\leqslant 1 \\ 0 & \text{if } 1<t \end{cases}$$

$$Q(0)=0, \quad Q'(0)=0$$

can now be obtained from equation (4.28):

$$z(t)=\int_0^t (e^{-(t-s)/2}-e^{-(t-s)})\begin{cases} c & \text{if } 0\leqslant s\leqslant 1 \\ 0 & \text{if } 1<s \end{cases} ds$$

$$=\begin{cases} \int_0^t c(e^{-(t-s)/2}-e^{-(t-s)})\,ds & \text{if } 0\leqslant t\leqslant 1 \\ \int_0^1 c(e^{-(t-s)/2}-e^{-(t-s)})\,ds & \text{if } 1<t \end{cases}$$

$$=\begin{cases} c(2e^{-(t-s)/2}-e^{-(t-s)})\big|_0^t & \text{if } 0\leqslant t\leqslant 1 \\ c(2e^{-(t-s)/2}-e^{-(t-s)})\big|_0^1 & \text{if } 1<t \end{cases}$$

$$=\begin{cases} c(1-2e^{-t/2}+e^{-t}) & \text{if } 0\leqslant t\leqslant 1 \\ c(2e^{-(t-1)/2}-e^{-(t-s)}-2e^{-t/2}+e^{-t}) & \text{if } 1<t \end{cases}$$

For the case $c=1$, the graph of z is given in Figure 4.1. ◀

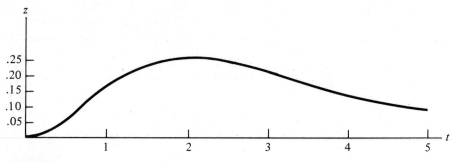

FIGURE 4.1

Exercises SECTION 4.5

Use Laplace transforms to solve the initial-value problems in Exercises 1–10.

1. $y'-2y=t$, $y(0)=3$
2. $y'+4y=\cos t$, $y(0)=5$
3. $y''+3y'+2y=t^2+1$, $y(0)=0$, $y'(0)=1$
4. $y''-5y'-6y=-2\cos 2t+e^t$, $y(0)=0$, $y'(0)=\frac{1}{10}$
5. $y''+y=e^{4t}$, $y(0)=0$, $y'(0)=0$

6. $y'' - y = 4$, $y(0) = 6$, $y'(0) = -3$

7. $y'' + 2y' + 2y = e^{-t}$, $y(0) = -1$, $y'(0) = 1$

8. $y''' + 2y'' + y' = t$, $y(0) = -4$, $y'(0) = 0$, $y''(0) = 0$

9. $y^{(4)} - y = e^{t}$, $y(0) = 0$, $y'(0) = 1$, $y''(0) = 0$, $y'''(0) = 0$

10. $y'' + 4y' + 4y = e^{t}$, $y(0) = 1$, $y'(0) = 1$

11. **(a)** Find the solution of

$$y'' + 3y' + 2y = \delta(t)$$

$$y(0) = 0, \quad y'(0) = 0$$

   (b) Use the solution found in part (a) to obtain the solution of

$$y'' + 3y' + 2y = e^{-3t}$$

$$y(0) = 0, \quad y'(0) = 0$$

12. **(a)** Find the solution of

$$y'' + 4y' + 4 = \delta(t)$$

$$y(0) = 0, \quad y'(0) = 0$$

   (b) Use the solution found in part (a) to obtain the solution of

$$y'' + 4y' + 4y = t$$

$$y(0) = 1, \quad y'(0) = -1$$

   Notice that the initial conditions in part (b) are different from those in part (a). Hint: See Exercise 8 of Section 2.2.

13. Find the solution of

$$y'' + y = F(t)$$

$$y(0) = 0, \quad y'(0) = 0$$

where $F(t) = \begin{cases} 1 & \text{if } 0 \leqslant t \leqslant 1 \\ 0 & \text{if } 1 < t \end{cases}$

14. In one model of foreign exchange speculation under floating exchange, it is assumed that the excess demand at time t caused by speculators is of the form $a_0 + a_1 R(t) + B \cos wt$, where $a_0 > 0$, $a_1 < 0$, and $B > 0$ are constants,

$R(t)$ is the rate of exchange at time t, and $B\cos wt$ represents external factors, such as seasonal influences. It is also assumed that the expected rate of exchange at time t (a value that is expected to be realized at some given time in the future) is of the form $b_0 R(t) + b_1 R'(t) + b_2 R''(t)$, where b_0, b_1, and b_2 are constants whose signs reflect the attitude of the speculators. Based upon these assumptions the following differential equation can be obtained [G. Gondolfo, *Mathematical Models in Economic Dynamics*, North Holland Publishing Company, Amsterdam, 1971, pages 227–235]:

$$b_2 R'' + b_1 R' + \left[a_1 - (1 - b_0)\right] R = -a_0 - B\cos wt$$

One case considered in the reference assigns the following values to the constants:

$$a_0 = 200 \qquad b_0 = 1 \qquad w = 4\pi$$
$$a_1 = -10 \qquad b_1 = -.4$$
$$B = 10 \qquad b_2 = -.2$$

Use Laplace transforms to find the solution for the case $R(0) = 0$ and $R'(0) = 1$.

15. The electrical circuit in Figure 4.2 is described by the differential equation

$$V' + \frac{V - E}{RC} = 0$$

If $E(t) \equiv 3$ volts, $RC = 1$ sec, and $V(0) = 2$ volts, use Laplace transforms to find the voltage V.

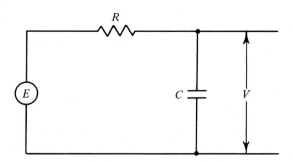

FIGURE 4.2

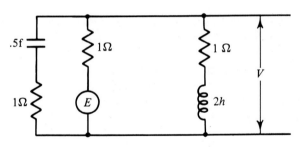

FIGURE 4.3

16. The electrical circuit in Figure 4.3 is described by the differential equation

$$4V'' + 11V' + 8V = 2E'' + 5E' + 2E$$

Use Laplace transforms to find the voltage V if $V(0)=0$, $V'(0)=0$, and

(a) $E(t)=3$ volts   **(b)** $E(t)=\sin t$ volts

17. The deflection $y(x)$ at a distance x from one end of a simply supported beam of length L due to a point load P at the beam's midpoint satisfies the differential equation

$$EIy^{(4)} = P\delta\left(x - \frac{L}{2}\right)$$

and the boundary conditions

$$y(0)=0, \quad y''(0)=0, \quad y(L)=0, \quad y''(L)=0$$

where E is Young's modulus and I is the moment of inertia of the cross section of the beam. (The beam's weight has been neglected.) Use Laplace transforms to find the deflection y. Hint: $y'(0)$ and $y'''(0)$ will have to be chosen so that the conditions at $x=L$ are satisfied.

18. Suppose that the beam in the preceding exercise has the end at $x=0$ restrained and that there is a uniform load of k lb/in² distributed on the beam. Then the equation for the deflection of the beam is

$$EIy^{(4)} = k$$

and the end conditions are

$$y(0)=0, \quad y'(0)=0, \quad y(L)=0, \quad y''(L)=0$$

Use Laplace transforms to find the deflection y.

5 Systems of Linear Differential Equations

5.1 Preliminaries

Throughout this chapter we will consider systems of first-order linear differential equations of the form

$$\frac{dx_1}{dt} = a_{11}(t)x_1 + a_{12}(t)x_2 + \cdots + a_{1n}(t)x_n + f_1(t)$$

$$\frac{dx_2}{dt} = a_{21}(t)x_1 + a_{22}(t)x_2 + \cdots + a_{2n}(t)x_n + f_2(t)$$

$$\cdots \cdots \cdots \cdots \cdots \cdots \cdots \cdots \cdots \cdots \cdots \cdots \quad \textbf{(5.1)}$$

$$\frac{dx_n}{dt} = a_{n1}(t)x_1 + a_{n2}(t)x_2 + \cdots + a_{nn}(t)x_n + f_n(t)$$

where a_{ij} and f_i are functions defined on an open interval I and each x_i is an unknown function of t. In the examples and exercises we will be interested only in the cases $n = 2, 3, 4$.

Such systems of differential equations arise frequently in applications. For example, consider the following sequence of chemical reactions in a well-stirred tank:

$$A \underset{k_4}{\overset{k_1}{\rightleftharpoons}} B \underset{k_3}{\overset{k_2}{\rightleftharpoons}} C$$

We assume that the rate at which each reaction occurs is proportional to the concentration of the chemical. The constants of proportionality are k_1, k_2, k_3, and k_4. If $C_1(t)$, $C_2(t)$, and $C_3(t)$ denote the concentrations at time t of A, B, and C, respectively, then

$$\frac{dC_1}{dt} = -k_1 C_1 + k_4 C_2$$

$$\frac{dC_2}{dt} = k_1 C_1 - (k_2 + k_4) C_2 + k_3 C_3 \tag{5.2}$$

$$\frac{dC_3}{dt} = k_2 C_2 - k_3 C_3$$

In order to simplify the notation we will introduce two concepts from linear algebra: vector and matrix.

DEFINITION 5.1

(i)   An **n×n matrix A** is a square array of numbers

$$\mathbf{A} = \begin{bmatrix} a_{11} & a_{12} & \cdots & a_{1n} \\ a_{21} & a_{22} & \cdots & a_{2n} \\ \cdots\cdots\cdots\cdots\cdots\cdots \\ a_{n1} & a_{n2} & \cdots & a_{nn} \end{bmatrix}$$

The numbers a_{ij} are called the **components** of the matrix $\mathbf{A}$. If each component of $\mathbf{A}$ is a function on an interval I, then $\mathbf{A}$ is called a **matrix function** on I. If each component of a matrix function $\mathbf{A}$ is continuous, then $\mathbf{A}$ is said to be continuous. Matrices will be denoted by boldface capital letters.

(ii)   An **n-vector x** is an ordered n-tuple of numbers

$$\mathbf{x} = \begin{bmatrix} x_1 \\ x_2 \\ \vdots \\ x_n \end{bmatrix}$$

The set of all n-vectors will be denoted by R^n. The n-vector all of whose components x_i are zero is called the **zero vector** and will be denoted by $\mathbf{0}$. The numbers x_i are called the **components** of the vector $\mathbf{x}$. If each component of $\mathbf{x}$ is a function on an interval I, then $\mathbf{x}$ is called a **vector function** on I. If each component of a vector function $\mathbf{x}$ is continuous or differentiable, then $\mathbf{x}$ is said to be continuous or differentiable, respectively. Vectors will be denoted by boldface capital or lower-case letters.

It is possible to define the arithmetic operations of addition and scalar multiplication of n-vectors as follows. If

$$\mathbf{x} = \begin{bmatrix} x_1 \\ x_2 \\ \vdots \\ x_n \end{bmatrix} \quad \text{and} \quad \mathbf{y} = \begin{bmatrix} y_1 \\ y_2 \\ \vdots \\ y_n \end{bmatrix}$$

are two n-vectors and c is any number, then

$$\mathbf{x} + \mathbf{y} = \begin{bmatrix} x_1 + y_1 \\ x_2 + y_2 \\ \vdots \\ x_n + y_n \end{bmatrix} \quad \text{and} \quad c\mathbf{x} = \begin{bmatrix} cx_1 \\ cx_2 \\ \vdots \\ cx_n \end{bmatrix}$$

► **Example 1**   Let $\mathbf{x} = \begin{bmatrix} 1 \\ 2 \end{bmatrix}$ and $\mathbf{y} = \begin{bmatrix} -3 \\ 4 \end{bmatrix}$ be 2-vectors and let $c=7$. Then

$$\mathbf{x} + \mathbf{y} = \begin{bmatrix} 1 \\ 2 \end{bmatrix} + \begin{bmatrix} -3 \\ 4 \end{bmatrix} = \begin{bmatrix} -2 \\ 6 \end{bmatrix}$$

and

$$c\mathbf{x} = 7\begin{bmatrix} 1 \\ 2 \end{bmatrix} = \begin{bmatrix} 7 \\ 14 \end{bmatrix}$$

Similarly,

$$3\mathbf{x} - 2\mathbf{y} = 3\begin{bmatrix} 1 \\ 2 \end{bmatrix} - 2\begin{bmatrix} -3 \\ 4 \end{bmatrix}$$

$$= \begin{bmatrix} 3 \\ 6 \end{bmatrix} + \begin{bmatrix} 6 \\ -8 \end{bmatrix}$$

$$= \begin{bmatrix} 9 \\ -2 \end{bmatrix} \qquad ◄$$

It is also possible to define the product of an $n \times n$ matrix $\mathbf{A} = [a_{ij}]$ and an n-vector

$$\mathbf{x} = \begin{bmatrix} x_1 \\ x_2 \\ \vdots \\ x_n \end{bmatrix}$$

as follows:

$$\mathbf{Ax} = \begin{bmatrix} \sum_{j=1}^{n} a_{1j}x_j \\ \sum_{j=1}^{n} a_{2j}x_j \\ \vdots \\ \sum_{j=1}^{n} a_{nj}x_j \end{bmatrix}$$

To find the ith component of the vector $\mathbf{Ax}$, we multiply each component of the ith row of the matrix $\mathbf{A}$ by the corresponding component of the vector $\mathbf{x}$:

$$\begin{bmatrix} \vdots & \vdots & & \vdots \\ a_{i1} & a_{i2} & \cdots & a_{in} \\ \vdots & \vdots & & \vdots \end{bmatrix} \begin{bmatrix} x_1 \\ x_2 \\ \vdots \\ x_n \end{bmatrix} = \begin{bmatrix} \vdots \\ a_{i1}x_1 + a_{i2}x_2 + \cdots + a_{in}x_n \\ \vdots \end{bmatrix}$$

▶ **Example 2** Let

$$\mathbf{A} = \begin{bmatrix} 1 & 2 & 3 \\ 4 & 5 & 6 \\ 7 & 8 & 9 \end{bmatrix} \quad \text{and} \quad \mathbf{x} = \begin{bmatrix} 3 \\ 2 \\ 1 \end{bmatrix}$$

Then

$$\mathbf{Ax} = \begin{bmatrix} 1 & 2 & 3 \\ 4 & 5 & 6 \\ 7 & 8 & 9 \end{bmatrix} \begin{bmatrix} 3 \\ 2 \\ 1 \end{bmatrix}$$

$$= \begin{bmatrix} 1\cdot3 + 2\cdot2 + 3\cdot1 \\ 4\cdot3 + 5\cdot2 + 6\cdot1 \\ 7\cdot3 + 8\cdot2 + 9\cdot1 \end{bmatrix}$$

$$= \begin{bmatrix} 10 \\ 28 \\ 46 \end{bmatrix} \qquad ◀$$

With these definitions the following theorem can be proved directly.

THEOREM 5.1   If x and y are n-vectors, c a constant, and A an $n \times n$ matrix, then

$$\text{(i)}\quad A(x+y)=Ax+Ay$$
$$\text{(ii)}\quad A(cx)=c(Ax)$$
$$\text{(iii)}\quad A0=0$$

▶ **Example 3**   Let $A = \begin{bmatrix} 1 & 2 \\ 3 & 4 \end{bmatrix}$, $x = \begin{bmatrix} -1 \\ 5 \end{bmatrix}$, $y = \begin{bmatrix} 7 \\ -3 \end{bmatrix}$, and $c=6$.

Then

$$A(x+y) = \begin{bmatrix} 1 & 2 \\ 3 & 4 \end{bmatrix}\left(\begin{bmatrix} -1 \\ 5 \end{bmatrix} + \begin{bmatrix} 7 \\ -3 \end{bmatrix}\right)$$

$$= \begin{bmatrix} 1 & 2 \\ 3 & 4 \end{bmatrix}\begin{bmatrix} 6 \\ 2 \end{bmatrix}$$

$$= \begin{bmatrix} 10 \\ 26 \end{bmatrix}$$

and

$$Ax+Ay = \begin{bmatrix} 1 & 2 \\ 3 & 4 \end{bmatrix}\begin{bmatrix} -1 \\ 5 \end{bmatrix} + \begin{bmatrix} 1 & 2 \\ 3 & 4 \end{bmatrix}\begin{bmatrix} 7 \\ -3 \end{bmatrix}$$

$$= \begin{bmatrix} 9 \\ 17 \end{bmatrix} + \begin{bmatrix} 1 \\ 9 \end{bmatrix}$$

$$= \begin{bmatrix} 10 \\ 26 \end{bmatrix}$$

so that $A(x+y)=Ax+Ay$. Also

$$A(cx) = \begin{bmatrix} 1 & 2 \\ 3 & 4 \end{bmatrix}\left(6\begin{bmatrix} -1 \\ 5 \end{bmatrix}\right)$$

$$= \begin{bmatrix} 1 & 2 \\ 3 & 4 \end{bmatrix}\begin{bmatrix} -6 \\ 30 \end{bmatrix}$$

$$= \begin{bmatrix} 54 \\ 102 \end{bmatrix}$$

and

$$c(\mathbf{Ax}) = 6 \begin{bmatrix} 1 & 2 \\ 3 & 4 \end{bmatrix} \begin{bmatrix} -1 \\ 5 \end{bmatrix}$$

$$= 6 \begin{bmatrix} 9 \\ 17 \end{bmatrix}$$

$$= \begin{bmatrix} 54 \\ 102 \end{bmatrix}$$

so that $\mathbf{A}(c\mathbf{x}) = c(\mathbf{Ax})$.   ◀

In Exercise 4 the reader is asked to prove Theorem 5.1 for the special case $n=2$.

The above notation is easily generalized to the case in which each a_{ij} and each x_i is a function. This is the case that occurs in differential equation theory.

DEFINITION 5.2   Let $\mathbf{x}(t)$ be a vector function each of whose components is defined on an open interval:

$$\mathbf{x}(t) = \begin{bmatrix} x_1(t) \\ x_2(t) \\ \vdots \\ x_n(t) \end{bmatrix}$$

Then the derivative $d\mathbf{x}/dt$ is given by

$$\frac{d\mathbf{x}}{dt} = \begin{bmatrix} \dfrac{dx_1}{dt} \\ \dfrac{dx_2}{dt} \\ \vdots \\ \dfrac{dx_n}{dt} \end{bmatrix}$$

For simplicity of notation $d\mathbf{x}/dt$ will also be denoted by $\mathbf{x}'$.

▶ **Example 4**   If

$$\mathbf{x}(t) = \begin{bmatrix} t^2 + t \\ e^t \\ \cos t \end{bmatrix}$$

then

$$\mathbf{x}'(t) = \begin{bmatrix} 2t + 1 \\ e^t \\ -\sin t \end{bmatrix}$$ ◀

We are now able to write the system of differential equations in (5.1) as

$$\begin{bmatrix} x_1'(t) \\ x_2'(t) \\ \vdots \\ x_n'(t) \end{bmatrix} = \begin{bmatrix} a_{11}(t) & a_{12}(t) & \cdots & a_{1n}(t) \\ a_{21}(t) & a_{22}(t) & \vdots & a_{2n}(t) \\ \cdots & \cdots & \cdots & \cdots \\ a_{n1}(t) & a_{n2}(t) & \cdots & a_{nn}(t) \end{bmatrix} \begin{bmatrix} x_1(t) \\ x_2(t) \\ \vdots \\ x_n(t) \end{bmatrix} + \begin{bmatrix} f_1(t) \\ f_2(t) \\ \vdots \\ f_n(t) \end{bmatrix}$$

If we set

$$\mathbf{x}(t) = \begin{bmatrix} x_1(t) \\ x_2(t) \\ \vdots \\ x_n(t) \end{bmatrix}, \qquad \mathbf{f}(t) = \begin{bmatrix} f_1(t) \\ f_2(t) \\ \vdots \\ f_n(t) \end{bmatrix}$$

and $\mathbf{A}(t) = [a_{ij}(t)]$, then the system of differential equations can be written as

$$\mathbf{x}'(t) = \mathbf{A}(t)\mathbf{x}(t) + \mathbf{f}(t)$$

▶ **Example 5**   Consider the system of differential equations in (5.2).
Setting

$$\mathbf{x}(t) = \begin{bmatrix} C_1(t) \\ C_2(t) \\ C_3(t) \end{bmatrix} \quad \text{and} \quad \mathbf{A} = \begin{bmatrix} -k_1 & k_4 & 0 \\ k_1 & -(k_2 + k_4) & k_3 \\ 0 & k_2 & -k_3 \end{bmatrix}$$

this system can be written as

$$\mathbf{x}'(t) = \mathbf{A}\mathbf{x}(t)$$ ◀

It is possible to use a change of variables to transform an nth-order linear differential equation

$$\left(D^n + a_{n-1}(t)D^{n-1} + \cdots + a_1(t)D + a_0(t)\right)y = f(t) \qquad \text{(5.3)}$$

into a system of first-order linear differential equations that may be written in the form

$$\frac{d\mathbf{x}}{dt} = \mathbf{A}\mathbf{x} + \mathbf{F}(t) \qquad \text{(5.4)}$$

As we will see in Section 5.5, it is more convenient to apply numerical methods to the system of differential equations in (5.4) than to the nth-order linear differential equation in (5.3). In order to transform equation (5.3) into equation (5.4), we change variables as follows:

$$x_1 = y$$
$$x_2 = Dy$$
$$\cdots \cdots$$
$$x_n = D^{n-1}y$$

With this notation we have

$$Dx_1 = Dy = x_2$$
$$Dx_2 = D^2y = x_3$$
$$\cdots \cdots \cdots \cdots \cdots \cdots \cdots \cdots$$
$$Dx_{n-1} = D^{n-1}y = x_n$$
$$Dx_n = D^n y$$
$$= -a_0(t)y - a_1(t)Dy - \cdots - a_{n-1}(t)D^{n-1}y + f(t)$$
$$= -a_0(t)x_1 - a_1(t)x_2 - \cdots - a_{n-1}(t)x_{n-1} + f(t)$$

If we now set

$$\mathbf{x} = \begin{bmatrix} x_1 \\ x_2 \\ \vdots \\ x_n \end{bmatrix} \qquad \mathbf{F}(t) = \begin{bmatrix} 0 \\ 0 \\ \vdots \\ f(t) \end{bmatrix}$$

$$\mathbf{A} = \begin{bmatrix} 0 & 1 & 0 & \cdots & 0 \\ 0 & 0 & 1 & \cdots & 0 \\ \cdots & \cdots & \cdots & \cdots & \cdots \\ 0 & 0 & \cdots & 0 & 1 \\ -a_0(t) & -a_1(t) & \cdots & -a_{n-2}(t) & -a_{n-1}(t) \end{bmatrix}$$

then equation (5.3) can be rewritten as

$$\frac{d\mathbf{x}}{dt} = \mathbf{A}(t)\mathbf{x} + \mathbf{F}(t)$$

▶ **Example 6**   Consider the differential equation

$$(D^2 + 3D - 5)y = \sin t \tag{5.5}$$

Using the change of variables $x_1 = y$, $x_2 = Dy$, we have

$$Dx_1 = Dy = x_2$$

$$Dx_2 = D^2y$$

$$= 5y - 3Dy + \sin t$$

$$= 5x_1 - 3x_2 + \sin t$$

Setting

$$\mathbf{x} = \begin{bmatrix} x_1 \\ x_2 \end{bmatrix}$$

we are able to rewrite equation (5.5) as

$$\frac{d\mathbf{x}}{dt} = \begin{bmatrix} 0 & 1 \\ 5 & -3 \end{bmatrix} \mathbf{x} + \begin{bmatrix} 0 \\ \sin t \end{bmatrix} \qquad\blacktriangleleft$$

Exercises   SECTION 5.1

1. Let $\mathbf{A} = \begin{bmatrix} 1 & 2 & 3 \\ 4 & 5 & 6 \\ 7 & 8 & 9 \end{bmatrix}$

Determine $\mathbf{Ax}$ if

(a) $\mathbf{x} = \begin{bmatrix} 1 \\ 0 \\ 1 \end{bmatrix}$ (b) $\mathbf{x} = \begin{bmatrix} 0 \\ 2 \\ 3 \end{bmatrix}$ (c) $\mathbf{x} = \begin{bmatrix} 0 \\ 0 \\ 0 \end{bmatrix}$ (d) $\mathbf{x} = \begin{bmatrix} 1 \\ 3 \\ 2 \end{bmatrix}$

2. Let

$$A = \begin{bmatrix} 2 & -5 & 6 & 7 \\ 1 & 3 & 5 & 4 \\ 2 & -1 & -3 & 1 \\ 4 & 5 & 7 & -3 \end{bmatrix}$$

Determine Ax if

(a) $x = \begin{bmatrix} 1 \\ 2 \\ 2 \\ 1 \end{bmatrix}$
(b) $x = \begin{bmatrix} 2 \\ 3 \\ 4 \\ 5 \end{bmatrix}$
(c) $x = \begin{bmatrix} 1 \\ 0 \\ 1 \\ 0 \end{bmatrix}$
(d) $x = \begin{bmatrix} 1 \\ 1 \\ 0 \\ 0 \end{bmatrix}$

3. Let $A = \begin{bmatrix} 1 & 2 \\ 2 & 4 \end{bmatrix}$ and $x = \begin{bmatrix} 2 \\ -1 \end{bmatrix}$. Show that $Ax = 0$.
This shows that $Ax = 0$ does not imply that $x = 0$ or that A consists only of zeros.

4. Prove Theorem 5.1 for the case $n = 2$.

5. Use Theorem 5.1 to prove that if x and y are n-vectors, c_1 and c_2 are constants, and A is an $n \times n$ matrix, then

$$A(c_1 x + c_2 y) = c_1 Ax + c_2 Ay$$

6. Let u and v be differentiable vector functions on an open interval I, and let c_1 and c_2 be constants. Show that

$$[c_1 u(t) + c_2 v(t)]' = c_1 u'(t) + c_2 v'(t)$$

7. Write the system of differential equations

$$w'(t) = 2w + 3y - z$$

$$y'(t) = w - 4y + 5z$$

$$z'(t) = y$$

in the form $x'(t) = A(t)x(t)$, where x is a 3-vector and A is a 3×3 matrix.

In Exercises 8-11 rewrite each equation in the form $x' = Ax + f$.

8. $(D^2 + D + 2)y = e^t$
9. $(D^2 + 1)y = e^t + t^2$
10. $(D^3 + 2D^2 - 5D + 4)y = \cos t$
11. $(D^4 + 3D^2 - 2D + 3)y = t$

5.2 Systems of Linear Differential Equations

In the previous section we showed that a system of linear differential equations having the form

$$\frac{dx_1}{dt} = a_{11}(t)x_1 + a_{12}(t)x_2 + \cdots + a_{1n}(t)x_n + f_1(t)$$

$$\frac{dx_2}{dt} = a_{21}(t)x_1 + a_{22}(t)x_2 + \cdots + a_{2n}(t)x_n + f_2(t)$$

$$\cdots\cdots\cdots\cdots\cdots\cdots\cdots\cdots\cdots\cdots\cdots\cdots$$

$$\frac{dx_n}{dt} = a_{n1}(t)x_1 + a_{n2}(t)x_2 + \cdots + a_{nn}(t)x_n + f_n(t)$$

can be written as

$$\frac{d\mathbf{x}}{dt} = \mathbf{A}(t)\mathbf{x} + \mathbf{F}(t) \tag{5.6}$$

where $\mathbf{A}(t) = [a_{ij}(t)]$ and

$$\mathbf{x}(t) = \begin{bmatrix} x_1(t) \\ x_2(t) \\ \vdots \\ x_n(t) \end{bmatrix}$$

Equation (5.6) is called **homogeneous** if $\mathbf{F}(t)$ is the zero vector. Otherwise equation (5.6) is called **nonhomogeneous**. For simplicity of notation, $\mathbf{A}(t)$, $d\mathbf{x}/dt$, and $\mathbf{F}(t)$ will be frequently denoted by $\mathbf{A}$, $\mathbf{x}'$, and $\mathbf{F}$, respectively.

DEFINITION 5.3 Let $\mathbf{A}$ and $\mathbf{F}$ be defined on an open interval I. A vector function $\mathbf{u}$ is a solution of equation (5.6) on I if

$$\mathbf{u}'(t) = \mathbf{A}(t)\mathbf{u}(t) + \mathbf{F}(t)$$

for every t in I.

▶ **Example 1** Consider the differential equation

$$\mathbf{x}' = \begin{bmatrix} 5 & -3 \\ -2 & 4 \end{bmatrix} \mathbf{x} \tag{5.7}$$

The vector function

$$\mathbf{u}(t)=\begin{bmatrix} 3e^{7t} \\ -2e^{7t} \end{bmatrix}$$

is a solution of equation (5.7) on $(-\infty, \infty)$ since

$$\mathbf{u}'(t)=\begin{bmatrix} 21e^{7t} \\ -14e^{7t} \end{bmatrix}$$

and

$$\begin{bmatrix} 5 & -3 \\ -2 & 4 \end{bmatrix}\begin{bmatrix} 3e^{7t} \\ -2e^{7t} \end{bmatrix}=\begin{bmatrix} 21e^{7t} \\ -14e^{7t} \end{bmatrix}$$

Similarly,

$$\mathbf{v}(t)=\begin{bmatrix} e^{2t} \\ e^{2t} \end{bmatrix}$$

is a solution of equation (5.7) on $(-\infty, \infty)$ since

$$\mathbf{v}'(t)=\begin{bmatrix} 2e^{2t} \\ 2e^{2t} \end{bmatrix}$$

and

$$\begin{bmatrix} 5 & -3 \\ -2 & 4 \end{bmatrix}\begin{bmatrix} e^{2t} \\ e^{2t} \end{bmatrix}=\begin{bmatrix} 2e^{2t} \\ 2e^{2t} \end{bmatrix} \qquad \blacktriangleleft$$

▶ **Example 2** Consider the equation

$$\mathbf{x}'=\begin{bmatrix} 1 & 1 & -1 \\ 2 & 3 & -4 \\ 4 & 1 & -4 \end{bmatrix}\mathbf{x} \qquad (5.8)$$

The vector functions

$$\mathbf{u}(t)=\begin{bmatrix} e^{-3t} \\ 7e^{-3t} \\ 11e^{-3t} \end{bmatrix} \quad \text{and} \quad \mathbf{v}(t)=\begin{bmatrix} e^{2t} \\ 2e^{2t} \\ e^{2t} \end{bmatrix}$$

are solutions on $(-\infty, \infty)$ of equation (5.8) since

$$\mathbf{u}'(t) = \begin{bmatrix} -3e^{-3t} \\ -21e^{-3t} \\ -33e^{-3t} \end{bmatrix} \qquad \mathbf{v}'(t) = \begin{bmatrix} 2e^{2t} \\ 4e^{2t} \\ 2e^{2t} \end{bmatrix}$$

and

$$\begin{bmatrix} 1 & 1 & -1 \\ 2 & 3 & -4 \\ 4 & 1 & -4 \end{bmatrix} \begin{bmatrix} e^{-3t} \\ 7e^{-3t} \\ 11e^{-3t} \end{bmatrix} = \begin{bmatrix} -3e^{-3t} \\ -21e^{-3t} \\ -33e^{-3t} \end{bmatrix}$$

$$\begin{bmatrix} 1 & 1 & -1 \\ 2 & 3 & -4 \\ 4 & 1 & -4 \end{bmatrix} \begin{bmatrix} e^{2t} \\ 2e^{2t} \\ e^{2t} \end{bmatrix} = \begin{bmatrix} 2e^{2t} \\ 4e^{2t} \\ 2e^{2t} \end{bmatrix}$$

It is left for the reader to verify that

$$\mathbf{w}(t) = \begin{bmatrix} e^t \\ e^t \\ e^t \end{bmatrix}$$

is also a solution on $(-\infty, \infty)$ of equation (5.8)   ◄

There is a linear property for solutions of the homogeneous equation $\mathbf{x}' = \mathbf{A}\mathbf{x}$, just as there is for the solutions of an nth-order linear differential equation.

THEOREM 5.2   If $\mathbf{u}$ and $\mathbf{v}$ are solutions on an open interval I of

$$\mathbf{x}' = \mathbf{A}(t)\mathbf{x}$$

and c_1, c_2 are any two constants, then

$$\mathbf{w}(t) = c_1\mathbf{u}(t) + c_2\mathbf{v}(t)$$

is also a solution on I.

Proof Since the derivative of a sum is the sum of the derivatives, we have $\mathbf{w}'(t)=c_1\mathbf{u}'(t)+c_2\mathbf{v}'(t)$. (See Exercise 6 of Section 5.1.) By Theorem 5.1

$$A(t)\mathbf{w}=A(t)(c_1\mathbf{u}+c_2\mathbf{v})$$

$$=c_1A(t)\mathbf{u}+c_2A(t)\mathbf{v}$$

$$=c_1\mathbf{u}'+c_2\mathbf{v}'$$

$$=\mathbf{w}'$$

Thus $\mathbf{w}$ is a solution of $\mathbf{x}'=A(t)\mathbf{x}$. ∎

▶ **Example 3** In Example 1 we found that

$$\mathbf{u}(t)=\begin{bmatrix} 3e^{7t} \\ -2e^{7t} \end{bmatrix} \quad \text{and} \quad \mathbf{v}(t)=\begin{bmatrix} e^{2t} \\ e^{2t} \end{bmatrix}$$

are solutions of

$$\mathbf{x}'=\begin{bmatrix} 5 & -3 \\ -2 & 4 \end{bmatrix}\mathbf{x}$$

By Theorem 5.2

$$c_1\begin{bmatrix} 3e^{7t} \\ -2e^{7t} \end{bmatrix}+c_2\begin{bmatrix} e^{2t} \\ e^{2t} \end{bmatrix}=\begin{bmatrix} 3c_1e^{7t}+c_2e^{2t} \\ -2c_1e^{7t}+c_2e^{2t} \end{bmatrix}$$

is also a solution for any choice of the constants c_1 and c_2. ◀

In applications it is frequently the goal to find one specific solution of the differential equation

$$\frac{d\mathbf{x}}{dt}=A(t)\mathbf{x}+\mathbf{F}(t)$$

If one specific solution is desired, then additional information must be given.
When we considered the second-order nonhomogeneous linear differential equation

$$\frac{d^2y}{dt^2}+a_1(t)\frac{dy}{dt}+a_0(t)y=f(t) \tag{5.9}$$

we found that the initial conditions

$$y(t_0)=a \qquad y'(t_0)=b \tag{5.10}$$

were sufficient to determine a unique solution (provided a_1, a_0, and f satisfy the hypotheses of Theorem 2.1). Proceeding as in the previous section, we can transform equation (5.9) into a matrix differential equation by setting $x_1=y$ and $x_2=y'$. Doing this enables us to rewrite the equation as

$$\frac{d\mathbf{x}}{dt}=\mathbf{A}(t)\mathbf{x}+\mathbf{F}(t) \tag{5.11}$$

where

$$\mathbf{x}(t)=\begin{bmatrix} x_1(t) \\ x_2(t) \end{bmatrix} \qquad \mathbf{A}(t)=\begin{bmatrix} 0 & 1 \\ -a_0(t) & -a_1(t) \end{bmatrix} \qquad \mathbf{F}(t)=\begin{bmatrix} 0 \\ f(t) \end{bmatrix}$$

The initial conditions in (5.10) allow us to determine an initial condition for equation (5.11):

$$\mathbf{x}(t_0)=\begin{bmatrix} x_1(t_0) \\ x_2(t_0) \end{bmatrix}$$

$$=\begin{bmatrix} y(t_0) \\ y'(t_0) \end{bmatrix}$$

$$=\begin{bmatrix} a \\ b \end{bmatrix} \tag{5.12}$$

Since the initial-value problem in (5.9) and (5.10) has a unique solution, the initial-value problem in (5.11) and (5.12) must also have a unique solution. This is a special case of the following theorem.

THEOREM 5.3   Let $\mathbf{A}$ be an $n \times n$ matrix function that is continuous on an open interval I, let $\mathbf{F}$ be a continuous n-vector function on I, let t_0 be any number in I, and let $\mathbf{c}$ be any vector in R^n. Then there is a unique solution on I of

$$\mathbf{x}'=\mathbf{A}(t)\mathbf{x}+\mathbf{F}(t), \quad \mathbf{x}(t_0)=\mathbf{c}$$

The problem of finding the solution on I of $\mathbf{x}'=\mathbf{A}(t)\mathbf{x}+\mathbf{F}(t)$ such that $\mathbf{x}(t_0)=\mathbf{c}$ is called the **initial-value problem** for $\mathbf{x}'=\mathbf{A}(t)\mathbf{x}+\mathbf{F}(t)$.

▶ **Example 4**   In Example 1 we found that the vector functions

$$\mathbf{u}(t)=\begin{bmatrix} 3e^{7t} \\ -2e^{7t} \end{bmatrix} \quad \text{and} \quad \mathbf{v}(t)=\begin{bmatrix} e^{2t} \\ e^{2t} \end{bmatrix}$$

are solutions of the differential equation

$$\mathbf{x}'=\begin{bmatrix} 5 & -3 \\ -2 & 4 \end{bmatrix}\mathbf{x} \qquad (5.13)$$

We will now show that the unique solution $\mathbf{x}$ of the differential equation satisfying the initial condition

$$\mathbf{x}(0)=\begin{bmatrix} 1 \\ 2 \end{bmatrix} \qquad (5.14)$$

has the form

$$\mathbf{x}(t)=c_1\mathbf{u}(t)+c_2\mathbf{v}(t)$$

for some choice of the constants c_1 and c_2.

By Theorem 5.2 the function $\mathbf{x}(t)$ is in fact a solution of equation (5.13). We will now choose the constants c_1 and c_2 so that

$$\begin{bmatrix} 1 \\ 2 \end{bmatrix}=\mathbf{x}(0)$$

$$=c_1\mathbf{u}(0)+c_2\mathbf{v}(0)$$

$$=c_1\begin{bmatrix} 3 \\ -2 \end{bmatrix}+c_2\begin{bmatrix} 1 \\ 1 \end{bmatrix}$$

$$=\begin{bmatrix} 3c_1+c_2 \\ -2c_1+c_2 \end{bmatrix}$$

Thus c_1 and c_2 satisfy the system of equations

$$3c_1+c_2=1$$

$$-2c_1+c_2=2$$

A short calculation shows that $c_1 = -\frac{1}{5}$ and $c_2 = \frac{8}{5}$, so

$$\mathbf{x}(t) = -\frac{1}{5}\begin{bmatrix} 3e^{7t} \\ -2e^{7t} \end{bmatrix} + \frac{8}{5}\begin{bmatrix} e^{2t} \\ e^{2t} \end{bmatrix}$$

$$= \begin{bmatrix} -\frac{3}{5}e^{7t} + \frac{8}{5}e^{2t} \\ \frac{2}{5}e^{7t} + \frac{8}{5}e^{2t} \end{bmatrix}$$

is the unique solution of the initial-value problem in (5.13) and (5.14). ◀

Exercises   SECTION 5.2

1. Show that

$$\mathbf{u}(t) = \begin{bmatrix} 3e^{-t} \\ -2e^{-t} \end{bmatrix} \quad \text{and} \quad \mathbf{v}(t) = \begin{bmatrix} e^{4t} \\ e^{4t} \end{bmatrix}$$

are solutions of

$$\mathbf{x}' = \begin{bmatrix} 1 & 3 \\ 2 & 2 \end{bmatrix}\mathbf{x}$$

2. Show that

$$\mathbf{u}(t) = \begin{bmatrix} 4e^t \\ e^t \end{bmatrix} \quad \text{and} \quad \mathbf{v}(t) = \begin{bmatrix} e^{-2t} \\ e^{-2t} \end{bmatrix}$$

are solutions of

$$\mathbf{x}' = \begin{bmatrix} 2 & -4 \\ 1 & -3 \end{bmatrix}\mathbf{x}$$

3. Show that

$$\mathbf{u}(t) = \begin{bmatrix} e^{5t} \\ e^{5t} \\ 0 \end{bmatrix}, \quad \mathbf{v}(t) = \begin{bmatrix} 3e^{3t} \\ e^{3t} \\ 0 \end{bmatrix}, \quad \text{and} \quad \mathbf{w}(t) = \begin{bmatrix} 0 \\ 0 \\ e^{-3t} \end{bmatrix}$$

are solutions of

$$\mathbf{x}' = \begin{bmatrix} 2 & 3 & 0 \\ -1 & 6 & 0 \\ 0 & 0 & -3 \end{bmatrix}\mathbf{x}$$

In Exercises 4 and 5 find numbers a, b, and r such that either a or b is not zero and

$$\mathbf{u}(t) = e^{rt}\begin{bmatrix} a \\ b \end{bmatrix}$$

is a solution of:

4. $\mathbf{x}' = \begin{bmatrix} 1 & 2 \\ 2 & 1 \end{bmatrix}\mathbf{x}$

5. $\mathbf{x}' = \begin{bmatrix} 2 & 4 \\ 5 & 3 \end{bmatrix}\mathbf{x}$

6. Use Exercise 2 to show that the solution of the initial-value problem

$$\mathbf{x}' = \begin{bmatrix} 2 & -4 \\ 1 & -3 \end{bmatrix}\mathbf{x}, \quad \mathbf{x}(0) = \begin{bmatrix} 3 \\ 0 \end{bmatrix}$$

can be written in the form $c_1\mathbf{u}(t) + c_2\mathbf{v}(t)$, where $\mathbf{u}(t)$ and $\mathbf{v}(t)$ are the vectors given in Exercise 2.

7. Use Exercise 1 to show that the solution of the initial-value problem

$$\mathbf{x}' = \begin{bmatrix} 1 & 3 \\ 2 & 2 \end{bmatrix}\mathbf{x}, \quad \mathbf{x}(0) = \begin{bmatrix} 1 \\ 1 \end{bmatrix}$$

can be written in the form $c_1\mathbf{u}(t) + c_2\mathbf{v}(t)$, where $\mathbf{u}(t)$ and $\mathbf{v}(t)$ are the vectors given in Exercise 1.

5.3 Flow Lines (OPTIONAL)

In many applications, the matrix $\mathbf{A}(t)$ in the differential equation $\mathbf{x}' = \mathbf{A}\mathbf{x}$ is a constant $n \times n$ matrix $\mathbf{A}$. It is sometimes useful to consider the graphs in R^n, called **flow lines**, of the solutions of the differential equation

$$\mathbf{x}' = \mathbf{A}\mathbf{x} \qquad (5.15)$$

It can be shown that each point in R^n lies on precisely one flow line of $\mathbf{x}' = \mathbf{A}\mathbf{x}$. In order to illustrate this property, we will consider the special case $n = 2$ in which the equations for the flow lines are easy to compute.

For simplicity of notation we will rewrite equation (5.15) as

$$\begin{bmatrix} x' \\ y' \end{bmatrix} = \begin{bmatrix} a & b \\ c & d \end{bmatrix}\begin{bmatrix} x \\ y \end{bmatrix}$$

or, equivalently,

$$\frac{dx}{dt} = ax + by$$

$$\frac{dy}{dt} = cx + dy \tag{5.16}$$

Using the chain rule and the equations in (5.16), we have

$$\frac{dy}{dx} = \frac{dy}{dt} \Big/ \frac{dx}{dt}$$

$$= \frac{cx + dy}{ax + by} \tag{5.17}$$

which is a first-order homogeneous differential equation that can be solved using the method presented in Section 1.5. The graphs of the (possibly implicit) solutions of equation (5.17) form the flow lines for equation (5.16).

For example, consider the differential equation

$$\begin{bmatrix} x' \\ y' \end{bmatrix} = \begin{bmatrix} 1 & 1 \\ 3 & -1 \end{bmatrix} \begin{bmatrix} x \\ y \end{bmatrix} \tag{5.18}$$

From equation (5.17) we have the first-order homogeneous differential equation

$$\frac{dy}{dx} = \frac{3x - y}{x + y} \tag{5.19}$$

It is left as an exercise for the reader to verify that all solutions of this equation are determined by

$$(y - x)(y + 3x) = c$$

where c is an arbitrary constant. The graphs (flow lines) of these solutions are shown in Figure 5.1.

In Example 1 of Section 5.2 we found that the vector function

$$\mathbf{v}(t) = \begin{bmatrix} x(t) \\ y(t) \end{bmatrix} = \begin{bmatrix} e^{2t} \\ e^{2t} \end{bmatrix}$$

is a solution of equation (5.18). The graph of $\mathbf{v}(t)$ for $-\infty < t < \infty$ is the graph

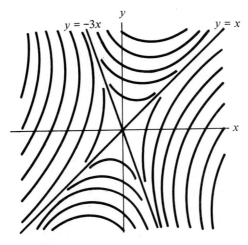

FIGURE 5.1

of $y=x$ for $x>0$. By Theorem 5.2 the vector function

$$\mathbf{w}(t)=-\mathbf{v}(t)=\begin{bmatrix} -e^{2t} \\ -e^{2t} \end{bmatrix}$$

is also a solution of equation (5.18). The graph of $\mathbf{w}(t)$ for $-\infty<t<\infty$ is the graph of $y=x$ for $x<0$. Similarly the solution

$$\mathbf{u}(t)=\begin{bmatrix} e^{-2t} \\ -3e^{-2t} \end{bmatrix}$$

determines the flow lines on the graph of $y=-3x$, with the origin excepted. The single point at the origin is the graph of the solution $\mathbf{x}(t)\equiv\begin{bmatrix} 0 \\ 0 \end{bmatrix}$.

Notice that the flow lines do not intersect each other. Moreover, every point in the xy-plane lies on a flow line. Thus, given any point P in the xy-plane, there is exactly one solution of equation (5.18) whose graph contains the point P. From this we conclude that the initial-value problem

$$\begin{bmatrix} x' \\ y' \end{bmatrix}=\begin{bmatrix} 1 & 1 \\ 3 & -1 \end{bmatrix}\begin{bmatrix} x \\ y \end{bmatrix}, \quad \begin{bmatrix} x(0) \\ y(0) \end{bmatrix}=\begin{bmatrix} a \\ b \end{bmatrix}$$

has a unique solution for every choice of a and b. This is exactly what is guaranteed by Theorem 5.3.

Exercises   SECTION 5.3

In Exercises 1–4 find equations for and sketch a few of the flow lines for the given differential equation.

1. $x' = \begin{bmatrix} 0 & 1 \\ -1 & 0 \end{bmatrix} x$

2. $x' = \begin{bmatrix} 1 & 2 \\ 2 & 1 \end{bmatrix} x$

3. $x' = \begin{bmatrix} -1 & -1 \\ 2 & -3 \end{bmatrix} x$

4. $x' = \begin{bmatrix} -3 & 2 \\ 2 & -3 \end{bmatrix} x$

5–8. The flow lines for the system of differential equations in (5.16) can be approximated by using the method of Section 1.10 on equation (5.17). Use this method to sketch the flow lines for each of the differential equations in Exercises 1–4.

9. The equations of motion for a two-dimensional steady flow of a fluid (the velocity of a particle depends only on its position (x, y) and not on time t) have the form

$$\frac{dx}{dt} = f_1(x, y) \qquad \frac{dy}{dt} = f_2(x, y)$$

Such a flow is called **irrotational** if the individual fluid particles do not rotate about those axes perpendicular to the direction of the particles' motion. A fluid is called **perfect** if:

   (i) it is incompressible, i.e., it has constant density, and
   (ii) there is no resistance to motion within the fluid.

A function $F(x, y)$ is called a **velocity potential** for the flow if

$$\frac{dx}{dt} = \frac{\partial F}{\partial x} \qquad \frac{dy}{dt} = \frac{\partial F}{\partial y}$$

It can be shown that if an irrotational fluid is perfect, then there exist velocity potentials for the flow, and every velocity potential satisfies Laplace's equation:

$$\frac{\partial^2 F}{\partial x^2} + \frac{\partial^2 F}{\partial y^2} = 0$$

(a) Show that the function $F(x, y) = x^2 - y^2$ satisfies Laplace's equation.
(b) Determine the flow lines for the planar flow of a perfect irrotational fluid having $F(x, y) = x^2 - y^2$ as a velocity potential.
(c) Graph the flow lines in the closed first quadrant. Having done this, you will see why this flow is sometimes referred to as a flow in a corner.
(d) Use Theorem 5.3 to explain why no two distinct fluid particles can simultaneously occupy the same position in the xy-plane.

5.4 General Solutions

A homogeneous nth-order linear differential equation has n linearly independent solutions. The same is true for the equation $\mathbf{x}' = \mathbf{A}\mathbf{x}$ whenever $\mathbf{A}$ is a continuous $n \times n$ matrix. We will first define what is meant by linearly independent solutions of $\mathbf{x}' = \mathbf{A}\mathbf{x}$ and then state an appropriate theorem.

DEFINITION 5.4 Let $\mathbf{g}_1, \mathbf{g}_2, \ldots, \mathbf{g}_k$ be n-vector functions defined on an open interval I. These functions are said to be **linearly dependent** if there are constants $c_1, c_2, \ldots, c_k$, at least one of which is nonzero, such that

$$c_1\mathbf{g}(t) + c_2\mathbf{g}_2(t) + \cdots + c_k\mathbf{g}_k(t) = \mathbf{0}$$

for every t in I. Otherwise, the functions are said to be **linearly independent**.

▶ **Example 1** Consider the vector functions

$$\mathbf{u}(t) = \begin{bmatrix} 3e^{7t} \\ -2e^{7t} \end{bmatrix} \quad \text{and} \quad \mathbf{v}(t) = \begin{bmatrix} e^{2t} \\ e^{2t} \end{bmatrix}$$

from Example 1 of Section 5.2. Let c_1 and c_2 be constants such that

$$\begin{bmatrix} 0 \\ 0 \end{bmatrix} = c_1 \begin{bmatrix} 3e^{7t} \\ -2e^{7t} \end{bmatrix} + c_2 \begin{bmatrix} e^{2t} \\ e^{2t} \end{bmatrix}$$

$$= \begin{bmatrix} 3c_1e^{7t} + c_2e^{2t} \\ -2c_1e^{7t} + c_2e^{2t} \end{bmatrix}$$

Thus

$$3c_1e^{7t} + c_2e^{2t} = 0 \tag{5.20}$$

and

$$-2c_1e^{7t} + c_2e^{2t} = 0 \tag{5.21}$$

for all values of t. Since e^{7t} and e^{2t} are linearly independent, we must have $c_1 = 0$ and $c_2 = 0$. Thus $c_1 = 0$, $c_2 = 0$ is the only choice of constants that satisfies equations (5.20) and (5.21). The vector functions $\mathbf{u}$ and $\mathbf{v}$ are linearly independent. ◀

▶ **Example 2**   Consider the vector functions

$$\mathbf{u}(t) = \begin{bmatrix} e^{-3t} \\ 7e^{-3t} \\ 11e^{-3t} \end{bmatrix} \qquad \mathbf{v}(t) = \begin{bmatrix} e^{2t} \\ 2e^{2t} \\ e^{2t} \end{bmatrix} \qquad \mathbf{w}(t) = \begin{bmatrix} e^{t} \\ e^{t} \\ e^{t} \end{bmatrix}$$

from Exercise 3 of Section 5.2. Let c_1, c_2, and c_3 be constants such that

$$\begin{bmatrix} 0 \\ 0 \\ 0 \end{bmatrix} = c_1 \begin{bmatrix} e^{-3t} \\ 7e^{-3t} \\ 11e^{-3t} \end{bmatrix} + c_2 \begin{bmatrix} e^{2t} \\ 2e^{2t} \\ e^{2t} \end{bmatrix} + c_3 \begin{bmatrix} e^{t} \\ e^{t} \\ e^{t} \end{bmatrix} = \begin{bmatrix} c_1 e^{-3t} + c_2 e^{2t} + c_3 e^{t} \\ 7c_1 e^{-3t} + 2c_2 e^{2t} + c_3 e^{t} \\ 11c_1 e^{-3t} + c_2 e^{2t} + c_3 e^{t} \end{bmatrix}$$

Thus

$$c_1 e^{-3t} + c_2 e^{2t} + c_3 e^{2t} + c_3 e^{t} = 0 \tag{5.22}$$

$$7c_1 e^{-3t} + 2c_2 e^{2t} + c_3 e^{t} = 0 \tag{5.23}$$

$$11c_1 e^{-3t} + c_2 e^{2t} + c_3 e^{t} = 0 \tag{5.24}$$

for all values of t. Since e^{-3t}, e^{2t}, and e^{t} are linearly independent, we must have $c_1 = 0$, $c_2 = 0$, and $c_3 = 0$. Hence the vector functions $\mathbf{u}$, $\mathbf{v}$, and $\mathbf{w}$ are linearly independent. ◀

It is now possible to state a theorem that will serve as the basis for determining all of the solutions of the homogeneous system of equations $\mathbf{x}' = \mathbf{A}\mathbf{x}$.

THEOREM 5.4   Let $\mathbf{A}$ be a continuous $n \times n$ matrix on an open interval I. Then there are n linearly independent solutions on I of $\mathbf{x}' = \mathbf{A}\mathbf{x}$. Moreover, if $u_1, u_2, \ldots, u_n$ are n linearly independent solutions on I of $\mathbf{x}' = \mathbf{A}\mathbf{x}$ and if $\mathbf{w}$ is any solution on I of $\mathbf{x}' = \mathbf{A}\mathbf{x}$, then there are constants $c_1, c_2, \ldots, c_n$ such that

$$\mathbf{w}(t) = c_1 \mathbf{u}_1(t) + c_2 \mathbf{u}_2(t) + \cdots + c_n \mathbf{u}_n(t)$$

for every t in I.

Proof   We will do the proof of the first part of the theorem for the special case $n=2$. Let t_0 be any number in the interval I. By Theorem 5.3 there are solutions $\mathbf{u}$ and $\mathbf{v}$ on I of $\mathbf{x}'=\mathbf{A}\mathbf{x}$ such that

$$\mathbf{u}(t_0)=\begin{bmatrix}1\\0\end{bmatrix} \quad \text{and} \quad \mathbf{v}(t_0)=\begin{bmatrix}0\\1\end{bmatrix}$$

Suppose that c_1 and c_2 are constants such that

$$c_1\mathbf{u}(t)+c_2\mathbf{v}(t)=\mathbf{0} \tag{5.25}$$

for every t in I. In particular, equation (5.25) holds when $t=t_0$. In such a case

$$\begin{bmatrix}0\\0\end{bmatrix}=c_1\begin{bmatrix}1\\0\end{bmatrix}+c_2\begin{bmatrix}0\\1\end{bmatrix}=\begin{bmatrix}c_1\\c_2\end{bmatrix}$$

so that $c_1=0$ and $c_2=0$. Thus the solutions $\mathbf{u}$ and $\mathbf{v}$ are linearly independent. The interested reader is referred to Exercises 8 and 9 for the proof of the second part of this theorem. ∎

   The preceding theorem tells us that if we can find n linearly independent solutions of $\mathbf{x}'=\mathbf{A}\mathbf{x}$, then we can find all of the solutions by considering sums of these n linearly independent solutions. This leads us to the following definition.

DEFINITION 5.5   Let $\mathbf{A}$ be a continuous $n\times n$ matrix on an open interval I, and let $\mathbf{u}_1, \mathbf{u}_2,\ldots,\mathbf{u}_n$ be n linearly independent solutions on I of $\mathbf{x}'=\mathbf{A}\mathbf{x}$. The expression

$$\mathbf{x}(t)=c_1\mathbf{u}_1(t)+c_2\mathbf{u}_2(t)+ \cdots +c_n\mathbf{u}_n(t)$$

where $c_1, c_2,\ldots, c_n$ are constants, is called a **general solution** on I of $\mathbf{x}'=\mathbf{A}\mathbf{x}$.

▶ **Example 3**   In Example 1 of Section 5.2 and Example 1 of this section we showed that

$$\mathbf{u}(t)=\begin{bmatrix}3e^{7t}\\-2e^{7t}\end{bmatrix} \quad \text{and} \quad \mathbf{v}(t)=\begin{bmatrix}e^{2t}\\e^{2t}\end{bmatrix}$$

are linearly independent solutions of

$$\mathbf{x}'=\begin{bmatrix}5&-3\\-2&4\end{bmatrix}\mathbf{x} \tag{5.26}$$

on $(-\infty, \infty)$. Therefore

$$\mathbf{x}(t) = c_1 \begin{bmatrix} 3e^{7t} \\ -2e^{7t} \end{bmatrix} + c_2 \begin{bmatrix} e^{2t} \\ e^{2t} \end{bmatrix}$$

is a general solution of equation (5.26) on $(-\infty, \infty)$. ◀

▶ **Example 4** In Example 2 of Section 5.2 and Example 2 of this section we showed that

$$\mathbf{u}(t) = \begin{bmatrix} e^{-3t} \\ 7e^{-3t} \\ 11e^{-3t} \end{bmatrix}, \qquad \mathbf{v}(t) = \begin{bmatrix} e^{2t} \\ 2e^{2t} \\ e^{2t} \end{bmatrix}, \qquad \text{and} \quad \mathbf{w}(t) = \begin{bmatrix} e^{t} \\ e^{t} \\ e^{t} \end{bmatrix}$$

are linearly independent solutions of

$$\mathbf{x}' = \begin{bmatrix} 1 & 1 & -1 \\ 2 & 3 & -4 \\ 4 & 1 & -4 \end{bmatrix} \mathbf{x} \qquad (5.27)$$

on $(-\infty, \infty)$. Therefore

$$\mathbf{x}(t) = c_1 \begin{bmatrix} e^{-3t} \\ 7e^{-3t} \\ 11e^{-3t} \end{bmatrix} + c_2 \begin{bmatrix} e^{2t} \\ 2e^{2t} \\ e^{2t} \end{bmatrix} + c_3 \begin{bmatrix} e^{t} \\ e^{t} \\ e^{t} \end{bmatrix}$$

is a general solution of equation (5.27). ◀

THEOREM 5.5 Let

$$\mathbf{u}_1(t) = \begin{bmatrix} u_{11}(t) \\ u_{21}(t) \\ \vdots \\ u_{n1}(t) \end{bmatrix}, \quad \mathbf{u}_2(t) = \begin{bmatrix} u_{12}(t) \\ u_{22}(t) \\ \vdots \\ u_{n2}(t) \end{bmatrix}, \dots, \mathbf{u}_n(t) = \begin{bmatrix} u_{1n}(t) \\ u_{2n}(t) \\ \vdots \\ u_{nn}(t) \end{bmatrix}$$

be n solutions on an open interval I of $\mathbf{x}' = \mathbf{A}\mathbf{x}$, where $\mathbf{A}$ is a continuous $n \times n$ matrix on I. These solutions are linearly independent if and only if the determinant (called the Wronskian of $\mathbf{u}_1, \mathbf{u}_2, \dots, \mathbf{u}_n$)

$$\det \begin{bmatrix} u_{11}(t) & u_{12}(t) & \cdots & u_{1n}(t) \\ u_{21}(t) & u_{22}(t) & \cdots & u_{2n}(t) \\ \vdots & \vdots & & \vdots \\ u_{n1}(t) & u_{n2}(t) & \cdots & u_{nn}(t) \end{bmatrix}$$

is not zero for at least one t in I.

The proof of this theorem relies on elementary concepts from linear algebra and will be omitted. In fact, it can be shown that the Wronskian is either zero for all t in I or is nonzero for all t in I.

▶ **Example 5** In Example 1 of Section 5.2 we found that the vector functions

$$\mathbf{u}(t)=\begin{bmatrix} 3e^{7t} \\ -2e^{7t} \end{bmatrix} \quad \text{and} \quad \mathbf{v}(t)=\begin{bmatrix} e^{2t} \\ e^{2t} \end{bmatrix}$$

are solutions on $(-\infty, \infty)$ of

$$\mathbf{x}'=\begin{bmatrix} 5 & -3 \\ -2 & 4 \end{bmatrix}\mathbf{x}$$

Since

$$\det\begin{bmatrix} 3e^{7t} & e^{2t} \\ -2e^{7t} & e^{2t} \end{bmatrix}=5e^{9t}\neq 0$$

these solutions are linearly independent. Compare this example with Example 1 of this section. ◀

▶ **Example 6** In Example 2 of Section 5.2 we found that the vector functions

$$\mathbf{u}(t)=\begin{bmatrix} e^{-3t} \\ 7e^{-3t} \\ 11e^{-3t} \end{bmatrix}, \quad \mathbf{v}(t)=\begin{bmatrix} e^{2t} \\ 2e^{2t} \\ e^{2t} \end{bmatrix}, \quad \text{and} \quad \mathbf{w}(t)=\begin{bmatrix} e^{t} \\ e^{t} \\ e^{t} \end{bmatrix}$$

are solutions on $(-\infty, \infty)$ of

$$\mathbf{x}'=\begin{bmatrix} 1 & 1 & -1 \\ 2 & 3 & -4 \\ 4 & 1 & -4 \end{bmatrix}\mathbf{x}$$

Since

$$\det\begin{bmatrix} e^{-3t} & e^{2t} & e^{t} \\ 7e^{-3t} & 2e^{2t} & e^{t} \\ 11e^{-3t} & e^{2t} & e^{t} \end{bmatrix}=-10$$

for any t, these solutions are linearly independent. Compare this example with Example 2 of this section. ◀

We are now in position to discuss the nonhomogeneous differential equation $x' = Ax + f$.

> **THEOREM 5.6**   Let A be a continuous $n \times n$ matrix function and f a continuous vector function on an open interval I. Let $u_1, u_2, \ldots, u_n$ be linearly independent solutions of $x' = Ax$, and let u be any solution on I of $x' = Ax + f$. Then every solution v on I can be written
>
> $$v(t) = c_1 u_1(t) = c_2 u_2(t) + \cdots + c_n u_n(t) + u(t)$$
>
> for an appropriate choice of the constants $c_1, c_2, \ldots, c_n$.

Proof   Consider the function $w = v - u$. Then

$$w' = v' - u'$$

$$= Av + f - (Au + f)$$

$$= Av - Au$$

$$= A(v - u) = Aw$$

so that w is a solution of the homogeneous equation $x' = Ax$. By Theorem 5.4 there exist constants $c_1, c_2, \ldots, c_n$ such that

$$w(t) = c_1 u_1(t) + c_2 u_2(t) + \cdots + c_n u_n(t)$$

for every t in I. It follows immediately that

$$v(t) = c_1 u_1(t) + c_2 u_2(t) + \cdots + c_n u_n(t) + u(t)$$

for every t in I.   ■

> **DEFINITION 5.6**   Let A be a continuous $n \times n$ matrix function, and let f be a continuous vector function on an open interval I. Let $u_1, u_2, \ldots, u_n$ be linearly independent solutions of $x' = Ax$, and let u be any solution of $x' = Ax + f$. Then
>
> $$x(t) = c_1 u_1(t) + c_2 u_2(t) + \cdots + c_n u_n(t) + u(t)$$
>
> where $c_1, c_2, \ldots, c_n$ are constants, is called a **general solution** of $x' = Ax + f$.

In other words, a general solution of $x' = Ax + f$ is the sum of a general solution of $x' = Ax$ and any particular solution of $x' = Ax + f$.

► **Example 7** The vector function

$$\mathbf{u}(t) = \begin{bmatrix} -\frac{7}{12}e^{-t} \\ -\frac{5}{6}e^{-t} \end{bmatrix}$$

is a solution of

$$\mathbf{x}' = \begin{bmatrix} 5 & -3 \\ -2 & 4 \end{bmatrix} \mathbf{x} + \begin{bmatrix} e^{-t} \\ 3e^{-t} \end{bmatrix}$$ (5.28)

since

$$\mathbf{u}'(t) = \begin{bmatrix} \frac{7}{12}e^{-t} \\ \frac{5}{6}e^{-t} \end{bmatrix}$$

and

$$\begin{bmatrix} 5 & -3 \\ -2 & 4 \end{bmatrix} \mathbf{u}(t) + \mathbf{f}(t) = \begin{bmatrix} 5 & -3 \\ -2 & 4 \end{bmatrix} \begin{bmatrix} -\frac{7}{12}e^{-t} \\ -\frac{5}{6}e^{-t} \end{bmatrix} + \begin{bmatrix} e^{-t} \\ 3e^{-t} \end{bmatrix}$$

$$= \begin{bmatrix} \frac{7}{12}e^{-t} \\ \frac{5}{6}e^{-t} \end{bmatrix}$$

Using Example 5, we find that

$$\mathbf{x}(t) = c_1 \begin{bmatrix} 3e^{7t} \\ -2e^{7t} \end{bmatrix} + c_2 \begin{bmatrix} e^t \\ e^t \end{bmatrix} - \begin{bmatrix} \frac{7}{12}e^{-t} \\ \frac{5}{6}e^{-t} \end{bmatrix}$$

is a general solution of equation (5.28). ◄

► **Example 8** The vector function

$$\mathbf{u}(t) = \begin{bmatrix} -\frac{11}{6}t - \frac{7}{4} \\ -\frac{4}{3}t - \frac{5}{2} \\ -\frac{13}{6}t - \frac{29}{12} \end{bmatrix}$$

is a solution of

$$\mathbf{x}' = \begin{bmatrix} 1 & 1 & -1 \\ 2 & 3 & -4 \\ 4 & 1 & -4 \end{bmatrix} \mathbf{x} + \begin{bmatrix} t \\ -t \\ -\frac{7}{3} \end{bmatrix}$$ (5.29)

since

$$\mathbf{u}'(t)=\begin{bmatrix} -\frac{11}{6} \\ -\frac{4}{3} \\ -\frac{13}{6} \end{bmatrix}$$

and

$$\begin{bmatrix} 1 & 1 & -1 \\ 2 & 3 & -4 \\ 4 & 1 & -4 \end{bmatrix}\mathbf{u}(t)+\begin{bmatrix} t \\ -t \\ 2 \end{bmatrix}=\begin{bmatrix} 1 & 1 & -1 \\ 2 & 3 & -4 \\ 4 & 1 & -4 \end{bmatrix}\begin{bmatrix} -\frac{11}{6}t-\frac{7}{4} \\ -\frac{4}{3}t-\frac{5}{2} \\ -\frac{13}{6}t-\frac{29}{12} \end{bmatrix}+\begin{bmatrix} t \\ -t \\ -\frac{7}{3} \end{bmatrix}$$

$$=\begin{bmatrix} -\frac{11}{6} \\ -\frac{4}{3} \\ -\frac{13}{6} \end{bmatrix}$$

Using Example 6, we find that

$$\mathbf{x}(t)=c_1\begin{bmatrix} e^{-3t} \\ 7e^{-3t} \\ 11e^{-3t} \end{bmatrix}+c_2\begin{bmatrix} e^{2t} \\ 2e^{2t} \\ e^{2t} \end{bmatrix}+c_3\begin{bmatrix} e^{t} \\ e^{t} \\ e^{t} \end{bmatrix}+\begin{bmatrix} -\frac{11}{6}t-\frac{7}{4} \\ -\frac{4}{3}t-\frac{5}{2} \\ -\frac{13}{6}t-\frac{29}{12} \end{bmatrix}$$

is a general solution of equation (5.29). ◀

Exercises SECTION 5.4

1. Show that $\mathbf{u}(t)=\begin{bmatrix} e^{-2t} \\ 2e^{-2t} \end{bmatrix}$ and $\mathbf{v}(t)=\begin{bmatrix} -3e^{5t} \\ e^{5t} \end{bmatrix}$ are linearly independent solutions of

$$\mathbf{x}'=\begin{bmatrix} 4 & -3 \\ -2 & -1 \end{bmatrix}\mathbf{x}$$

Find a general solution.

2. Show that $\mathbf{u}(t)=\begin{bmatrix} 2e^{4t} \\ 3e^{4t} \end{bmatrix}$ and $\mathbf{v}(t)=\begin{bmatrix} e^{-t} \\ -e^{-t} \end{bmatrix}$ are linearly independent solutions of

$$\mathbf{x}'=\begin{bmatrix} 1 & 2 \\ 3 & 2 \end{bmatrix}\mathbf{x}$$

Find a general solution.

3. Show that $\mathbf{u}(t) = \begin{bmatrix} 2e^{-2t} \\ -e^{-2t} \\ 0 \end{bmatrix}$, $\mathbf{v}(t) = \begin{bmatrix} e^{-3t} \\ 0 \\ -e^{-3t} \end{bmatrix}$, and $\mathbf{w}(t) = \begin{bmatrix} 0 \\ 1 \\ -1 \end{bmatrix}$ are linearly independent solutions of

$$\mathbf{x}' = \begin{bmatrix} -1 & 2 & 2 \\ 2 & 2 & 2 \\ -3 & -6 & -6 \end{bmatrix}$$

Find a general solution.

4. Show that each of the following vector functions is a solution of

$$\mathbf{x}' = \begin{bmatrix} 1 & 0 & -1 \\ 1 & 2 & 1 \\ 2 & 2 & 3 \end{bmatrix} \mathbf{x}$$

(a) $\mathbf{v}(t) = \begin{bmatrix} e^t + 2e^{2t} \\ -e^t - e^{2t} \\ -2e^{2t} \end{bmatrix}$ **(b)** $\mathbf{w}(t) = \begin{bmatrix} e^t + e^{3t} \\ -e^t - e^{3t} \\ -2e^{3t} \end{bmatrix}$

(c) $\mathbf{y}(t) = \begin{bmatrix} e^{3t} - 2e^{2t} \\ -e^{3t} + e^{2t} \\ -2e^{3t} + 2e^{2t} \end{bmatrix}$ **(d)** $\mathbf{z}(t) = \begin{bmatrix} e^t \\ -e^t \\ 0 \end{bmatrix}$

Find a general solution.

5. Find numbers r, c, and d so that either c or d is not zero and

$$\mathbf{u}(t) = e^{rt} \begin{bmatrix} c \\ d \end{bmatrix}$$

is a solution of

$$\mathbf{x}' = \begin{bmatrix} 2 & 3 \\ 4 & 3 \end{bmatrix} \mathbf{x}$$

There will be two choices for r and the corresponding values of c and d. Are the solutions obtained from these two choices linearly independent?

6. Find numbers A and B so that $\mathbf{z}(t) = e^t \begin{bmatrix} A \\ B \end{bmatrix}$ is a solution of

$$\mathbf{x}' = \begin{bmatrix} 4 & -3 \\ -2 & -1 \end{bmatrix} \mathbf{x} + \begin{bmatrix} e^t \\ 3e^t \end{bmatrix}$$

Use Exercise 1 to find a general solution of this system of differential equations.

7. Find numbers a, b, c, d so that $\mathbf{z}(t) = \begin{bmatrix} at+b \\ ct+d \end{bmatrix}$ is a solution of

$$\mathbf{x}' = \begin{bmatrix} 1 & 2 \\ 3 & 2 \end{bmatrix} \mathbf{x} + \begin{bmatrix} t+1 \\ 2t-1 \end{bmatrix}$$

Use Exercise 2 to find a general solution of this system of differential equations.

8. Let $\mathbf{A}$ be a 2×2 matrix function on an open interval I, and let $\mathbf{u}_1, \mathbf{u}_2$ be linearly independent solutions on I of $\mathbf{x}' = \mathbf{A}\mathbf{x}$. Show that the Wronskian

$$W[\mathbf{u}_1, \mathbf{u}_2](t) = \det \begin{bmatrix} u_{11}(t) & u_{12}(t) \\ u_{21}(t) & u_{22}(t) \end{bmatrix}$$

is *not* zero for every t in I. Hint: Suppose that $W[\mathbf{u}_1, \mathbf{u}_2](t_0) = 0$ for some t_0 in I. Use Exercise 2 of Appendix 2 to show that there are constants c_1 and c_2, not both zero, such that $c_1 \mathbf{u}_1(t_0) + c_2 \mathbf{u}_2(t_0) = \mathbf{0}$ (the zero vector). Consider the initial-value problem $\mathbf{x}' = \mathbf{A}\mathbf{x}$, $\mathbf{x}(t_0) = \mathbf{0}$.

9. Let $\mathbf{A}$, $\mathbf{u}_1$, and $\mathbf{u}_2$ be as in Exercise 8. Let $\mathbf{u}$ be any solution on I of $\mathbf{x}' = \mathbf{A}\mathbf{x}$. Show that there are constants c_1 and c_2 such that $\mathbf{u}(t) = c_1 \mathbf{u}_1(t) + c_2 \mathbf{u}_2(t)$ on I. Hint: Let t_0 be any number in I. Show that there are constants c_1 and c_2 such that $c_1 \mathbf{u}_1(t) + c_2 \mathbf{u}_2(t)$ is the solution on I of the initial-value problem $\mathbf{x}' = \mathbf{A}\mathbf{x}$, $\mathbf{x}(t_0) = \mathbf{u}(t_0)$.

5.5 Matrix Methods: Real Eigenvalues

In the next two sections we will present a method for finding a general solution of the differential equation

$$\mathbf{x}' = \mathbf{A}\mathbf{x} \tag{5.30}$$

where $\mathbf{A}$ is an $n \times n$ matrix.

Recall that the ordinary differential equation

$$(D^2 + aD + b)y = 0$$

has solutions of the form

$$y(t) = e^{rt}c$$

where c is an arbitrary constant. We will show that equation (5.30) has analogous functions for solutions. That is, we will show that equation (5.30) has solutions of the form

$$\mathbf{x}(t) = e^{\lambda t}\mathbf{v}$$

where λ is a constant (possibly a complex number) and $\mathbf{v}$ is a constant vector. To begin, we note that if

$$\mathbf{v} = \begin{bmatrix} v_1 \\ v_2 \\ \vdots \\ v_n \end{bmatrix}$$

then

$$\frac{d}{dt}(e^{\lambda t}\mathbf{v}) = \frac{d}{dt}\begin{bmatrix} e^{\lambda t}v_1 \\ e^{\lambda t}v_2 \\ \vdots \\ e^{\lambda t}v_n \end{bmatrix}$$

$$= \begin{bmatrix} \lambda e^{\lambda t}v_1 \\ \lambda e^{\lambda t}v_2 \\ \vdots \\ \lambda e^{\lambda t}v_n \end{bmatrix}$$

$$= \lambda e^{\lambda t}\begin{bmatrix} v_1 \\ v_2 \\ \vdots \\ v_n \end{bmatrix} = \lambda e^{\lambda t}\mathbf{v}$$

Thus $\mathbf{x}(t) = e^{\lambda t}\mathbf{v}$ is a solution of $\mathbf{x}' = \mathbf{A}\mathbf{x}$ if and only if

$$\lambda e^{\lambda t}\mathbf{v} = e^{\lambda t}\mathbf{A}\mathbf{v}$$

or, equivalently, if and only if

$$\mathbf{A}\mathbf{v} = \lambda\mathbf{v}$$

DEFINITION 5.7 Let $\mathbf{A}$ be an $n \times n$ matrix. A number λ is called an **eigenvalue** of $\mathbf{A}$ if there is a nonzero vector, called an **eigenvector** of $\mathbf{A}$ associated with λ, such that $\mathbf{Av} = \lambda \mathbf{v}$.

We have already shown the following theorem to be true.

THEOREM 5.7 If λ is an eigenvalue of the $n \times n$ matrix $\mathbf{A}$ and $\mathbf{v}$ is an eigenvector of $\mathbf{A}$ associated with λ, then

$$\mathbf{x}(t) = e^{\lambda t}\mathbf{v}$$

is a solution of

$$\mathbf{x}' = \mathbf{A}\mathbf{x}$$

The problem now is to find the eigenvalues and eigenvectors of an $n \times n$ matrix. We begin by noting that the $n \times n$ matrix

$$\mathbf{I} = \begin{bmatrix} 1 & 0 & 0 & \cdots & 0 \\ 0 & 1 & 0 & \cdots & 0 \\ & & \cdots & & \\ 0 & 0 & 0 & \cdots & 1 \end{bmatrix}$$

called the **$n \times n$ identity matrix**, has the property that

$$\mathbf{Ix} = \mathbf{x}$$

for every n-vector $\mathbf{x}$. The equation $\mathbf{Av} = \lambda \mathbf{v}$ can now be rewritten as $\mathbf{Av} - \lambda \mathbf{Iv} = \mathbf{0}$, so that

$$\mathbf{0} = \mathbf{Av} - \lambda \mathbf{Iv} = (\mathbf{A} - \lambda \mathbf{I})\mathbf{v} \qquad (5.31)$$

Thus the eigenvalues and eigenvectors of $\mathbf{A}$ are determined by the equation $(\mathbf{A} - \lambda \mathbf{I})\mathbf{v} = \mathbf{0}$ or, equivalently, by the system of equations

$$(a_{11} - \lambda)v_1 + a_{12}v_2 + \cdots + a_{1n}v_n = 0$$
$$a_{21}v_1 + (a_{22} - \lambda)v_2 + \cdots + a_{2n}v_n = 0$$
$$\cdots \cdots \cdots \cdots \cdots \cdots \cdots \cdots \cdots$$
$$a_{n1}v_1 + a_{n2}v_2 + \cdots + (a_{nn} - \lambda)v_n = 0$$

where

$$\mathbf{A} = \begin{bmatrix} a_{11} & a_{12} & \cdots & a_{1n} \\ a_{21} & a_{22} & \cdots & a_{2n} \\ & \cdots \cdots \cdots & \\ a_{n1} & a_{n2} & & a_{nn} \end{bmatrix} \quad \text{and} \quad \mathbf{v} = \begin{bmatrix} v_1 \\ v_2 \\ \vdots \\ v_n \end{bmatrix}$$

For each choice of λ this is a system of n linear equations in the n unknowns $v_1, v_2, \ldots, v_n$. Evidently $v_1 = v_2 = \cdots = v_n = 0$ is a solution of this system for every choice of λ. For certain choices of λ some of the equations are consequences of the other equations. In such a case there are solutions of the system other than $v_1 = v_2 = \cdots = v_n = 0$. It is precisely these cases that interest us.

A famous theorem in linear algebra states that a system of n linear equations in n unknowns

$$b_{11}x_1 + b_{12}x_2 + \cdots + b_{1n}x_n = 0$$
$$b_{21}x_1 + b_{22}x_2 + \cdots + b_{2n}x_n = 0$$
$$\cdots\cdots\cdots\cdots\cdots\cdots\cdots\cdots\cdots$$
$$b_{n1}x_1 + b_{n2}x_2 + \cdots + b_{nn}x_n = 0$$

has a solution other than $x_1 = x_2 = \cdots = x_n = 0$ if and only if

$$\det \begin{bmatrix} b_{11} & b_{12} & \cdots & b_{1n} \\ b_{21} & b_{22} & \cdots & b_{2n} \\ \cdots\cdots\cdots\cdots\cdots \\ b_{n1} & b_{n2} & \cdots & b_{nn} \end{bmatrix} = 0$$

Since we wish the vector $\mathbf{v}$ in equation (5.31) to be different from the zero vector, we must have

$$\det \begin{bmatrix} a_{11}-\lambda & a_{12} & \cdots & a_{1n} \\ a_{21} & a_{22}-\lambda & \cdots & a_{2n} \\ \cdots\cdots\cdots\cdots\cdots\cdots\cdots \\ a_{n1} & a_{n2} & \cdots & a_{nn}-\lambda \end{bmatrix} = 0$$

or, equivalently, $\det(\mathbf{A}-\lambda\mathbf{I})=0$. It can be shown that $\det(\mathbf{A}-\lambda\mathbf{I})$ is an nth-degree polynomial in λ, called the **characteristic polynomial** of $\mathbf{A}$, with $(-1)^n$ as the coefficient of λ^n. Thus the eigenvalues of $\mathbf{A}$ are roots of the characteristic polynomial of $\mathbf{A}$. The above discussion is summarized in the following theorem.

THEOREM 5.8   The eigenvalues of an $n \times n$ matrix $\mathbf{A}$ are the roots of the polynomial equation

$$\det(\mathbf{A}-\lambda\mathbf{I})=0$$

Thus, there are at most n distinct eigenvalues of $\mathbf{A}$.

▶ **Example 1** The eigenvalues of the matrix

$$A = \begin{bmatrix} 5 & -3 \\ -2 & 4 \end{bmatrix}$$

are the roots of the equation

$$0 = \det(A - \lambda I) = \det \begin{bmatrix} 5-\lambda & -3 \\ -2 & 4-\lambda \end{bmatrix} = \lambda^2 - 9\lambda + 14$$

Hence $\lambda = 2$ and $\lambda = 7$ are the eigenvalues of A. In order to find an eigenvector v of A associated with $\lambda = 2$, we need to find a nonzero vector v such that $(A - 2I)v = 0$. Then

$$\begin{bmatrix} 0 \\ 0 \end{bmatrix} = \begin{bmatrix} 5-2 & -3 \\ -2 & 4-2 \end{bmatrix} \begin{bmatrix} v_1 \\ v_2 \end{bmatrix} = \begin{bmatrix} 3v_1 - 3v_2 \\ -2v_1 + 2v_2 \end{bmatrix}$$

Thus v is an eigenvector of A if and only if

$$-v_1 + v_2 = 0$$

The eigenvectors of A associated with $\lambda = 2$ are

$$v = \begin{bmatrix} c_1 \\ c_1 \end{bmatrix} = c_1 \begin{bmatrix} 1 \\ 1 \end{bmatrix}$$

for every number $c_1 \neq 0$. The number c_1 cannot equal zero since the zero vector cannot be an eigenvector. Note that c_1 can be any nonzero number whatsoever. Hence there are infinitely many eigenvectors associated with the eigenvalue $\lambda = 2$. This will be the case in general; each eigenvector of an $n \times n$ matrix has infinitely many associated eigenvectors. However, there are at most n that are linearly independent. We will now find the eigenvalues of A associated with $\lambda = 7$. If w is an eigenvector of A associated with $\lambda = 7$, then

$$\begin{bmatrix} 0 \\ 0 \end{bmatrix} = \begin{bmatrix} 5-7 & -3 \\ -2 & 4-7 \end{bmatrix} \begin{bmatrix} w_1 \\ w_2 \end{bmatrix} = \begin{bmatrix} -2w_1 & -3w_2 \\ -2w_1 & -3w_2 \end{bmatrix}$$

Thus w is an eigenvector of A associated with $\lambda = 7$ if and only if

$$-2w_1 - 3w_2 = 0$$

The eigenvectors of $\mathbf{A}$ associated with $\lambda=7$ are

$$\mathbf{w}=\begin{bmatrix} 3c_2 \\ -2c_2 \end{bmatrix}=c_2\begin{bmatrix} 3 \\ -2 \end{bmatrix}$$

for every number $c_2\neq0$. Hence the vector functions

$$\mathbf{u}_1(t)=c_1e^{2t}\begin{bmatrix} 1 \\ 1 \end{bmatrix} \qquad \mathbf{u}_2(t)=c_2e^{7t}\begin{bmatrix} 3 \\ -2 \end{bmatrix}$$

are solutions of

$$\mathbf{x}'=\begin{bmatrix} 5 & -3 \\ -2 & 4 \end{bmatrix}\mathbf{x}$$

for every choice of the numbers c_1 and c_2. ◄

▶ **Example 2**   The eigenvalues of the matrix

$$\mathbf{A}=\begin{bmatrix} 1 & 1 & -1 \\ 2 & 3 & -4 \\ 4 & 1 & -4 \end{bmatrix}$$

are the roots of the equation

$$0=\det(\mathbf{A}-\lambda\mathbf{I})=\det\begin{bmatrix} 1-\lambda & 1 & -1 \\ 2 & 3-\lambda & -4 \\ 4 & 1 & -4-\lambda \end{bmatrix}=-\lambda^3+7\lambda-6$$

$$=-(\lambda-1)(\lambda-2)(\lambda+3)$$

Hence the eigenvalues of $\mathbf{A}$ are $\lambda=1$, $\lambda=2$, and $\lambda=-3$. When $\lambda=1$, the equation $(\mathbf{A}-\lambda\mathbf{I})\mathbf{v}=\mathbf{0}$ becomes

$$\begin{bmatrix} 0 \\ 0 \\ 0 \end{bmatrix}=\begin{bmatrix} 0 & 1 & -1 \\ 2 & 2 & -4 \\ 4 & 1 & -5 \end{bmatrix}\begin{bmatrix} v_1 \\ v_2 \\ v_3 \end{bmatrix}=\begin{bmatrix} v_2-v_3 \\ 2v_1+2v_2-4v_3 \\ 4v_1+v_2-5v_3 \end{bmatrix}$$

so that

$$v_2-v_3=0$$
$$2v_1+2v_2-4v_3=0$$
$$4v_1+v_2-5v_3=0$$

A short calculation shows that $v_3 = v_2$ and $v_1 = v_2$. The eigenvectors of $\mathbf{A}$ associated with $\lambda = 1$ are

$$\mathbf{v} = \begin{bmatrix} c_1 \\ c_1 \\ c_1 \end{bmatrix} = c_1 \begin{bmatrix} 1 \\ 1 \\ 1 \end{bmatrix}$$

for every $c_1 \neq 0$. When $\lambda = 2$, the equation $(\mathbf{A} - \lambda \mathbf{I})\mathbf{v} = \mathbf{0}$ becomes

$$\begin{bmatrix} 0 \\ 0 \\ 0 \end{bmatrix} = \begin{bmatrix} -1 & 1 & -1 \\ 2 & 1 & -4 \\ 4 & 1 & -6 \end{bmatrix} \begin{bmatrix} v_1 \\ v_2 \\ v_3 \end{bmatrix} = \begin{bmatrix} -v_1 + v_2 - v_3 \\ 2v_1 + v_2 - 4v_3 \\ 4v_1 + v_2 - 6v_3 \end{bmatrix}$$

so that

$$-v_1 + v_2 - v_3 = 0$$

$$2v_1 + v_2 - 4v_3 = 0$$

$$4v_1 + v_2 - 6v_3 = 0$$

A short calculation shows that $v_1 = v_3$ and $v_2 = 2v_3$. The eigenvectors of $\mathbf{A}$ associated with $\lambda = 2$ are

$$\mathbf{v} = \begin{bmatrix} c_2 \\ 2c_2 \\ c_2 \end{bmatrix} = c_2 \begin{bmatrix} 1 \\ 2 \\ 1 \end{bmatrix}$$

for every number $c_2 \neq 0$. When $\lambda = -3$, the equation $(\mathbf{A} - \lambda \mathbf{I})\mathbf{v} = \mathbf{0}$ becomes

$$\begin{bmatrix} 0 \\ 0 \\ 0 \end{bmatrix} = \begin{bmatrix} 4 & 1 & -1 \\ 2 & 6 & -4 \\ 4 & 1 & -1 \end{bmatrix} \begin{bmatrix} v_1 \\ v_2 \\ v_3 \end{bmatrix} = \begin{bmatrix} 4v_1 + v_2 - v_3 \\ 2v_1 + 6v_2 - 4v_3 \\ 4v_1 + v_2 - v_3 \end{bmatrix}$$

so that

$$4v_1 + v_2 - v_3 = 0$$

$$2v_1 + 6v_2 - 4v_3 = 0$$

$$4v_1 + v_2 - v_3 = 0$$

A short calculation shows that $v_2 = 7v_1$ and $v_3 = 11v_1$. The eigenvectors of **A** associated with $\lambda = -3$ are

$$
\mathbf{v} = \begin{bmatrix} c_3 \\ 7c_3 \\ 11c_3 \end{bmatrix} = c_3 \begin{bmatrix} 1 \\ 7 \\ 11 \end{bmatrix}
$$

for every number $c_3 \neq 0$. Hence the vector functions

$$
\mathbf{u}_1(t) = c_1 e^t \begin{bmatrix} 1 \\ 1 \\ 1 \end{bmatrix} \qquad \mathbf{u}_2(t) = c_2 e^{2t} \begin{bmatrix} 1 \\ 2 \\ 1 \end{bmatrix} \qquad \mathbf{u}_3(t) = c_3 e^{-3t} \begin{bmatrix} 1 \\ 7 \\ 11 \end{bmatrix}
$$

are solutions of

$$
\mathbf{x}' = \begin{bmatrix} 1 & 1 & -1 \\ 2 & 3 & -4 \\ 4 & 1 & -4 \end{bmatrix} \mathbf{x}
$$

for every choice of the numbers c_1, c_2, and c_3. ◀

Eigenvalues and eigenvectors provide us with a means of finding solutions of the differential equation

$$
\frac{d\mathbf{x}}{dt} = \mathbf{A}\mathbf{x}
$$

However, we need more than that to construct a general solution: we need n linearly independent solutions when **A** is an $n \times n$ matrix. The following theorem is stated without proof.

THEOREM 5.9   Let $\mathbf{v}_1, \mathbf{v}_2, \dots, \mathbf{v}_k$ be eigenvectors of an $n \times n$ matrix **A** associated with the distinct eigenvalues $\lambda_1, \lambda_2, \dots, \lambda_k$, respectively. Then $\mathbf{u}_1(t) = e^{\lambda_1 t}\mathbf{v}_1$, $\mathbf{u}_2(t) = e^{\lambda_2 t}\mathbf{v}_2, \dots,$ $\mathbf{u}_k(t) = e^{\lambda_k t}\mathbf{v}_k$ are linearly independent solutions of $\mathbf{x}' = \mathbf{A}\mathbf{x}$.

▶ **Example 3**   Let

$$
\mathbf{A} = \begin{bmatrix} 5 & -3 \\ -2 & 4 \end{bmatrix}
$$

In Example 1 we found that $\begin{bmatrix} 1 \\ 1 \end{bmatrix}$ and $\begin{bmatrix} 3 \\ -2 \end{bmatrix}$ are eigenvectors of **A** associated with the eigenvalues $\lambda = 2$ and $\lambda = 7$, respectively. Hence a general solution (see Definition 5.5) of $\mathbf{x}' = \mathbf{A}\mathbf{x}$ is

$$\mathbf{x}(t) = c_1 e^{2t} \begin{bmatrix} 1 \\ 1 \end{bmatrix} + c_2 e^{7t} \begin{bmatrix} 3 \\ -2 \end{bmatrix}$$ ◄

▶ **Example 4**   Let

$$\mathbf{A} = \begin{bmatrix} 1 & 1 & -1 \\ 2 & 3 & -4 \\ 4 & 1 & -4 \end{bmatrix}$$

In Example 2 we showed that

$$\begin{bmatrix} 1 \\ 1 \\ 1 \end{bmatrix}, \quad \begin{bmatrix} 1 \\ 2 \\ 1 \end{bmatrix}, \quad \text{and} \quad \begin{bmatrix} 1 \\ 7 \\ 11 \end{bmatrix}$$

are eigenvectors associated with the eigenvalues $\lambda = 1$, $\lambda = 2$, and $\lambda = -3$, respectively. Hence a general solution of $\mathbf{x}' = \mathbf{A}\mathbf{x}$ is

$$\mathbf{x}(t) = c_1 e^{t} \begin{bmatrix} 1 \\ 1 \\ 1 \end{bmatrix} + c_2 e^{2t} \begin{bmatrix} 1 \\ 2 \\ 1 \end{bmatrix} + c_3 e^{-3t} \begin{bmatrix} 1 \\ 7 \\ 11 \end{bmatrix}$$ ◄

It is important to note that the eigenvalues of an $n \times n$ matrix $\mathbf{A}$ need not be real numbers. For example, the eigenvalues of

$$\mathbf{B} = \begin{bmatrix} 0 & 1 \\ -1 & 0 \end{bmatrix}$$

are $\lambda = i$ and $\lambda = -i$. Solutions of $\mathbf{x}' = \mathbf{A}\mathbf{x}$ that correspond to complex eigenvalues will be discussed in the next section. It is also important to note that an eigenvalue may be a root of the characteristic polynomial of multiplicity greater than one. Such matrices rarely arise in elementary applications and we will not single them out for special consideration. The methods described in this and the following sections can be used to find a general solution even when the matrix $\mathbf{A}$ has an eigenvalue of multiplicity greater than one, provided that the matrix $\mathbf{A}$ has n linearly independent eigenvectors. If the matrix has fewer linearly independent eigenvectors, then the situation is more complicated. Nonetheless, a generalization of the method discussed above will determine a general solution. In Section 5.8 we will present a method of finding a general solution which is applicable even when the matrix $\mathbf{A}$ has an eigenvalue of multiplicity greater than one.

Exercises   SECTION 5.5

In Exercises 1–7 find a general solution of $x'=Ax$ given A.

1. $\begin{bmatrix} 0 & 1 \\ 1 & 0 \end{bmatrix}$

2. $\begin{bmatrix} 6 & 8 \\ 6 & -2 \end{bmatrix}$

3. $\begin{bmatrix} -2 & 1 \\ 1 & -1 \end{bmatrix}$

4. $\begin{bmatrix} 1 & -1 & 0 \\ 2 & 3 & 2 \\ 1 & 1 & 2 \end{bmatrix}$

5. $\begin{bmatrix} -1 & 2 & 2 \\ 2 & 2 & 2 \\ -3 & -6 & -6 \end{bmatrix}$

6. $\begin{bmatrix} 1 & 2 & -4 \\ 0 & 0 & 1 \\ 0 & 4 & 0 \end{bmatrix}$

7. $\begin{bmatrix} 2 & 1 & 1 & 0 \\ 0 & 1 & -1 & 0 \\ 0 & 0 & -2 & 0 \\ 0 & -2 & -1 & -1 \end{bmatrix}$

8. $\begin{bmatrix} -2 & 1 & 0 & 0 \\ 1 & -1 & 0 & 0 \\ 0 & 0 & 6 & 8 \\ 0 & 0 & 6 & -2 \end{bmatrix}$

In Exercises 9–16 find the solution of $dx/dt=Ax$, $x(0)=x_0$, where:

9. A is as in Exercise 1 and $x_0 = \begin{bmatrix} 2 \\ 0 \end{bmatrix}$

10. A is as in Exercise 2 and $x_0 = \begin{bmatrix} 0 \\ 1 \end{bmatrix}$

11. A is as in Exercise 3 and $x_0 = \begin{bmatrix} 1 \\ 0 \end{bmatrix}$

12. A is as in Exercise 4 and $x_0 = \begin{bmatrix} 1 \\ 0 \\ 0 \end{bmatrix}$

13. **A** is as in Exercise 5 and $\mathbf{x}_0 = \begin{bmatrix} 0 \\ 1 \\ 0 \end{bmatrix}$

14. **A** is as in Exercise 6 and $\mathbf{x}_0 = \begin{bmatrix} 0 \\ 0 \\ 1 \end{bmatrix}$

15. **A** is as in Exercise 7 and $\mathbf{x}_0 = \begin{bmatrix} 1 \\ 0 \\ 3 \\ 0 \end{bmatrix}$

16. **A** is as in Exercise 8 and $\mathbf{x}_0 = \begin{bmatrix} 0 \\ 1 \\ 1 \\ 0 \end{bmatrix}$

5.6 Matrix Methods: Complex Eigenvalues

The method described in the previous section can be slightly modified to obtain solutions of $d\mathbf{x}/dt = \mathbf{A}\mathbf{x}$ even when **A** has complex eigenvalues. The basis of our calculations in such a case is the following theorem.

THEOREM 5.10   Let $\mathbf{x} = \mathbf{y} + i\mathbf{z}$, where **y** and **z** are real-valued vector functions, be a solution of $d\mathbf{x}/dt = \mathbf{A}\mathbf{x}$. Then both **y** and **z** are solutions of $d\mathbf{x}/dt = \mathbf{A}\mathbf{x}$.

Proof   If $\mathbf{x} = \mathbf{y} + i\mathbf{z}$ is a solution of $d\mathbf{x}/dt = \mathbf{A}\mathbf{x}$, then

$$\frac{d\mathbf{y}}{dt} + i\frac{d\mathbf{z}}{dt} = \frac{d\mathbf{x}}{dt} = \mathbf{A}\mathbf{x} = \mathbf{A}(\mathbf{y} + i\mathbf{z}) = \mathbf{A}\mathbf{y} + i\mathbf{A}\mathbf{z} \qquad (5.32)$$

Equating the real and imaginary parts of each side of equation (5.32) gives $d\mathbf{y}/dt = \mathbf{A}\mathbf{y}$ and $d\mathbf{z}/dt = \mathbf{A}\mathbf{z}$. Hence both **y** and **z** are solutions of $d\mathbf{x}/dt = \mathbf{A}\mathbf{x}$. ∎

If $\lambda = a + bi$, $b \neq 0$, is a complex eigenvalue of an $n \times n$ real matrix **A** with $\mathbf{v} = \mathbf{u} + i\mathbf{w}$ as an associated eigenvector (**u** and **w** have real components), then $e^{\lambda t}\mathbf{v}$ is a solution of $d\mathbf{x}/dt = \mathbf{A}\mathbf{x}$. We will now use Theorem 5.10 to obtain two real solutions from this complex solution. The complex-valued vector function

$e^{(a+bi)t}(\mathbf{u}+i\mathbf{w})$ can be written as

$$e^{(a+bi)t}(\mathbf{u}+i\mathbf{w})=e^{at}e^{ibt}(\mathbf{u}+i\mathbf{w})$$

$$=e^{at}(\cos bt+i\sin bt)(\mathbf{u}+i\mathbf{w})$$

$$=e^{at}(\mathbf{u}\cos bt-\mathbf{w}\sin bt)+ie^{at}(\mathbf{u}\sin bt+\mathbf{w}\cos bt)$$

Hence, if $\lambda=a+bi$, $b\neq0$, is an eigenvalue of $\mathbf{A}$ with associated eigenvector $\mathbf{v}=\mathbf{u}+i\mathbf{w}$, then

$$\mathbf{y}(t)=e^{at}(\mathbf{u}\cos bt-\mathbf{w}\sin bt) \tag{5.33}$$

and

$$\mathbf{z}(t)=e^{at}(\mathbf{u}\sin bt+\mathbf{w}\cos bt) \tag{5.34}$$

are solutions of $d\mathbf{x}/dt=\mathbf{A}\mathbf{x}$. In fact, they are linearly independent solutions of $d\mathbf{x}/dt=\mathbf{A}\mathbf{x}$. This can be verified as follows: Suppose there are constants c_1 and c_2 such that

$$c_1\mathbf{y}(t)+c_2\mathbf{z}(t)=0 \tag{5.35}$$

for all t. Setting $t=0$ and $t=\pi/2b$ in equation (5.35) we obtain

$$c_1\mathbf{u}+c_2\mathbf{w}=0$$

$$e^{a\pi/2b}(-c_1\mathbf{w}+c_2\mathbf{u})=0$$

Then $c_1\mathbf{u}=-c_2\mathbf{w}$ and $c_1\mathbf{w}=c_2\mathbf{u}$, so

$$c_1^2\mathbf{u}=-c_1c_2\mathbf{w}=-c_2c_1\mathbf{w}=-c_2^2\mathbf{u} \tag{5.36}$$

and

$$c_1^2\mathbf{w}=c_1c_2\mathbf{u}=c_2c_1\mathbf{u}=-c_2^2\mathbf{w} \tag{5.37}$$

Since $\mathbf{v}=\mathbf{u}+i\mathbf{w}$ is not the zero vector, either $\mathbf{u}$ or $\mathbf{w}$ (or both) is not the zero vector. If $\mathbf{u}\neq0$, then from equation (5.36) we obtain

$$c_1^2=-c_2^2$$

The left side of this equation is nonnegative, while the right side is nonpositive.

Hence $c_1 = 0$ and $c_2 = 0$ are the only values that satisfy the equation. If $\mathbf{w} \neq \mathbf{0}$, then a similar argument using equation (5.37) shows that $c_1 = 0$ and $c_2 = 0$. The functions $\mathbf{y}$ and $\mathbf{z}$ are linearly independent.

If $\lambda = a + bi$, $b \neq 0$, is a complex eigenvalue of the real matrix $\mathbf{A}$ with associated eigenvector $\mathbf{v} = \mathbf{u} + i\mathbf{w}$, then $\mu = a - bi$ is also an eigenvalue of $\mathbf{A}$ and has $\bar{\mathbf{v}} = \mathbf{u} - i\mathbf{w}$ as an associated eigenvalue. Then

$$e^{(a-bi)t}(\mathbf{u} - i\mathbf{w}) = e^{at}e^{-bi}(\mathbf{u} - i\mathbf{w})$$

$$= e^{at}(\cos bt - i \sin bt)(\mathbf{u} - i\mathbf{w})$$

$$= \mathbf{y}(t) - i\mathbf{z}(t)$$

where $\mathbf{y}(t)$ and $\mathbf{z}(t)$ are as in equations (5.33) and (5.34). Thus the eigenvalue $a - bi$ determines (up to a multiple of -1) the same real-valued solutions of $d\mathbf{x}/dt = \mathbf{A}\mathbf{x}$ as does the eigenvalue $a + bi$. This enables us to associate with each pair $a + bi$, $b \neq 0$, of complex eigenvalues of $\mathbf{A}$ a pair of real-valued solutions of $d\mathbf{x}/dt = \mathbf{A}\mathbf{x}$.

▶ **Example 1** The eigenvalues of

$$\mathbf{A} = \begin{bmatrix} 1 & -2 \\ 3 & 3 \end{bmatrix}$$

are easily found to be $\lambda = 2 \pm \sqrt{5}\, i$. When $\lambda = 2 + \sqrt{5}\, i$, the equation $(\mathbf{A} - \lambda \mathbf{I})\mathbf{v} = \mathbf{0}$ becomes

$$\begin{bmatrix} 0 \\ 0 \end{bmatrix} = \begin{bmatrix} 1 - (2 + \sqrt{5}\, i) & -2 \\ 3 & 3 - (2 + \sqrt{5}\, i) \end{bmatrix} \begin{bmatrix} v_1 \\ v_2 \end{bmatrix}$$

$$= \begin{bmatrix} (-1 - \sqrt{5}\, i)v_1 - 2v_2 \\ 3v_1 + (1 - \sqrt{5}\, i)v_2 \end{bmatrix}$$

so that

$$(-1 - \sqrt{5}\, i)v_1 - 2v_2 = 0$$

and we have

$$v_2 = \left(-\frac{1}{2} - \frac{\sqrt{5}}{2}\, i \right) v_1$$

Thus

$$\mathbf{v}=\begin{bmatrix} 2 \\ -1-\sqrt{5}\,i \end{bmatrix}$$

is an eigenvector of $\mathbf{A}$ associated with $\lambda=2+\sqrt{5}\,i$. Then

$$e^{(2+\sqrt{5}\,i)t}\begin{bmatrix} 2 \\ -1-\sqrt{5}\,i \end{bmatrix}=e^{2t}\left(\cos\sqrt{5}\,t+i\sin\sqrt{5}\,t\right)\begin{bmatrix} 2 \\ -1-\sqrt{5}\,i \end{bmatrix}$$

$$=e^{2t}\begin{bmatrix} 2\cos\sqrt{5}\,t+i2\sin\sqrt{5}\,t \\ -\cos\sqrt{5}\,t+\sqrt{5}\,\sin\sqrt{5}\,t+i\left(-\sqrt{5}\,\cos\sqrt{5}\,t-\sin\sqrt{5}\,t\right) \end{bmatrix}$$

$$=e^{2t}\begin{bmatrix} 2\cos\sqrt{5}\,t \\ -\cos\sqrt{5}\,t+\sqrt{5}\,\sin\sqrt{5}\,t \end{bmatrix}+ie^{2t}\begin{bmatrix} 2\sin\sqrt{5}\,t \\ -\sqrt{5}\,\cos\sqrt{5}\,t-\sin\sqrt{5}\,t \end{bmatrix}$$

The vector functions

$$\mathbf{u}_1(t)=e^{2t}\begin{bmatrix} 2\cos\sqrt{5}\,t \\ -\cos\sqrt{5}\,t+\sqrt{5}\,\sin\sqrt{5}\,t \end{bmatrix} \qquad \mathbf{u}_2(t)=e^{2t}\begin{bmatrix} 2\sin\sqrt{5}\,t \\ -\sqrt{5}\,\cos\sqrt{5}\,t-\sin\sqrt{5}\,t \end{bmatrix}$$

are linearly independent solutions of $d\mathbf{x}/dt=\mathbf{A}\mathbf{x}$. Hence

$$\mathbf{x}(t)=c_1e^{2t}\begin{bmatrix} 2\cos\sqrt{5}\,t \\ -\cos\sqrt{5}\,t+\sqrt{5}\,\sin\sqrt{5}\,t \end{bmatrix}+c_2e^{2t}\begin{bmatrix} 2\sin\sqrt{5}\,t \\ -\sqrt{5}\,\cos\sqrt{5}\,t-\sin\sqrt{5}\,t \end{bmatrix}$$

is a general solution of $d\mathbf{x}/dt=\mathbf{A}\mathbf{x}$. ◄

▶ **Example 2** The eigenvalues of

$$\mathbf{A}=\begin{bmatrix} 1 & -1 & -1 \\ 1 & -1 & 0 \\ 1 & 0 & -1 \end{bmatrix}$$

are $\lambda=-1$, $\lambda=i$, and $\lambda=-i$. When $\lambda=-1$, the equation $(\mathbf{A}-\lambda\mathbf{I})\mathbf{v}=\mathbf{0}$ becomes

$$\begin{bmatrix} 0 \\ 0 \\ 0 \end{bmatrix}=\begin{bmatrix} 2 & -1 & -1 \\ 1 & 0 & 0 \\ 1 & 0 & 0 \end{bmatrix}\begin{bmatrix} v_1 \\ v_2 \\ v_3 \end{bmatrix}$$

$$\begin{bmatrix} 0 \\ 0 \\ 0 \end{bmatrix}=\begin{bmatrix} 2v_1-v_2-v_3 \\ v_1 \\ v_1 \end{bmatrix}$$

so that

$$2v_1 - v_2 - v_3 = 0 \quad \text{and} \quad v_1 = 0$$

A short calculation shows that $v_1 = 0$ and $v_3 = -v_2$. Thus

$$\mathbf{v} = \begin{bmatrix} 0 \\ 1 \\ -1 \end{bmatrix}$$

is an eigenvector of $\mathbf{A}$ associated with $\lambda = -1$. When $\lambda = i$, the equation $(\mathbf{A} - \lambda \mathbf{I})\mathbf{v} = \mathbf{0}$ becomes

$$\begin{bmatrix} 0 \\ 0 \\ 0 \end{bmatrix} = \begin{bmatrix} 1-i & -1 & -1 \\ 1 & -1-i & 0 \\ 1 & 0 & -1-i \end{bmatrix} \begin{bmatrix} v_1 \\ v_2 \\ v_3 \end{bmatrix}$$

$$= \begin{bmatrix} (1-i)v_1 - v_2 - v_3 \\ v_1 + (-1-i)v_2 \\ v_1 + (-1-i)v_3 \end{bmatrix}$$

so that

$$(1-i)v_1 - v_2 - v_3 = 0$$

$$v_1 + (-1-i)v_2 = 0$$

$$v_1 + (-1-i)v_3 = 0$$

Another calculation shows that $v_1 = (1+i)v_2$ and $v_3 = v_2$. Thus

$$\mathbf{v} = \begin{bmatrix} 1+i \\ 1 \\ 1 \end{bmatrix}$$

is an eigenvector of $\mathbf{A}$ associated with $\lambda = i$. Then

$$e^{it} \begin{bmatrix} 1+i \\ 1 \\ 1 \end{bmatrix} = (\cos t + i \sin t) \begin{bmatrix} 1+i \\ 1 \\ 1 \end{bmatrix}$$

$$= \begin{bmatrix} \cos t - \sin t + i(\cos t + \sin t) \\ \cos t + i \sin t \\ \cos t + i \sin t \end{bmatrix}$$

$$= \begin{bmatrix} \cos t - \sin t \\ \cos t \\ \cos t \end{bmatrix} + i \begin{bmatrix} \cos t + \sin t \\ \sin t \\ \sin t \end{bmatrix}$$

Thus

$$\mathbf{u}_1(t)=e^{-t}\begin{bmatrix} 0 \\ 1 \\ -1 \end{bmatrix} \quad \mathbf{u}_2(t)=\begin{bmatrix} \cos t-\sin t \\ \cos t \\ \cos t \end{bmatrix} \quad \mathbf{u}_3(t)=\begin{bmatrix} \cos t+\sin t \\ \sin t \\ \sin t \end{bmatrix}$$

are three solutions of $d\mathbf{x}/dt=\mathbf{Ax}$. When $t=0$

$$\mathbf{u}_1(0)=\begin{bmatrix} 0 \\ 1 \\ -1 \end{bmatrix} \quad \mathbf{u}_2(0)=\begin{bmatrix} 1 \\ 1 \\ 1 \end{bmatrix} \quad \mathbf{u}_3(0)=\begin{bmatrix} 1 \\ 0 \\ 0 \end{bmatrix}$$

Since

$$\det\begin{bmatrix} 0 & 1 & 1 \\ 1 & 1 & 0 \\ -1 & 1 & 0 \end{bmatrix}=2$$

the vectors $\mathbf{u}_1(0)$, $\mathbf{u}_2(0)$, and $\mathbf{u}_3(0)$ are linearly independent. By Theorem 5.5 the vector functions $\mathbf{u}_1$, $\mathbf{u}_2$, and $\mathbf{u}_3$ are linearly independent. A general solution of $d\mathbf{x}/dt=\mathbf{Ax}$ is

$$\mathbf{x}(t)=c_1e^{-t}\begin{bmatrix} 0 \\ 1 \\ -1 \end{bmatrix}+c_2\begin{bmatrix} \cos t-\sin t \\ \cos t \\ \cos t \end{bmatrix}+c_3\begin{bmatrix} \cos t+\sin t \\ \sin t \\ \sin t \end{bmatrix} \qquad \blacktriangleleft$$

Exercises   SECTION 5.6

In Exercises 1–6 find a general solution of $d\mathbf{x}/dt=\mathbf{Ax}$, where $\mathbf{A}$ is:

1. $\begin{bmatrix} 0 & -1 \\ 1 & 0 \end{bmatrix}$

2. $\begin{bmatrix} 3 & 1 \\ -1 & 3 \end{bmatrix}$

3. $\begin{bmatrix} -1 & -2 \\ 3 & 1 \end{bmatrix}$

4. $\begin{bmatrix} 1 & 0 & 0 \\ 1 & 0 & 1 \\ 2 & -2 & -2 \end{bmatrix}$

5. $\begin{bmatrix} 0 & 2 & 1 \\ 1 & 0 & 2 \\ 2 & 1 & 0 \end{bmatrix}$

6. $\begin{bmatrix} 1 & 0 & -2 \\ 0 & 2 & 1 \\ 1 & 0 & -1 \end{bmatrix}$

In Exercises 7–12 find a general solution of $d\mathbf{x}/dt = \mathbf{A}\mathbf{x}$, $\mathbf{x}(0) = \mathbf{x}_0$, where:

7. $\mathbf{A}$ is as in Exercise 1 and $\mathbf{x}_0 = \begin{bmatrix} 2 \\ 2 \end{bmatrix}$

8. $\mathbf{A}$ is as in Exercise 2 and $\mathbf{x}_0 = \begin{bmatrix} 1 \\ 0 \end{bmatrix}$

9. $\mathbf{A}$ is as in Exercise 3 and $\mathbf{x}_0 = \begin{bmatrix} 2 \\ 1 \end{bmatrix}$

10. $\mathbf{A}$ is as in Exercise 4 and $\mathbf{x}_0 = \begin{bmatrix} 1 \\ 0 \\ 1 \end{bmatrix}$

11. $\mathbf{A}$ is as in Exercise 5 and $\mathbf{x}_0 = \begin{bmatrix} 1 \\ 1 \\ 1 \end{bmatrix}$

12. $\mathbf{A}$ is as in Exercise 6 and $\mathbf{x}_0 = \begin{bmatrix} 1 \\ 2 \\ 1 \end{bmatrix}$

5.7 Applications

▶ **Example 1**   As we have seen, differential equations can be used to describe a wide range of phenomena. They have even been used to describe warfare. Suppose that there are two armies, called the Blue army and the Red army, and that each army is composed of men of equal fighting value, which in general will not be the same for the Blue as for the Red. In a given time each man will score, on an average, a certain number of hits. Consequently, the number of men knocked out of action per unit time will be directly proportional to the numerical strength of the opposing army. If $b(t)$ and $r(t)$ represent the numerical strengths of the Blue army and Red army, respectively, then the differential equations

$$\frac{db}{dt} = -cr \qquad (c > 0)$$

$$\frac{dr}{dt} = -kb \qquad (k > 0) \tag{5.38}$$

describe the situation. [F. W. Lancaster, Mathematics in warfare, in *The World of Mathematics*, Volume 4, edited by J. R. Newman, Simon and Schuster, New York, 1956, page 2140.] The equations in (5.38) may be rewritten as

$$\begin{bmatrix} b' \\ r' \end{bmatrix} = \begin{bmatrix} 0 & -c \\ -k & 0 \end{bmatrix} \begin{bmatrix} b \\ r \end{bmatrix} \qquad (5.39)$$

The eigenvalues of the above matrix are easily found to be the roots of $\lambda^2 - ck = 0$. That is, the eigenvalues are $\lambda = \pm\sqrt{ck}$. When $\lambda = \sqrt{ck}$, the equation for the associated eigenvectors is

$$\begin{bmatrix} 0 \\ 0 \end{bmatrix} = \begin{bmatrix} -\sqrt{ck} & -c \\ -c & -\sqrt{ck} \end{bmatrix} \begin{bmatrix} v_1 \\ v_2 \end{bmatrix} = \begin{bmatrix} -\sqrt{ck}\, v_1 - cv_2 \\ -kv_1 - \sqrt{ck}\, v_2 \end{bmatrix}$$

Thus

$$\begin{aligned} -\sqrt{ck}\, v_1 - \quad cv_2 &= 0 \\ -kv_1 - \sqrt{ck}\, v_2 &= 0 \end{aligned}$$

A short calculation shows that $v_1 = -\sqrt{c/k}\, v_2$, so the eigenvectors associated with $\lambda = \sqrt{ck}$ are

$$\mathbf{v} = d_1 \begin{bmatrix} \sqrt{c} \\ -\sqrt{k} \end{bmatrix}$$

where d_1 is any nonzero constant. A similar argument shows that the eigenvectors associated with $\lambda = -\sqrt{ck}$ are

$$\mathbf{v} = d_2 \begin{bmatrix} \sqrt{c} \\ \sqrt{k} \end{bmatrix}$$

where d_2 is any nonzero constant. Hence

$$\begin{bmatrix} b(t) \\ r(t) \end{bmatrix} = c_1 e^{\sqrt{ck}\, t} \begin{bmatrix} \sqrt{c} \\ -\sqrt{k} \end{bmatrix} + c_2 e^{-\sqrt{ck}\, t} \begin{bmatrix} \sqrt{c} \\ \sqrt{k} \end{bmatrix} \qquad (5.40)$$

is a general solution of equation (5.39).

Suppose that the Red army is twice as effective as the Blue army, i.e., $c = 2k$. If the Blue army is not to be eliminated, initially it must be significantly larger than the Red army. We will determine the relative sizes of $b(0)$ and $r(0)$ to assure that $b(t)>0$ for all $t>0$. For simplicity of notation we will set $R=r(0)$ and $B=b(0)$. Since R and B represent the initial sizes of the armies, they are both positive numbers.

We begin by choosing the constants c_1 and c_2 in equation (5.40) so that the resulting solution satisfies the initial conditions given above. From equation (5.40) we have

$$\begin{bmatrix} B \\ R \end{bmatrix} = \begin{bmatrix} b(0) \\ r(0) \end{bmatrix} = c_1 \begin{bmatrix} \sqrt{2k} \\ -\sqrt{k} \end{bmatrix} + c_2 \begin{bmatrix} \sqrt{2k} \\ -\sqrt{k} \end{bmatrix}$$

so that

$$c_1\sqrt{2k} + c_2\sqrt{2k} = B$$

$$-c_1\sqrt{k} + c_2\sqrt{k} = R$$

Solving this pair of equations for c_1 and c_2 yields

$$c_1 = \frac{\sqrt{2}\,B - 2R}{4\sqrt{k}} \qquad c_2 = \frac{\sqrt{2}\,B + 2R}{4\sqrt{k}}$$

Hence

$$r(t) = \frac{2R - \sqrt{2}\,B}{4} e^{\sqrt{2}\,kt} + \frac{2R + \sqrt{2}\,B}{4} e^{-\sqrt{2}\,kt} \tag{5.41}$$

$$b(t) = -\sqrt{2}\left(\frac{2R - \sqrt{2}\,B}{4} e^{\sqrt{2}\,kt} - \frac{2R + \sqrt{2}\,B}{4} e^{-\sqrt{2}\,kt} \right) \tag{5.42}$$

Evidently $b(t)>0$ for all $t>0$ if and only if

$$\left(2R - \sqrt{2}\,B\right)e^{\sqrt{2}\,kt} - \left(2R + \sqrt{2}\,B\right)e^{-\sqrt{2}\,kt} < 0 \tag{5.43}$$

for all $t>0$. If we multiply each side of this inequality by $e^{\sqrt{2}\,kt}$ and rearrange the terms, we find that the inequality in (5.43) is equivalent to

$$\left(2R - \sqrt{2}\,B\right)e^{2\sqrt{2}\,kt} < 2R + \sqrt{2}\,B \tag{5.44}$$

for all $t>0$. Since $k>0$, we must have

$$\lim_{t\to\infty} e^{2\sqrt{2}\,kt} = \infty$$

Therefore the inequality in (5.44) holds for all $t>0$ if and only if $2R-\sqrt{2}\,B\leqslant0$. We have shown that $b(t)>0$ for all $t>0$ if and only if $\sqrt{2}\,R\leqslant B$. Thus the Blue army is not annihilated by the Red army if and only if the initial size of the Blue army is greater than or equal to $\sqrt{2}$ times the initial size of the Red army.

Let us now consider what happens to the Red army whenever the Blue army is not annihilated. If $\sqrt{2}\,R<B$, then $r(t)=0$ if and only if

$$\left(\sqrt{2}\,B-2R\right)e^{\sqrt{2}\,kt}=\left(2R+\sqrt{2}\,B\right)e^{-\sqrt{2}\,kt}$$

If we multiply each side of this equation by $e^{\sqrt{2}\,kt}$ and rearrange the terms, we find that

$$e^{2\sqrt{2}\,kt}=\frac{2R+\sqrt{2}\,B}{\sqrt{2}\,B-2R}$$

so that

$$t=\frac{1}{2\sqrt{2}\,k}\ln\frac{2R+\sqrt{2}\,B}{\sqrt{2}\,B-2R}$$

is the time at which the Red army vanishes. If $\sqrt{2}\,R=B$, then

$$r(t)=Re^{-\sqrt{2}\,kt}\quad\text{and}\quad b(t)=Be^{-\sqrt{2}\,kt}$$

so that both forces decrease monotonically to zero as $t\to\infty$.

We have shown that if $c=2k$, then:

(i) The Blue army is annihilated if $B<\sqrt{2}\,R$.
(ii) The Red army is annihilated if $\sqrt{2}\,R<B$.
(iii) Neither army is annihilated if $\sqrt{2}\,R=B$, but both forces decrease exponentially to zero as $t\to\infty$.

Graphs illustrating these three cases are shown in Figures 5.2–5.4. ◀

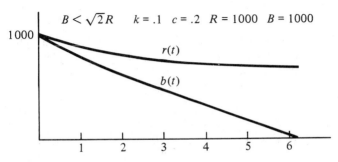

$B<\sqrt{2}\,R$ $k=.1$ $c=.2$ $R=1000$ $B=1000$

FIGURE 5.2

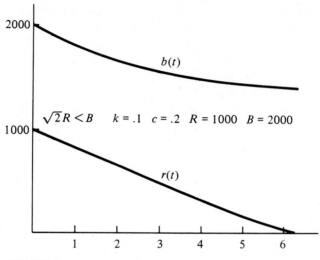

FIGURE 5.3

▶ **Example 2** When a radioactive substance decays, it often goes through a number of stages before a stable state is attained. If

$$S_1 \rightarrow S_2 \rightarrow \cdots \rightarrow S_{n-1} \rightarrow S_n$$

is a chain of radioactive decompositions, and if x_i denotes the mass of the substance S_i, then a system of differential equations is obtained:

$$\frac{dx_1}{dt} = k_1 x_1$$

$$\frac{dx_2}{dt} = -k_1 x_1 + k_2 x_2$$

$$\cdots\cdots\cdots\cdots\cdots\cdots\cdots\cdots\cdots \qquad \textbf{(5.45)}$$

$$\frac{dx_{n-1}}{dt} = -k_{n-2} x_{n-2} + k_{n-1} x_{n-1}$$

$$\frac{dx_n}{dt} = k_{n-1} x_{n-1}$$

where k_i is the constant of proportionality between dx_i/dt and x_i. An example of such a reaction is

$$_{92}U^{234} \rightarrow {}_{90}Th^{230} \rightarrow {}_{88}Ra^{226} \rightarrow {}_{86}Em^{222} \rightarrow {}_{84}Po^{218}$$

in which the half-lives T of the decompositions are $(2.48)10^5$ years,

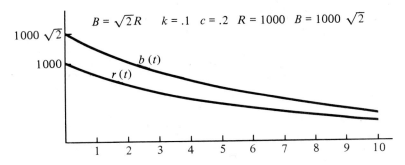

FIGURE 5.4

$(8)10^4$ years, 1622 years, and 3.83 days, respectively. Since the half-life of $_{86}Em^{222}$ is so small relative to the other half-lives, we will assume that the amounts of $_{86}Em^{222}$ and $_{84}Po^{218}$ present at a given time are identical. In Section 1.3 we found that $k = -T^{-1}\ln 2$ [equation (1.18)]. Thus we have

$$k_1 = -(2.48)^{-1}10^{-5}\ln 2 \approx -(2.79)10^{-6}$$

$$k_2 = -(8)^{-1}10^{-4}\ln 2 \approx -(8.66)10^{-6} \qquad (5.46)$$

$$k_3 = -(1622)^{-1}\ln 2 \approx -(4.27)10^{-4}$$

We will determine how many milligrams of $_{84}Po^{218}$ are present after 100,000 years if there are 206 milligrams of $_{92}U^{234}$ present initially. In terms of differential equations this means that we will determine the value when $t = 100,000$ of the fourth component of the solution of the initial-value problem

$$\mathbf{x}' = \mathbf{Ax}, \quad \mathbf{x}(0) = \begin{bmatrix} 206 \\ 0 \\ 0 \\ 0 \end{bmatrix}$$

where

$$\mathbf{x} = \begin{bmatrix} x_1 \\ x_2 \\ x_3 \\ x_4 \end{bmatrix} \qquad \mathbf{A} = \begin{bmatrix} k_1 & 0 & 0 & 0 \\ -k_1 & k_2 & 0 & 0 \\ 0 & -k_2 & k_3 & 0 \\ 0 & 0 & -k_3 & 0 \end{bmatrix}$$

and k_1, k_2, k_3 are as in (5.46). Straightforward calculations show that the eigenvalues of A are k_1, k_2, k_3, and 0. The equation for the eigenvectors associated with $\lambda = k_1$ is

$$\begin{bmatrix} 0 \\ 0 \\ 0 \\ 0 \end{bmatrix} = \begin{bmatrix} 0 & 0 & 0 & 0 \\ -k_1 & k_2-k_1 & 0 & 0 \\ 0 & -k_2 & k_3-k_1 & 0 \\ 0 & 0 & -k_3 & -k_1 \end{bmatrix} \begin{bmatrix} v_1 \\ v_2 \\ v_3 \\ v_4 \end{bmatrix}$$

$$= \begin{bmatrix} 0 \\ -k_1 v_1 + (k_2-k_1)v_2 \\ -k_2 v_2 + (k_3-k_1)v_3 \\ -k_3 v_3 - k_1 v_4 \end{bmatrix}$$

Thus

$$v_2 = k_1(k_2-k_1)^{-1} v_1$$

$$v_3 = k_2(k_3-k_1)^{-1} v_2 = k_1 k_2 (k_2-k_1)^{-1}(k_3-k_1)^{-1} v_1$$

$$v_4 = -k_3 k_1^{-1} v_3 = -k_2 k_3 (k_2-k_1)^{-1}(k_3-k_1)^{-1} v_1$$

so that

$$v = c_1 \begin{bmatrix} 1 \\ k_1(k_2-k_1)^{-1} \\ k_1 k_2 (k_2-k_1)^{-1}(k_3-k_1)^{-1} \\ -k_2 k_3 (k_2-k_1)^{-1}(k_3-k_1)^{-1} \end{bmatrix}$$

is an eigenvector of A associated with $\lambda = k_1$ for every nonzero constant c_1. Using the values for k_1, k_2, and k_3 in (5.46) and choosing $c_1 = 206$, the components of v are nearly integers. In order to simplify the calculations we will round off these components. Doing this we find that the vector

$$v_1 = \begin{bmatrix} 206 \\ 98 \\ 2 \\ -306 \end{bmatrix}$$

is approximately an eigenvector of A associated with $\lambda = k_1$.

Similar calculations show that the vector

$$\mathbf{v}_2 = \begin{bmatrix} 0 \\ 97 \\ 2 \\ -99 \end{bmatrix}$$

is approximately an eigenvector of $\mathbf{A}$ associated with $\lambda = k_2$. Straightforward calculations show that

$$\mathbf{v}_3 = \begin{bmatrix} 0 \\ 0 \\ 1 \\ -1 \end{bmatrix} \quad \text{and} \quad \mathbf{v}_4 = \begin{bmatrix} 0 \\ 0 \\ 0 \\ 1 \end{bmatrix}$$

are eigenvectors of $\mathbf{A}$ associated with $\lambda = k_3$ and $\lambda = 0$, respectively. A general solution has the form (neglecting the rounding errors in $\mathbf{v}_1$ and $\mathbf{v}_2$)

$$\mathbf{x}(t) = c_1 e^{k_1 t} \begin{bmatrix} 206 \\ 98 \\ 2 \\ -306 \end{bmatrix} + c_2 e^{k_2 t} \begin{bmatrix} 0 \\ 97 \\ 2 \\ -99 \end{bmatrix} + c_3 e^{k_3 t} \begin{bmatrix} 0 \\ 0 \\ 1 \\ -1 \end{bmatrix} + c_4 \begin{bmatrix} 0 \\ 0 \\ 0 \\ 1 \end{bmatrix}$$

We now choose c_1, c_2, c_3, and c_4 so that

$$\begin{bmatrix} 206 \\ 0 \\ 0 \\ 0 \end{bmatrix} = \mathbf{x}(0) = \begin{bmatrix} 206c_1 \\ 98c_1 + 97c_2 \\ 2c_1 + 2c_2 + c_3 \\ -306c_1 - 99c_2 - c_3 + c_4 \end{bmatrix}$$

which leads directly to a system of equations for the coefficients c_1, c_2, and c_3. Solving this system gives

$$c_1 = 1, \quad c_2 = -\tfrac{98}{97} \approx -1.01, \quad c_3 \approx .02, \quad c_4 \approx 206.03$$

so that

$$\mathbf{x}(t) \approx \begin{bmatrix} 206 e^{-(2.79)10^{-6}t} \\ 98 e^{-(2.79)10^{-6}t} - 98 e^{-(8.66)10^{-6}t} \\ 2 e^{-(2.79)10^{-6}t} - 2.02 e^{-(8.66)10^{-6}t} + .02 e^{-(4.27)10^{-4}t} \\ -306 e^{-(2.79)10^{-6}t} + 99.99 e^{-(8.66)10^{-6}t} - .02 e^{-(4.27)10^{-4}t} + 206.03 \end{bmatrix}$$

The fourth component, $x_4(t)$, of $\mathbf{x}(t)$ represents the mass of $_{86}\text{Em}^{222}$ and, by our initial assumption, the mass of $_{84}\text{Po}^{218}$. Hence

$$x_4(10^5) = -306e^{-.279} + 99.99e^{-.866} - .02e^{-42.7} + 206.03$$

$$\approx 16.59 \text{ milligrams}$$

Hence after 100,000 years there will be 16.59 milligrams of $_{84}\text{Po}^{218}$. Note that $x_4(t) \rightarrow 206.03$ as $t \rightarrow \infty$. The discrepancy between 206.03 and our initial mass of 206 milligrams is due to rounding off.   ◀

Exercises   SECTION 5.7

1. In Example 1 of this section suppose that $c = k$, $r(0) = 500$, and $b(0) = 1000$. What is the number of Blue forces when the Red forces are annihilated?

2. In Figures 5.2 and 5.3 it appears that the time at which $B(t) = 0$ when $k = .1$, $c = .2$, $R = 1000$, and $B = 1000$ is the same time at which $R(t) = 0$ when $k = .1$, $c = .2$, $R = 1000$, and $B = 2000$. Is this in fact the case?

3. Consider the two compartment system indicated in Figure 5.5, where

   $I_i =$ rate at which material enters the compartment i

   $f_{ij} =$ rate at which material passes from compartment j to compartment i

   $f_{0i} =$ rate at which material is removed from compartment i

   $q_i =$ concentration of material in compartment i

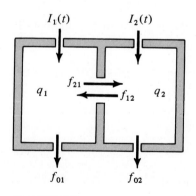

FIGURE 5.5

and $0 \leqslant f_{01}, f_{02}, f_{12}, f_{21}$. Then

$$\frac{dq_1}{dt} = -(f_{01} + f_{21})q_1 + f_{12}q_2 + I_1(t)$$

$$\frac{dq_2}{dt} = f_{21}q_1 - (f_{02} + f_{12})q_2 + I_2(t)$$

Let the initial concentrations of q_1 and q_2 be A and B, respectively.

(a) The system is called closed if $I_1(t) = I_2(t) = f_{01} = f_{02} = 0$. Find a general solution in this case.

(b) The system is called open without intakes if $I_1 = I_2 = 0$. Find a general solution for the special case $f_{01} = f_{02} = f_{12} = f_{21} = a$.
[J. A. Jacquez, *Compartmental Analysis in Biology and Medicine*, Elsevier Publishing Company, Amsterdam, 1972, pages 62–65.]

4. Consider the two compartment system given in Exercise 3.

(a) The system is called open with intakes if $I_1(t) = k_1$, $I_2(t) = k_2$, and $f_{01} \neq 0 \neq f_{02}$. Find a general solution for the special case $f_{12} = f_{21} = f_{02} = f_{01} = a$. Hint: Use Exercise 3(b) and find a solution of the form $q_1(t) \equiv c$, $q_2(t) \equiv d$ for some numbers c and d.

(b) The system is called a mamillary system with intake if $I_1(t) = p_1$, $I_2(t) = 0$, $f_{01} = 0$, but $f_{02} \neq 0$. Find a general solution for the special case $f_{12} = f_{21} = f_{02} = a$.

5. The schematic diagram in Figure 5.6 represents one model for the movement of iodine through a dog. If f_i is the amount of iodine in compartment

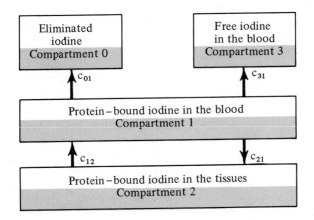

FIGURE 5.6

i, then the differential equations corresponding to this model are

$$\frac{df_1}{dt} = -(c_{21}+c_{01}+c_{31})f_1+c_{12}f_2$$

$$\frac{df_2}{dt} = c_{21}f_1-c_{12}f_2 \qquad\qquad (5.47)$$

$$\frac{df_3}{dt} = c_{31}f_1-c_{03}f_3$$

where $0 < c_{12}, c_{21}, c_{01}, c_{31}, c_{03}$. [Hirsh Cohen, Mathematics and the biological sciences, in *The Mathematical Sciences*, edited by COSRIMS, MIT Press, Cambridge, Mass., 1969, pages 227–228.] The objective of this exercise is to show that $f_1(t) \to 0$, $f_2(t) \to 0$, and $f_3(t) \to 0$ as $t \to \infty$, regardless of the initial values $f_1(0)$, $f_2(0)$, and $f_3(0)$. This can be done as follows:

(a) Show that the eigenvalues of the matrix

$$\mathbf{A} = \begin{bmatrix} -(c_{21}+c_{01}+c_{31}) & c_{12} & 0 \\ c_{21} & -c_{12} & 0 \\ c_{31} & 0 & -c_{03} \end{bmatrix}$$

are $-c_{03}$ and the roots r_1 and r_2 of the polynomial $x^2+(c_{01}+c_{12}+c_{21}+c_{31})x+c_{12}(c_{01}+c_{31})$.

(b) Show that r_1 and r_2 are negative real numbers. Hint: Show that $(c_{01}+c_{12}+c_{21}+c_{31})^2-4c_{12}(c_{01}+c_{31})=(c_{01}-c_{12}+c_{21}+c_{31})^2+4c_{12}c_{21}$.

(c) Assuming that $r_1 \neq -c_{03}$ and $r_2 \neq -c_{03}$, argue that a general solution of $\mathbf{x}' = \mathbf{A}\mathbf{x}$ has the form

$$\mathbf{x}(t) = c_1 e^{-c_{03}t}\mathbf{v}_1 + c_2 e^{r_1 t}\mathbf{v}_2 + c_3 e^{r_2 t}\mathbf{v}_3$$

for some vectors $\mathbf{v}_1, \mathbf{v}_2, \mathbf{v}_3$.

(d) Argue that any solution of the system of differential equations in (5.47) tends to the zero vector as $t \to \infty$.

6. Consider the mechanical system consisting of two masses m_1 and m_2, each of which is restrained by an elastic spring attached to a fixed base and by a third elastic spring that connects the two masses as shown in Figure 5.7.

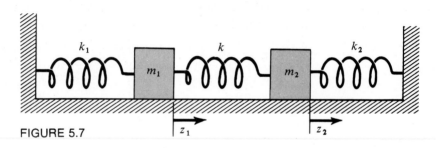

FIGURE 5.7

Let z_1 and z_2 measure the displacement from equilibrium of the masses m_1 and m_2, respectively. Then

$$m_1 \frac{d^2 z_1}{dt^2} = -(k_1 + k)z_1 + kz_2$$

$$m_2 \frac{d^2 z_2}{dt^2} = kz_1 - (k_2 + k)z_2 \qquad \text{(5.48)}$$

where k, k_1, and k_2 are spring constants.

(a) Use the change of variables $x_1 = z_1$, $x_2 = dz_1/dt$, $x_3 = z_2$, $x_4 = dz_2/dt$ to transform the equations in (5.48) into a matrix equation of the form $dx/dt = Ax$ where

$$A = \begin{bmatrix} 0 & 1 & 0 & 0 \\ \dfrac{-(k_1+k)}{m_1} & 0 & \dfrac{k}{m_1} & 0 \\ 0 & 0 & 0 & 1 \\ \dfrac{k}{m_2} & 0 & \dfrac{-(k_2+k)}{m_2} & 0 \end{bmatrix} \quad \text{and} \quad x = \begin{bmatrix} x_1 \\ x_2 \\ x_3 \\ x_4 \end{bmatrix}$$

(b) Show that the characteristic equation of A is

$$\left(\lambda^2 + \frac{k_1+k}{m_1} \right)\left(\lambda^2 + \frac{k_2+k}{m_2} \right) - \frac{k^2}{m_1 m_2} = 0$$

(c) Suppose that the mass m_2 is initially held fixed, and that the mass m_1 is displaced by 1 unit in the positive direction, held fixed, and then released. In such a case

$$x(0) = \begin{bmatrix} 1 \\ 0 \\ 0 \\ 0 \end{bmatrix}$$

For the special case $m_1 = m_2$, $k_1 = k_2 = k$, find the solution to $dx/dt = Ax$ subject to this initial condition.

7. Suppose that in the previous problem there is a frictional force acting on each mass proportional to the velocity of the mass. The equations in (5.48) become modified to

$$m_1 \frac{d^2 z_1}{dt^2} = -r_1 \frac{dz_1}{dt} - (k_1+k)z_1 + kz_2$$

$$m_2 \frac{d^2 z_2}{dt^2} = r_2 \frac{dz_2}{dt} + kz_1 - (k_2+k)z_2$$

Use the change of variables given above to write this system of equations in matrix form.

8. Consider a uniform shaft with three identical disks equally spaced along it (Figure 5.8). At the $x=0$ end the shaft is fixed while the $x=3L$ end is free.

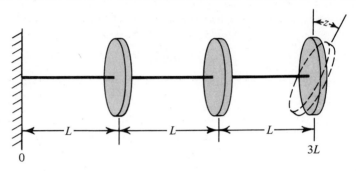

FIGURE 5.8

If z_1, z_2, and z_3 denote the angular deflections of the three disks, then the torsional oscillations of the disks are described by the equations

$$\frac{d^2z_1}{dt^2}=k(-2z_1+z_2)$$

$$\frac{d^2z_2}{dt}=k(+z_1-2z_2+z_3) \qquad \text{(5.49)}$$

$$\frac{d^2z_3}{dt}=k(+z_2-z_3)$$

where k is a constant that depends upon the torsional stiffness of the shaft, the moment of inertia of the disks, and the length of the shaft. For simplicity assume that the units have been chosen so that $k=1$.

(a) Use the change of variables $x_1=z_1$, $x_2=dz_1/dt$, $x_3=z_2$, $x_4=dz_2/dt$, $x_5=z_3$, $x_6=dz_3/dt$ to transform the equations in (5.49) into a matrix equation of the form $d\mathbf{x}/dt=\mathbf{Ax}$ where

$$\mathbf{A}=\begin{bmatrix} 0 & 1 & 0 & 0 & 0 & 0 \\ -2 & 0 & 1 & 0 & 0 & 0 \\ 0 & 0 & 0 & 1 & 0 & 0 \\ 1 & 0 & -2 & 0 & 1 & 0 \\ 0 & 0 & 0 & 0 & 0 & 1 \\ 0 & 0 & 1 & 0 & -1 & 0 \end{bmatrix}$$

(b) Show that the characteristic equation of **A** is

$$\lambda^6 + 5\lambda^4 + 6\lambda^2 + 1 = 0$$

(c) Show that all of the eigenvalues are complex numbers with real parts equal to zero. Hint: Use the change of variables $z = -\lambda^2$ and show that the equation $z^3 - 5z^2 + 6z - 1 = 0$ has three positive real roots. This may be done using the Intermediate-Value Theorem and the numbers $z = 0$, $z = .5$, $z = 2$, and $z = 4$.

(d) Use part (c) to describe the motion of the disks without computing z_1, z_2, and z_3.

9. Consider the electrical circuit shown in Figure 5.9, where E is an electromotive force, c_1, c_2, c_3 are capacitances, I_1 and I_2 currents, and $L_1 L_2$ induc-

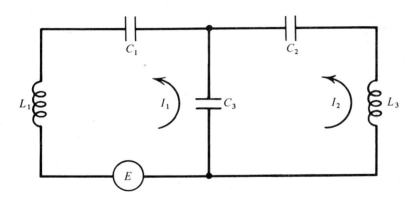

FIGURE 5.9

tances. Using Kirchhoff's laws, it can be shown that

$$\left[L_1 D^2 + \left(\frac{1}{c_1} + \frac{1}{c_3} \right) \right] I_1 - \frac{1}{c_3} I_2 = E$$

$$\left[L_2 D^2 + \left(\frac{1}{c_2} + \frac{1}{c_3} \right) \right] I_2 - \frac{1}{c_3} I_1 = 0$$

(a) For the case $L_1 = L_2 = 1$ henry and $c_1 = c_2 = c_3 = 1$ farad, write this system of equations in the form $d\mathbf{x}/dt = \mathbf{A}\mathbf{x} + \mathbf{f}(t)$.

(b) Find a general solution of the homogeneous equation.

(c) Suppose that E is constant. Find a constant vector that is a solution of the nonhomogeneous equation found in part (a).

(d) Find a general solution of the nonhomogeneous equation in part (a).

(e) Find the solution of the nonhomogeneous equation in part (a) that satisfies the initial condition

$$\mathbf{x}(0) = \begin{bmatrix} 0 \\ 0 \\ 0 \\ 0 \end{bmatrix}$$

5.8 Elimination

In this section we will present an alternate method to the one presented in Sections 5.5 and 5.6. This method has the advantage that it can be applied to a system of nonhomogeneous linear differential equations

$$\frac{dx_1}{dt} = a_{11}x_1 + a_{12}x_2 + \cdots + a_{1n}x_n + f_1(t)$$

$$\frac{dx_2}{dt} = a_{21}x_1 + a_{22}x_2 + \cdots + a_{2n}x_n + f_2(t)$$

$$\cdots \cdots \cdots \cdots \cdots \cdots \cdots \cdots \cdots \cdots \cdots$$

$$\frac{dx_3}{dt} = a_{n1}x_1 + a_{n2}x_2 + \cdots + a_{nn}x_n + f_n(t)$$

where the a_{ij} are constants. Equivalently (where $D = d/dt$),

$$(D - a_{11})x_1 - a_{12}x_2 - \cdots - a_{1n}x_n = f_1(t)$$
$$-a_{21}x_1 + (D - a_{22})x_2 - \cdots - a_{2n}x_n = f_2(t)$$
$$\cdots \cdots \cdots \cdots \cdots \cdots \cdots \cdots \cdots \cdots \cdots$$
$$-a_{n1}x_1 - a_{n2}x_2 - \cdots + (D - a_{nn})x_n = f_n(t)$$

The method of elimination consists of treating this system of equations as though it were a linear system of equations and solving for the functions x_i ($i = 1, 2, \ldots, n$) in turn by eliminating the other functions.

▶ **Example 1**   Consider the system of differential equations

$$(D - 5)x_1 + 3x_2 = 0$$
$$2x_1 + (D - 4)x_2 = 0 \qquad \text{(5.50)}$$

which is the system encountered in Example 1 of Section 5.2 and Example 1 of Section 5.5, but written in a slightly different form. If we multiply the first of these equations by $D-4$ and the second by -3, we obtain

$$(D-4)(D-5)x_1+3(D-4)x_2=0$$

$$-6x_1-3(D-4)x_2=0$$

Adding these equations gives a second-order linear differential equation for the function x_1:

$$[(D-4)(D-5)-6]x_1=0$$

so that

$$(D-2)(D-7)x_1=0 \qquad\qquad \textbf{(5.51)}$$

Since 2 and 7 are the roots of the auxiliary equation $(r-2)(r-7)=0$, a general solution of equation (5.51) is

$$x_1(t)=d_1e^{2t}+d_2e^{7t}$$

The function x_2 can now be obtained from the first equation in (5.50), which may be rewritten as

$$x_2(t)=-\tfrac{1}{3}(D-5)x_1$$

Then

$$x_2(t)=-\tfrac{1}{3}(D-5)\left(d_1e^{2t}+d_2e^{7t}\right)$$

$$=d_1e^{2t}-\tfrac{2}{3}d_2e^{7t}$$

If we now set $c_2=d_1$ and $c_1=\tfrac{1}{3}d_2$, the functions x_1 and x_2 can be written as

$$x_1(t)=3c_1e^{7t}+c_2e^{2t}$$

$$x_2(t)=-2c_1e^{7t}+c_2e^{2t}$$

which is the general solution for the system of differential equations found in Example 3 of Section 5.4. The reader should compare the above calculations with those in Example 1 of Section 5.5. In particular, notice that the auxiliary equation for equation (5.51) is identical with the characteristic equation for the matrix **A** in that example. ◀

▶ **Example 2**   Consider the system of differential equations

$$(D-1)x_1 - x_2 + x_3 = 0 \tag{5.52}$$

$$-2x_1 + (D-3)x_2 + 4x_3 = 0 \tag{5.53}$$

$$-4x_1 - x_2 + (D+4)x_3 = 0 \tag{5.54}$$

If we multiply equation (5.52) by 2, multiply equation (5.53) by $D-1$, and add, we obtain

$$[(D-1)(D-3)-2]x_2 + [4(D-1)+2]x_3 = 0$$

which simplifies to

$$(D^2 - 4D + 1)x_2 + (4D - 2)x_3 = 0 \tag{5.55}$$

If we multiply equation (5.53) by -2 and add the result to equation (5.54), we obtain

$$[-2(D-3)-1]x_2 + [(D+4)-8]x_3 = 0$$

which simplifies to

$$-(2D-5)x_2 + (D-4)x_3 = 0 \tag{5.56}$$

We will now solve equations (5.55) and (5.56) for the functions x_2 and x_3. If we multiply equation (5.55) by $D-4$, multiply equation (5.56) by $-(4D-2)$, and add, we obtain

$$[(D-4)(D^2 - 4D + 1) + (4D - 2)(2D - 5)]x_2 = 0$$

which simplifies to

$$(D+3)(D-2)(D-1)x_2 = 0 \tag{5.57}$$

Therefore

$$x_2(t) = c_1 e^{-3t} + c_2 e^{2t} + c_3 e^t \tag{5.58}$$

where c_1, c_2, c_3 are arbitrary constants. Multiplying equation (5.55) by $2D-5$, multiplying equation (5.56) by $D^2 - 4D + 1$, and adding gives

$$[(2D-5)(4D-2) + (D-4)(D^2 - 4D + 1)]x_3 = 0$$

which simplifies to

$$(D+3)(D-2)(D-1)x_3=0$$

Therefore

$$x_3(t)=d_1e^{-3t}+d_2e^{2t}+d_3e^{t} \tag{5.59}$$

where d_1, d_2, d_3 are constants that must be chosen so that the functions in (5.58) and (5.59) form a solution for the simultaneous equations (5.55) and (5.56). Inserting these choices for x_2 and x_3 into equation (5.55) yields

$$0=(D^2-4D+1)(c_1e^{-3t}+c_2e^{2t}+c_3e^{t})$$

$$+(4D-2)(d_1e^{-3t}+d_2e^{2t}+d_3e^{t})$$

$$=(22c_1-14d_1)e^{-3t}+(-3c_2+6d_2)e^{2t}+(-2c_3+2d_3)e^{t}$$

for every t. Since the functions e^{-3t}, e^{2t}, and e^{t} are linearly independent, we must have

$$22c_1-14d_1=0$$

$$-3c_2+6d_2=0$$

$$-2c_3+2d_3=0$$

Hence

$$d_1=\tfrac{11}{7}c_1, d_2=\tfrac{1}{2}c_2, d_3=c_3 \tag{5.60}$$

so that

$$x_3(t)=\tfrac{11}{7}c_1e^{-3t}+\tfrac{1}{2}c_2e^{2t}+c_3e^{t}$$

If equation (5.56) is used instead of equation (5.55), then the equalities in (5.60) are again obtained. The reader should verify this. The function x_1 can now be determined from equation (5.53), which can be rewritten as

$$x_1=\tfrac{1}{2}(D-3)x_2+2x_3$$

Then

$$x_1(t)=\tfrac{1}{2}(D-3)(c_1e^{-3t}+c_2e^{2t}+c_3e^{t})+2(\tfrac{11}{7}c_1e^{-3t}+\tfrac{1}{2}c_2e^{2t}+c_3e^{t})$$

$$=\tfrac{1}{7}c_1e^{-3t}+\tfrac{1}{2}c_2e^{2t}+c_3e^{t}$$

If we set $C_1 = \frac{1}{7}c_1$, $C_2 = \frac{1}{2}c_2$, and $C_3 = c_3$, then x_1, x_2, and x_3 can be written as

$$x_1(t) = C_1 e^{-3t} + C_2 e^{2t} + C_3 e^t$$

$$x_2(t) = 7C_1 e^{-3t} + 2C_2 e^{2t} + C_3 e^t$$

$$x_3(t) = 11C_1 e^{-3t} + C_2 e^{2t} + C_3 e^t$$

which is the general solution for the system of equations found in Example 4 of Section 5.4. The reader should compare the above calculations with those in Example 2 of Section 5.5. In particular, notice that the auxiliary equation for equation (5.57) is identical with the characteristic equation for the matrix **A** in that example. ◀

▶ **Example 3** Consider the system of differential equations

$$\frac{dx_1}{dt} = 7x_1 - 2x_2 + e^{-t}$$

$$\frac{dx_2}{dt} = 3x_1 + 2x_2 + t$$

(5.61)

which can be rewritten as

$$(D-7)x_1 + 2x_2 = e^{-t}$$

$$-3x_1 + (D-2)x_2 = t$$

If we multiply the first equation by 3, multiply the second equation by $D-7$, and add, we obtain

$$[(D-7)(D-2)+6]x_2 = 3e^{-t} + (D-7)t$$

which simplifies to

$$(D-5)(D-4)x_2 = 3e^{-t} + 1 - 7t$$

(5.62)

Using the method of undetermined coefficients, we find that

$$z(t) = \frac{1}{10}e^{-t} - \frac{43}{400} - \frac{7}{20}t$$

is a solution of equation (5.62). Since 5 and 4 are the roots of the auxiliary equation $(r-s)(r-4)=0$, a general solution of equation (5.62) is

$$x_2(t)=c_1e^{5t}+c_2e^{4t}+\tfrac{1}{10}e^{-t}-\tfrac{43}{400}-\tfrac{7}{20}t \qquad \textbf{(5.63)}$$

Using the second equation in (5.61), we have

$$x_1(t)=\tfrac{1}{3}(D-2)x_2-\tfrac{1}{3}t$$

$$=\tfrac{1}{3}(D-2)\left(c_1e^{5t}+c_2e^{4t}+\tfrac{1}{10}e^{-t}-\tfrac{43}{400}-\tfrac{7}{20}t\right)-\tfrac{1}{3}t$$

$$=c_1e^{5t}+\tfrac{2}{3}c_2e^{4t}-\tfrac{1}{10}e^{-t}-\tfrac{9}{200}-\tfrac{1}{10}t \qquad \textbf{(5.64)}$$

The functions x_1 and x_2 given in (5.63) and (5.64), respectively, form a general solution for the system of differential equations in (5.61).   ◀

Exercises   SECTION 5.8

In Exercises 1–9 find the general solution of the given system of differential equations.

1. $\dfrac{dx_1}{dt}=x_2$

   $\dfrac{dx_2}{dt}=x_1$

2. $\dfrac{dx_1}{dt}=-x_2$

   $\dfrac{dx_2}{dt}=x_1$

3. $\dfrac{dx_1}{dt}=6x_1+8x_2$

   $\dfrac{dx_2}{dt}=6x_1-2x_2$

4. $\dfrac{dx_1}{dt}=-2x_1+x_2$

   $\dfrac{dx_2}{dt}=x_1-x_2$

5. $\dfrac{dx_1}{dt}=3x_1+2x_2+t$

   $\dfrac{dx_2}{dt}=2x_1+3x_2+3$

6. $\dfrac{dx_1}{dt} = 6x_1 + 8x_2$

$\dfrac{dx_2}{dt} = 6x_1 - 2x_2 + e^{-t}$

7. $\dfrac{dx_1}{dt} = 3x_1 + 2x_2 + 2x_3$

$\dfrac{dx_2}{dt} = x_1 + 4x_2 + x_3$

$\dfrac{dx_3}{dt} = -2x_1 - 4x_2 - x_3$

8. $\dfrac{dx_1}{dt} = 3x_1 + 2x_2$

$\dfrac{dx_2}{dt} = 3x_1 - 2x_2$

$\dfrac{dx_3}{dt} = x_1 + x_2 + x_3$

9. $\dfrac{dx_1}{dt} = x_1 - x_2 - x_3 + e^t$

$\dfrac{dx_2}{dt} = x_1 - x_2 + t$

$\dfrac{dx_3}{dt} = x_1 - x_3 + 1$

10. Find the solution of the system of differential equations in Exercise 1 such that $x_1(0) = 1$, $x_2(0) = 2$.

11. Find the solution of the system of differential equations in Exercise 5 such that $x_1(0) = 3$, $x_2(0) = 1$.

12. Find the solution of the system of differential equations in Exercise 7 such that $x_1(0) = 2$, $x_2(0) = 1$, $x_3(0) = 0$.

In Exercises 13–16 use the method of elimination to solve these problems from Section 5.7 (pages 280–81):

13. Exercise 1
14. Exercise 3
15. Exercise 4(a)
16. Exercise 4(b)

17. Suppose that a substance S_1 is produced in a cell at a rate proportional to its own concentration c_1. Let a second substance S_2 catalyze the decomposition of S_1, so that S_1 is decomposed at a rate proportional to the concentration c_2 of S_2. We assume that the production of S_2 is catalyzed by S_1 at a rate proportional to c_1, while at the same time S_2 decomposes at a rate proportional to its own concentration c_2. Under these conditions it can be shown that

$$\frac{dc_1}{dt} = A_1c_1 + B_1c_2 + C_1$$

$$\frac{dc_2}{dt} = A_2c_1 + B_2c_2 + C_2$$

[N. Rashevsky, *Mathematical Biophysics*, The University of Chicago Press, Chicago, 1938, pages 54–57.] Find a general solution to this problem for the special case considered in the book cited above:

$$A_1 = .1 \qquad B_1 = -.1 \qquad C_1 = 2(10^{-5})$$
$$A_2 = .3 \qquad B_2 = -.1 \qquad C_2 = 10^{-5}$$

18. One model for an oral glucose tolerance test is

$$\frac{dg}{dt} = -m_1g - m_2h + J(t)$$

$$\frac{dh}{dt} = -m_3g - m_4h + k(t)$$

where $0 < m_1, m_2, m_3, m_4$ and

$g = $ concentration of glucose above a fasting state level

$h = $ effective hormonal concentration above fasting state level

$J, k = $ exogenous inputs into the system via intestines or intravenously

At $t=0$ we assume that $g=h=0$. [E. Ackerman, L. C. Gatewood, J. W. Rosevear, and G. D. Molnar, Blood glucose regulation and diabetes, in *Concepts and Models of Biomathematics*, edited by F. Heinmets, Marcel Dekker, New York, 1969, pages 135–137.]

(a) Let r_1 and r_2 be the roots of the polynomial $x^2 + (m_1 + m_4)x + m_1m_4 - m_2m_3$. Show that r_1 and r_2 are negative real numbers.
(b) Find the form of a general solution for the system of differential equations in terms of r_1 and r_2 for the special case $J(t) \equiv A$, $k(t) \equiv B$, $m_1m_4 - m_2m_3 \neq 0$, and $(m_1 - m_4)^2 - 4m_2m_3 > 0$.
(c) Compute $\lim_{t \to \infty} g(t)$ and $\lim_{t \to \infty} h(t)$. Interpret these values in relation to the model.

19. One model for the cooling of a bare homogeneous metal ingot with a tubular shell is

$$\frac{dT_1}{dt} = -a_1 T_1 + a_1 T_2 + A_1\left(T_1^4 - T_0^4\right)$$

$$\frac{dT_2}{dt} = a_1 T_1 - (a_1 + a_2) T_2 + A_2\left(T_2^4 - T_0^4\right) + c$$

where $0 < a_1, a_2$ and

T_1 = temperature of the ingot

T_2 = temperature of a tubular shell that surrounds the ingot

T_0 = temperature of the surrounding atmosphere (assumed constant)

[K. D. Tocher, Mathematical models in the steel industry, in *Mathematical Models in Metallurgical Process Development*, The Iron and Steel Institute, London, 1970, page 13.] If we ignore the radiation terms (that is, assume $A_1 = A_2 = 0$), the system of equations may be written as

$$\frac{dT_1}{dt} = -a_1 T_1 + a_1 T_2$$

$$\frac{dT_2}{dt} = a_1 T_1 - (a_1 + a_2) T_2 + c$$

(a) Show that the roots r_1 and r_2 of the polynomial $f(x) = x^2 + (2a_1 + a_2)\lambda + a_1 a_2$ are negative real numbers.
(b) Let $T_1(0) = a$ and $T_2(0) = b$. Find the solution of the system of differential equations in terms of a, b, r_1, and r_2.
(c) Compute $\lim_{t \to \infty} T_1(t)$ and $\lim_{t \to \infty} T_2(t)$. Interpret these values in relation to the model.

5.9 Laplace Transforms

Consider again the system of nonhomogeneous linear differential equations

$$\frac{dx_1}{dt} = a_{11}x_1 + a_{12}x_2 + \cdots + a_{1n}x_n + f_1(t)$$

$$\frac{dx_2}{dt} = a_{21}x_1 + a_{22}x_2 + \cdots + a_{2n}x_n + f_2(t)$$

$$\cdots\cdots\cdots\cdots\cdots\cdots\cdots\cdots\cdots$$

$$\frac{dx_n}{dt} = a_{n1}x_1 + a_{n2}x_2 + \cdots + a_{nn}x_n + f_n(t)$$

where the functions $f_1, f_2, \ldots, f_n$ are defined on $[0, \infty)$. If we take the Laplace transform of each of the above equations, use the identity $L[dz/dt] = sL[z] - z(0)$, and rearrange the resulting equations, we obtain a system of n equations in n unknowns:

$$(a_{11} - s)L[x_1] + a_{12}L[x_2] + \cdots + a_{1n}L[x_n] = -L[f_1] - x_1(0)$$
$$a_{21}L[x_1] + (a_{22} - s)L[x_2] + \cdots + a_{2n}L[x_n] = -L[f_2] - x_2(0)$$
$$\cdots\cdots\cdots\cdots\cdots\cdots\cdots\cdots\cdots\cdots\cdots\cdots\cdots\cdots$$
$$a_{n1}L[x_1] + a_{n2}L[x_2] + \cdots + (a_{nn} - s)L[x_n] = -L[f_n] - x_n(0)$$

If we now solve this system for $L[x_1], L[x_2], \ldots, L[x_n]$, then, in all cases of interest here, the functions $x_1, x_2, \ldots, x_n$ may be found by using a table of Laplace transforms. In order to compare this method with the methods of Sections 5.5, 5.6, and 5.8, we will redo some of the examples found there.

▶ **Example 1**   Consider the system of differential equations

$$\frac{dx_1}{dt} = 5x_1 - 3x_2$$
$$\frac{dx_2}{dt} = -2x_1 + 4x_2$$

(5.65)

Taking the Laplace transform of each side of each equation and using the identity $L[dz/dt] = sL[z] - z(0)$, we obtain

$$sL[x_1] - x_1(0) = 5L[x_1] - 3L[x_2]$$

$$sL[x_2] - x_2(0) = -2L[x_1] + 4L[x_2]$$

We could carry $x_1(0)$ and $x_2(0)$ through all of our calculations as arbitrary constants, but this is an algebraic nuisance. Hence, for the sake of simplicity, we will take $x_1(0) = 3$ and $x_2(0) = 4$ as the initial conditions in this example. The system of equations now becomes

$$(s - 5)L[x_1] + 3L[x_2] = 3$$
$$2L[x_1] + (s - 4)L[x_2] = 4$$

(5.66)

If we multiply the first equation by $s - 4$, multiply the second equation by -3, and add, we obtain

$$((s - 5)(s - 4) - 6)L[x_1] = (s - 4)(3) - 12$$

which simplifies to

$$L[x_1] = \frac{3s - 24}{(s-2)(s-7)}$$

We will now use the method of partial fractions to find numbers A and B so that

$$\frac{3s - 24}{(s-2)(s-7)} = \frac{A}{s-2} + \frac{B}{s-7}$$

Multiplying each side of this equation by $(s-2)(s-7)$ yields

$$3s - 24 = A(s-7) + B(s-2)$$

Setting $s=2$ and then setting $s=7$, we find that $-18 = -5A$ and $-3 = 5B$, respectively. Therefore $A = \frac{18}{5}$ and $B = -\frac{3}{5}$. We now have

$$L[x_1] = \frac{18}{5} \frac{1}{s-2} - \frac{3}{5} \frac{1}{s-7} \qquad (5.67)$$

Since $L[e^{at}] = 1/(s-a)$, we have

$$x_1(t) = \tfrac{18}{5} e^{2t} - \tfrac{3}{5} e^{7t} \qquad (5.68)$$

The reader should note the similarity between the calculations done here and those in Example 1 of Section 5.5 and Example 1 of Section 5.8. In particular, notice that the characteristic polynomial of the matrix in Example 1 of Section 5.5 and the auxiliary polynomial for equation (5.57) appear here in the Laplace transform of x_1 as $(s-2)(s-7)$. The function x_2 may now be computed in two ways: (1) use the second equation in (5.65) and x_1 from (5.68), or (2) use the second equation in (5.66) and $L[x_1]$ from (5.67) to compute $L[x_2]$, and then find x_2 knowing $L[x_2]$. The latter method requires the use of partial fractions and therefore is more cumbersome. Using the second equation in (5.65) and x_1 from (5.68), we have

$$x_2(t) = \tfrac{5}{3} x_1 - \tfrac{1}{3} \frac{dx_1}{dt}$$

$$= \tfrac{1}{3}(5 - D)x_1$$

$$= \tfrac{1}{3}(5 - D)\left(\tfrac{18}{5} e^{2t} - \tfrac{3}{5} e^{7t}\right)$$

$$= \tfrac{18}{5} e^{2t} + \tfrac{2}{5} e^{7t}$$

In Example 3 of Section 5.4 we found that a general solution of the system of differential equations (5.65) is

$$x_1(t)=c_1e^{2t}+3c_2e^{7t}$$

$$x_2(t)=c_1e^{2t}-2c_2e^{7t}$$

The solution computed above is obtained from this general solution by setting $c_1=\frac{18}{5}$ and $c_2=-\frac{1}{5}$.   ◄

▶ **Example 2**   Consider the system of differential equations

$$\frac{dx_1}{dt}=x_1+x_2-x_3$$

$$\frac{dx_2}{dt}=2x_1+3x_2-4x_3 \qquad\qquad \text{(5.69)}$$

$$\frac{dx_3}{dt}=4x_1+x_2-4x_3$$

with the initial conditions

$$x_1(0)=0, \quad x_2(0)=5, \quad x_3(0)=1$$

Taking the Laplace transform of each side of each equation in (5.69) and using the identity $L[dz/dt]=sL[z]-z(0)$, we obtain

$$sL[x_1]-x_1(0)=L[x_1]+L[x_2]-L[x_3]$$

$$sL[x_2]-x_2(0)=2L[x_1]+3L[x_2]-4L[x_3]$$

$$sL[x_3]-x_3(0)=4L[x_1]+L[x_2]-4L[x_3]$$

or, equivalently,

$$(s-1)L[x_1]-L[x_2]+L[x_3]=0 \qquad\qquad \text{(5.70)}$$

$$-2L[x_1]+(s-3)L[x_2]+4L[x_3]=5 \qquad\qquad \text{(5.71)}$$

$$-4L[x_1]-L[x_2]+(s+4)L[x_3]=1 \qquad\qquad \text{(5.72)}$$

which is a system of three linear equations in three unknowns: $L[x_1]$, $L[x_2]$, and $L[x_3]$. If we multiply equation (5.70) by 2, multiply equation

(5.71) by $s-1$, and add, we obtain

$$((s-1)(s-3)-2)L[x_2]+(4(s-1)+2)L[x_3]=5(s-1)$$

which simplifies to

$$(s^2-4s+1)L[x_2]+(4s-2)L[x_3]=5s-5 \qquad \text{(5.73)}$$

If we multiply equation (5.71) by -2 and add the result to equation (5.72), we obtain

$$(-2(s-3)-1)L[x_2]+((s+4)-8)L[x_3]=-9$$

which simplifies to

$$-(2s-5)L[x_2]+(s-4)L[x_3]=-9 \qquad \text{(5.74)}$$

Equations (5.73) and (5.74) give us two equations that we can solve for $L[x_1]$ and $L[x_2]$. If we multiply equation (5.73) by $s-4$, multiply equation (5.74) by $-(4s-2)$, and add, we obtain

$$((s-4)(s^2-4s+1)+(4s-2)(2s-5))L[x_2]=(5s-5)(s-4)+9(4s-2)$$

which simplifies to

$$(s+3)(s-2)(s-1)L[x_2]=5s^2+11s+2$$

Hence

$$L[x_2]=\frac{5s^2+11s+2}{(s+3)(s-2)(s-1)}$$

Using the method of partial fractions, $L[x_2]$ can be written in the more convenient form

$$L[x_2]=\frac{7}{10}\frac{1}{s+3}+\frac{44}{5}\frac{1}{s-2}-\frac{9}{2}\frac{1}{s-1} \qquad \text{(5.75)}$$

From equation (5.74) we find that

$$L[x_3]=\frac{1}{s-4}(-9+(2s-5)L[x_2])$$

$$=-\frac{9}{s-4}+\frac{7}{10}\frac{2s-5}{(s-4)(s+3)}+\frac{44}{5}\frac{2s-5}{(s-4)(s-2)}$$

$$-\frac{9}{2}\frac{2s-5}{(s-4)(s-1)}$$

Using the method of partial fractions, we find that

$$\frac{2s-5}{(s-4)(s+3)} = \frac{3}{7}\frac{1}{s-4} + \frac{11}{7}\frac{1}{s+3}$$

$$\frac{2s-5}{(s-4)(s-2)} = \frac{3}{2}\frac{1}{s-4} + \frac{1}{2}\frac{1}{s-2}$$

$$\frac{2s-5}{(s-4)(s-1)} = \frac{1}{s-4} + \frac{1}{s-1}$$

Using these identities, $L[x_3]$ can now be rewritten as

$$L[x_3] = -\frac{9}{s-4} + \frac{7}{10}\left(\frac{3}{7}\frac{1}{s-4} + \frac{11}{7}\frac{1}{s+3}\right)$$

$$+ \frac{44}{5}\left(\frac{3}{2}\frac{1}{s-4} + \frac{1}{2}\frac{1}{s-2}\right) - \frac{9}{2}\left(\frac{1}{s-4} + \frac{1}{s-1}\right)$$

$$= \frac{11}{10}\frac{1}{s+3} + \frac{22}{5}\frac{1}{s-2} - \frac{9}{2}\frac{1}{s-1} \qquad \textbf{(5.76)}$$

From equations (5.75) and (5.76) we find that

$$x_2(t) = \tfrac{7}{10}e^{-3t} + \tfrac{44}{5}e^{2t} - \tfrac{9}{2}e^{t}$$

$$x_3(t) = \tfrac{11}{10}e^{-3t} + \tfrac{22}{5}e^{2t} - \tfrac{9}{2}e^{t}$$

The function x_1 can now be computed from the second equation in (5.69):

$$x_1(t) = \tfrac{1}{2}\left(\frac{dx_2}{dt} - 3x_2\right) + 2x_3$$

$$= \tfrac{1}{2}(D-3)x_2 + 2x_3$$

$$= \tfrac{1}{2}(D-3)\left(\tfrac{7}{10}e^{-3t} + \tfrac{44}{5}e^{2t} - \tfrac{9}{2}e^{t}\right) + 2\left(\tfrac{11}{20}e^{-3t} + \tfrac{22}{5}e^{2t} - \tfrac{9}{2}e^{t}\right)$$

$$= \tfrac{1}{10}e^{-3t} + \tfrac{22}{5}e^{2t} - \tfrac{9}{2}e^{t}$$

In Example 4 of Section 5.4 we found that a general solution of the system of differential equations in (5.69) is

$$x_1(t)=c_1e^{-3t}+c_2e^{2t}+c_3e^t$$

$$x_2(t)=7c_1e^{-3t}+2c_2e^{2t}+c_3e^t$$

$$x_3(t)=11c_1e^{-3t}+c_2e^{2t}+c_3e^t$$

The solution computed above is obtained from this general solution by setting $c_1=\frac{1}{10}$, $c_2=\frac{22}{5}$, and $c_3=-\frac{9}{2}$.

The reader should note the similarity between the above calculations and those done in Example 2 of Section 5.5 and Example 2 of Section 5.8. In particular, notice that the characteristic polynomial of the matrix in Example 2, Section 5.5, and the auxiliary polynomial for equation (5.51) appear here in the Laplace transform for x_2 as $(s+3)(s-2)(s-1)$.

◀

▶ **Example 3**   Consider the system of differential equations

$$\frac{dx_1}{dt}=7x_1-2x_2+e^{-t}$$

(5.77)

$$\frac{dx_2}{dt}=3x_1+2x_2+t$$

with the initial conditions $x_1(0)=0$ and $x_2(0)=0$. Taking the Laplace transform of each side of each equation and using the identity $L[dz/dt]$ $=sL[z]-z(0)$, we obtain

$$sL[x_1]-x_1(0)=7L[x_1]-2L[x_2]+L[e^{-t}]$$

$$sL[x_2]-x_2(0)=3L[x_1]+2L[x_2]+L[t]$$

which simplifies to

$$(s-7)L[x_1]+2L[x_2]=\frac{1}{s+1}$$

$$-3L[x_1]+(s-2)L[x_2]=\frac{1}{s^2}$$

If we multiply the first equation by 3 and the second equation by $s-7$, and add, we obtain

$$((s-7)(s-2)+6)\,L[x_2]=\frac{3}{s+1}+\frac{s-7}{s^2}$$

which simplifies to

$$(s-5)(s-4)L[x_2]=\frac{3}{s+1}+\frac{s-7}{s^2}$$

Hence

$$L[x_2]=\frac{3}{(s-5)(s-4)(s+1)}+\frac{s-7}{s^2(s-5)(s-4)}$$

Using the method of partial fractions, we find that

$$\frac{3}{(s-5)(s-4)(s+1)}=\frac{1}{2}\frac{1}{s-5}-\frac{3}{5}\frac{1}{s-4}+\frac{1}{10}\frac{1}{s+1}$$

$$\frac{s-7}{s^2(s-5)(s-4)}=-\frac{43}{400}\frac{1}{s}-\frac{7}{20}\frac{1}{s^2}-\frac{2}{25}\frac{1}{s-5}+\frac{3}{16}\frac{1}{s-4}$$

So

$$L[x_2]=\left(\frac{1}{2}\frac{1}{s-5}-\frac{3}{5}\frac{1}{s-4}+\frac{1}{10}\frac{1}{s+1}\right)$$

$$+\left(-\frac{43}{400}\frac{1}{s}-\frac{7}{20}\frac{1}{s^2}-\frac{2}{25}\frac{1}{s-5}+\frac{3}{16}\frac{1}{s-4}\right)$$

$$=\frac{21}{50}\frac{1}{s-5}-\frac{33}{80}\frac{1}{s-4}+\frac{1}{10}\frac{1}{s+1}-\frac{43}{400}\frac{1}{s}-\frac{7}{20}\frac{1}{s^2}$$

so that

$$x_2(t)=\tfrac{21}{50}e^{5t}-\tfrac{33}{80}e^{4t}+\tfrac{1}{10}e^{-t}-\tfrac{43}{400}-\tfrac{7}{20}t$$

From the second equation in (5.77) we have

$$x_1(t)=\tfrac{1}{3}\left(\frac{dx_2}{dt}-2\right)x_2-\tfrac{1}{3}t$$

$$=\tfrac{1}{3}(D-2)x_2-\tfrac{1}{3}t$$

$$=\tfrac{1}{3}(D-2)\left(\tfrac{21}{50}e^{5t}-\tfrac{33}{80}e^{4t}+\tfrac{1}{10}e^{-t}-\tfrac{43}{400}-\tfrac{7}{20}t\right)-\tfrac{1}{3}t$$

$$=\tfrac{21}{50}e^{5t}-\tfrac{11}{40}e^{4t}-\tfrac{1}{10}e^{-t}-\tfrac{9}{200}-\tfrac{1}{10}t$$

In Example 3 of Section 5.8 we found that a general solution of the system of differential equations in (5.77) is

$$x_1(t) = c_1 e^{5t} + \tfrac{2}{3} c_2 e^{4t} - \tfrac{1}{10} e^{-t} - \tfrac{9}{200} - \tfrac{1}{10} t$$

$$x_2(t) = c_1 e^{5t} + c_2 e^{4t} + \tfrac{1}{10} e^{-t} - \tfrac{43}{400} - \tfrac{7}{20} t$$

The solution computed above is obtained from this general solution by setting $c_1 = \tfrac{21}{50}$ and $c_2 = -\tfrac{33}{80}$. ◀

Exercises　SECTION 5.9

Use Laplace transforms to solve each of the following problems. Note that the systems of differential equations in Exercises 1–9 are the same as those in Exercises 1–9 of Section 5.8. Moreover, Exercise 2 is Exercise 1 of Section 5.6, while Exercises 1, 3, and 4 are Exercises 1, 2, and 3 of Section 5.5. Compare the methods of solution of the corresponding problems.

1. $\dfrac{dx_1}{dt} = x_2, \quad x_1(0) = 0$

$\dfrac{dx_2}{dt} = x_1, \quad x_2(0) = 1$

2. $\dfrac{dx_1}{dt} = -x_2, \quad x_1(0) = 1$

$\dfrac{dx_2}{dt} = x_1, \quad x_2(0) = 0$

3. $\dfrac{dx_1}{dt} = 6x_1 + 8x_2, \quad x_1(0) = 3$

$\dfrac{dx_2}{dt} = 6x_1 - 2x_2, \quad x_2(0) = 5$

4. $\dfrac{dx_1}{dt} = -2x_1 + x_2, \quad x_1(0) = 2$

$\dfrac{dx_2}{dt} = x_1 - x_2, \quad x_2(0) = -1$

5. $\dfrac{dx_1}{dt} = 3x_1 + 2x_2 + t, \quad x_1(0) = -1$

$\dfrac{dx_2}{dt} = 2x_1 + 3x_2 + 3, \quad x_2(0) = 1$

6. $\dfrac{dx_1}{dt} = 6x_1 + 8x_2, \quad x_1(0) = 0$

$\dfrac{dx_2}{dt} = 6x_1 - 2x_2 + e^{-t}, \quad x_2(0) = 0$

7. $\dfrac{dx_1}{dt} = 3x_1 + 2x_2 + 2x_3, \quad x_1(0) = 0$

$\dfrac{dx_2}{dt} = x_1 + 4x_2 + x_3, \quad x_2(0) = 1$

$\dfrac{dx_3}{dt} = -2x_1 - 4x_2 - x_3, \quad x_3(0) = 2$

8. $\dfrac{dx_1}{dt} = 3x_1 + 2x_2, \quad x_1(0) = 3$

$\dfrac{dx_2}{dt} = 3x_1 - 2x_2, \quad x_2(0) = -2$

$\dfrac{dx_3}{dt} = x_1 + x_2 + x_3, \quad x_3(0) = 3$

9. $\dfrac{dx_1}{dt} = x_1 - x_2 - x_3 + e^t, \quad x_1(0) = 0$

$\dfrac{dx_2}{dt} = x_1 - x_2 + t, \quad x_2(0) = 0$

$\dfrac{dx_3}{dt} = x_1 - x_3 + 1, \quad x_3(0) = 0$

10. Solve the warfare problem presented in Example 1 of Section 5.7 using Laplace transforms.

11. Do Exercise 1 of Section 5.7 using Laplace transforms.

12. Do Example 1 of Section 5.7 using Laplace transforms.

13. Do Exercise 17 of Section 5.8 using Laplace transforms.

14. Do Exercise 19 of Section 5.8 using Laplace transforms.

15. Systems of equations which contain derivatives of order greater than one commonly arise in engineering. For example, consider the mechanical system consisting of two masses m_1 and m_2, each of which is restrained by an elastic spring attached to a fixed base and by a third elastic spring that connects the two masses as shown in Figure 5.10. Let x_1 and x_2 measure the displacement from equilibrium of the masses m_1 and m_2, respectively.

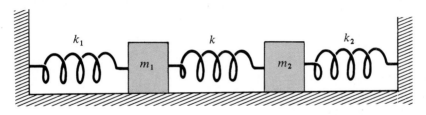

FIGURE 5.10

Then

$$m_1 \frac{d^2 x_1}{dt^2} = -(k_1 + k)x_1 + kx_2$$

$$m_2 \frac{d^2 x_2}{dt^2} = kx_1 - (k_2 + k)x_2$$

where k, k_1, and k_2 are the spring constants. Suppose that m_2 is initially held fixed and that m_1 is displaced by one unit in the positive direction, held fixed, and then released. In such a case

$$x_1(0) = 1, \quad x_1'(0) = 0, \quad x_2(0) = 0, \quad x_2'(0) = 0$$

Use Laplace transforms to find the displacements x_1 and x_2 of the two masses for the special case $m_1 = m_2$ and $k_1 = k = k_2$. Compare your method of solution to that used to solve Exercise 6 of Section 5.7.

16. Consider the electrical circuit shown in Figure 5.11, where C_1, C_2, and C_3 are capacitances, E is an electromotive force, I_1 and I_2 are currents, and

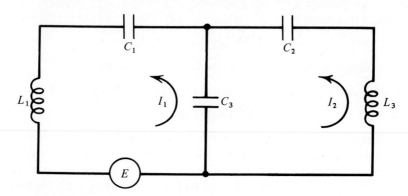

FIGURE 5.11

L_1, L_2 are inductances. Using Kirchhoff's laws, it can be shown that

$$\left[L_1 D^2 + \left(\frac{1}{C_1} + \frac{1}{C_2} \right) \right] I_1 - \frac{1}{C_3} I_2 = E$$

$$-\frac{1}{C_3} I_1 + \left[L_2 D^2 + \left(\frac{1}{C_2} + \frac{1}{C_3} \right) \right] I_2 = 0$$

Use Laplace transforms to find the currents I_1 and I_2 for the special case $L_1=L_2=1$ henry, $C_1=C_2=C_3=1$ farad, $E=110$ volts, and $I_1(0)=I_1'(0)=I_2(0)=I_2'(0)=0$.

5.10 Numerical Methods

The numerical methods discussed in Sections 1.8 and 1.9 can be extended to cover systems of differential equations. For simplicity we will only consider the initial-value problem for a system of two differential equations of the form

$$\frac{dx}{dt}=f(t,x,y), \quad x(t_0)=x_0$$

$$\frac{dy}{dt}=g(t,x,y), \quad y(t_0)=y_0$$

(5.78)

For small values of h we have

$$\frac{x(t+h)-x(t)}{h}\simeq f(t,x(t),y(t))$$

$$\frac{y(t+h)-y(t)}{h}\simeq g(t,x(t),y(t))$$

so that

$$x(t+h)\simeq x(t)+hf(t,x(t),y(t))$$

$$y(t+h)\simeq x(t)+hg(t,x(t),y(t))$$

If we choose $h=1/n$ and $t_i=t_0+ih$ for $i=1,2,\dots$, then

$$x(t_{i+1})=x(t_i+h)\simeq x(t_i)+hf(t_i,x(t_i),y(t_i))$$

$$y(t_{i+1})=y(t_i+h)\simeq y(t_i)+hg(t_i,x(t_i),y(t_i))$$

Euler's method computes approximations x_i and y_i to $x(t_i)$ and $y(t_i)$, respectively, according to the iterative procedure

$$x_{i+1}=x_i+hf(t_i,x_i,y_i)$$

$$y_{i+1}=y_i+hg(t_i,x_i,y_i)$$

As in the case for a single differential equation, at each step (after the first) we compute the values of x and y required in the following step.

▶ **Example 1**   One of the classical problems of physics is to analyze the motion of a simple pendulum. Ideally, a pendulum consists of a bob of mass m suspended from a frictionless pivot by a massless rod of length L as shown in Figure 5.12.

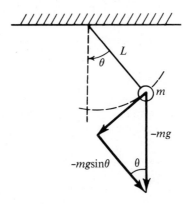

FIGURE 5.12

When the bob is displaced from a vertical position by an angle θ, measured in radians, the bob travels along an arc of length $s = L\theta$. Hence the angular acceleration a is given by

$$a = \frac{d^2s}{dt^2} = L\frac{d^2\theta}{dt^2}$$

The gravitational force on the bob $-mg$ acts downward. The component of this force acting along the path of motion of the bob and toward the equilibrium position of the bob is $-mg\sin\theta$. If there are no other forces acting on the pendulum, then, by Newton's second law of motion ($F = ma$), we have

$$-mg\sin\theta = mL\frac{d^2\theta}{dt^2}$$

so that

$$\frac{d^2\theta}{dt^2} + \frac{g}{L}\sin\theta = 0$$

For simplicity we will assume that the numerical value of g/L is 1. We will also assume that the pendulum is displaced by an angle of .1 radian, held stationary, and then released. In such a case $\theta(0) = .1$ and $\theta'(0) = 0$.

If we rename the variables according to $x=\theta$, $y=d\theta/dt$, then we have

$$\frac{dx}{dt}=\frac{d\theta}{dt}=y, \quad x(0)=\theta(0)=.1$$

$$\frac{dy}{dt}=\frac{d^2\theta}{dt^2}=-\sin\theta=-\sin x, \quad y(0)=\theta'(0)=0$$

Thus the motion of the pendulum is described by the initial-value problem

$$\frac{dx}{dt}=y, \quad x(0)=.1$$

$$\frac{dy}{dt}=-\sin x, \quad y(0)=0 \tag{5.79}$$

Euler's method for this problem is

$$x_{i+1}=x_i+hy_i$$

$$y_{i+1}=y_i-h\sin x_i$$

Choosing $n=10$ and $h=1/n$, we have

$$x_1=x_0+hy_0=.1$$

$$y_1=y_0-h\sin x_0=-10^{-1}\sin(.1)\approx-.0100$$

Continuing in this fashion, we have

t_i	x_i $(\approx\theta(t_i))$	y_i $(\approx\theta'(t_i))$
0.0	.1000	.0000
.1	.1000	$-.0100$
.2	.0990	$-.0200$
.3	.0970	$-.0299$
.4	.0940	$-.0395$
.5	.0901	$-.0489$
1.4	.0188	$-.1055$
1.5	.0083	$-.1074$
1.6	$-.0025$	$-.1082$
1.7	$-.0133$	$-.1080$
3.0	$-.1147$	$-.0177$
3.1	$-.1165$	$-.0063$
3.2	$-.1171$	.0053
3.3	$-.1166$	.0171

Shortly after being released, the pendulum swings from right to left. Hence we would expect that $x_{i+1} \leqslant x_i$ and $y_i < 0$ for small values of i. Looking at the table, we see that this is true for $i < 32$. When the pendulum is vertical, $\theta = 0$. From the table we see that $\theta(t) = 0$ when t is approximately 1.6. Hence the pendulum is vertical approximately 1.6 seconds after it is released. Notice that y_i, the approximate angular velocity, changes from negative to positive when i changes from 31 to 32. This means that the pendulum begins to swing from left to right approximately 3.1 seconds after being released. It should also be noted that the data in the table seems to indicate that the pendulum swings farther to the left and, hence, higher than its initial position. Since $\theta'(0) = 0$, this is impossible. This apparent absurdity is due to the errors inherent in any numerical method, especially Euler's method. Next we will consider a Runge-Kutta method, in which these errors are usually not so large.

The Runge-Kutta method can also be adapted to initial-value problems for systems of differential equations of the form in (5.78):

$$x_{i+1} = x_i + \frac{h}{6}(k_{1i} + 2k_{2i} + 2k_{3i} + k_{4i})$$

$$y_{i+1} = y_i + \frac{h}{6}(j_{1i} + 2j_{2i} + 2j_{3i} + j_{4i})$$

where

$$k_{1i} = f(t_i, x_i, y_i)$$

$$k_{2i} = f\left(t_i + \frac{h}{2}, x_i + \frac{h}{2}k_{1i}, y_i + \frac{h}{2}j_{1i}\right)$$

$$k_{3i} = f\left(t_i + \frac{h}{2}, x_i + \frac{h}{2}k_{2i}, y_i + \frac{h}{2}j_{2i}\right)$$

$$k_{4i} = f(t_{i+1}, x_i + hk_{3i}, y_i + hj_{3i})$$

$$j_{1i} = g(t_i, x_i, y_i)$$

$$j_{2i} = g\left(t_i + \frac{h}{2}, x_i + \frac{h}{2}k_{1i}, y_i + \frac{h}{2}j_{1i}\right)$$

$$j_{3i} = g\left(t_i + \frac{h}{2}, x_i + \frac{h}{2}k_{2i}, y_i + \frac{h}{2}j_{2i}\right)$$

$$j_{4i} = g(t_{i+1}, x_i + hk_{3i}, y_i + hj_{3i})$$

▶ **Example 2**   For the pendulum problem described by the initial-value problem in (5.79), we have

$$f(t, x, y)=y$$

$$g(t, x, y)=-\sin x$$

so that

$$k_{1i}=y_i \qquad\qquad j_{1i}=-\sin x_i$$

$$k_{2i}=y_i+\frac{h}{2}j_{1i} \qquad\qquad j_{2i}=-\sin\left(x_i+\frac{h}{2}k_{1i}\right)$$

$$=y_i-\frac{h}{2}\sin x_i \qquad\qquad =-\sin\left(x_i+\frac{h}{2}y_i\right)$$

$$k_{3i}=y_i+\frac{h}{2}j_{2i} \qquad\qquad j_{3i}=-\sin\left(x_i+\frac{h}{2}k_{2i}\right)$$

$$=y_i-\frac{h}{2}\sin\left(x_i+\frac{h}{2}y_i\right) \qquad =-\sin\left(x_i+\frac{h}{2}\left[y_i-\frac{h}{2}\sin x_i\right]\right)$$

$$k_{4i}=y_i+hj_{3i}=y_i-h\sin\left(x_i+\frac{h}{2}\left[y_i-\frac{h}{2}\sin x_i\right]\right)$$

$$j_{4i}=-\sin(x_i+hk_{3i})=-\sin\left(x_i+h\left[y_i-\frac{h}{2}\sin\left(x_i+\frac{h}{2}y_i\right)\right]\right)$$

With $n=10$ and $h=1/n$, we have

$$k_{10}=0 \qquad\qquad\qquad j_{10}=-\sin(.1)$$

$$\simeq .09983$$

$$k_{20}=-(.05)\sin(.1) \qquad\qquad j_{20}=-\sin(.1)$$

$$\simeq -.00499 \qquad\qquad\qquad \simeq -.09983$$

$$k_{30}=-(.05)\sin(.1) \qquad\qquad j_{30}=-\sin[.1-(.05)^2\sin(.1)]$$

$$\simeq -.00499 \qquad\qquad\qquad \simeq -.09975$$

$$k_{40}=-(.1)\sin[.1-(.05)^2\sin(.1)] \qquad j_{40}=-\sin[.1-(.1)(.05)\sin(.1)]$$

$$\simeq -.09975 \qquad\qquad\qquad \simeq -.09934$$

so that

$$x_1 = .1 + \frac{.1}{6}(k_{10} + 2k_{20} + 2k_{30} + k_{40}) \approx .0995$$

$$y_1 = 0 + \frac{.1}{6}(j_{10} + 2j_{20} + 2j_{30} + j_{40}) \approx -.0010$$

Continuing with the Runge-Kutta method, we have

t_i	x_i $(\approx \theta(t_i))$	y_i $(\approx \theta'(t_i))$
0.0	.1000	.0000
.1	.0995	−.0100
.2	.0980	−.0198
.3	.0955	−.0295
.4	.0921	−.0389
.5	.0878	−.0479
1.4	.0170	−.0985
1.5	.0072	−.0997
1.6	−.0028	−.0999
1.7	−.0128	−.0991
3.0	−.0990	−.0143
3.1	−.0999	−.0043
3.2	−.0998	.0056
3.3	−.0988	.0156

Again we see that the pendulum is vertical approximately 1.6 seconds after it is released and that it starts to swing from left to right approximately 3.1 seconds after it is released. However, this time the data indicates that the pendulum swings essentially the same distance to the left as its initial displacement. Theoretically, the pendulum swings from an angle of .1 radian to an angle of −.1 radian and back again. ◀

▶ **Example 3** In his book *The Prevention of Malaria*, Sir Ronald Ross (1857–1932) developed a system of differential equations to represent, in certain circumstances, the way malaria affects a community. (He received the 1902 Nobel Prize in Physiology and Medicine for his work on malaria.) The model he formulated is one of the earliest attempts to apply differential equations to epidemiology. To describe this model, we adopt the following notation:

$p(t)$=human population at time t
$H(t)$=human population affected with malaria at time t
f=infective rate, i.e., fraction of affected human
    population that is infective
r=recovery rate, i.e., fraction of affected human
    population that becomes healthy per unit of time
m=human death rate

p', M, f', r', m' will denote the corresponding quantities relative to the mosquito population.

If a mosquito bites a human on an average b' times per unit of time, then $f'M$ infective mosquitos will inflict $b'f'M$ bites per unit of time, and a fraction of these bites $(p-H)/p$ occur on healthy people. We will assume that once bitten, the person becomes affected. Then the number of new infections per unit of time is $b'f'M(p-H)/p$. Similarly, if a human is bitten, on the average, b times per unit of time, then the number of new infections among the mosquitos is $bfH(p'-M)/p'$.

Obviously the rate at which humans are bitten, bp, must coincide with the rate at which the mosquitos bite, $b'p'$. Hence $b=b'p'/p$, so the number of new infections per unit of time among the mosquitos is $b'fH(p'-M)/p$.

If we assume that immigration and emigration in the area under consideration are negligible for both people and mosquitos, then

$$\left\{ \begin{array}{c} \text{rate of affection} \\ \text{among humans} \end{array} \right\} = \left\{ \begin{array}{c} \text{rate of new} \\ \text{infections} \end{array} \right\} - \{\text{death rate}\} - \left\{ \begin{array}{c} \text{recovery} \\ \text{rate} \end{array} \right\}$$

That is,

$$\frac{dH}{dt} = \frac{b'f'M}{p}(p-H) - mH - rH$$

A similar argument gives an equation for the rate of affection among the mosquitos:

$$\frac{dM}{dt} = \frac{b'fH}{p}(p'-M) - m'M - r'M$$

We further assume that the birth rate is equal to the death rate for both the humans and the mosquitos, so the populations p and p' are constant. To simplify the notation slightly, we assume that the humans have one unit of population and that the mosquitos have A units of population, i.e., $p=1$ and $p'=A$.

Ross analyzed the probable values of m, m', r, r', and concluded that m is negligible with respect to r, while r' is negligible with respect to m'. We are now able to write the differential equations as

$$\frac{dH}{dt}=b'f'M(1-H)-rH$$

$$\frac{dM}{dt}=b'fH(A-M)-m'M$$

A detailed discussion of these equations is given by A. Lotka in "Analysis of Malaria Epidemiology" (American Journal of Hygiene, January Supplement, 1923). This article gives the values

$$A=19.418 \qquad r=.231046 \qquad f'=.33333$$
$$m'=3.2958 \qquad f=.25 \qquad b'=.82396$$

and time is measured in months. With these constants the differential equations become (where we have rounded to three decimal places for simplicity)

$$\frac{dH}{dt}=.275M-.231H-.275HM$$

$$\frac{dM}{dt}=4.000H-3.296M-.206HM$$

Due to the presence of HM in each of these equations, they are nonlinear, and therefore the methods of the previous sections are not applicable to this system of differential equations. The Runge-Kutta method for this system is

$$H_{i+1}=H_i+\frac{h}{6}(k_{1i}+2k_{2i}+2k_{3i}+k_{4i})$$

$$M_{i+1}=M_i+\frac{h}{6}(j_{1i}+2j_{2i}+2j_{3i}+j_{4i})$$

where

$$k_{1i} = .275 M_i - .231 H_i - .275 H_i M_i$$

$$j_{1i} = 4.00 H_i - 3.296 M_i - .206 H_i M_i$$

$$k_{2i} = .275 \left(M_i + \frac{h}{2} j_{1i} \right) - .231 \left(H_i + \frac{h}{2} k_{1i} \right)$$

$$- .275 \left(H_i + \frac{h}{2} k_{1i} \right) \left(M_i + \frac{h}{2} j_{1i} \right)$$

$$j_{2i} = 4.00 \left(H_i + \frac{h}{2} k_{1i} \right) - 3.296 \left(M_i + \frac{h}{2} j_{1i} \right)$$

$$- .206 \left(H_i + \frac{h}{2} k_{1i} \right) \left(M_i + \frac{h}{2} j_{1i} \right)$$

$$k_{3i} = .275 \left(M_i + \frac{h}{2} j_{2i} \right) - .231 \left(H_i + \frac{h}{2} k_{2i} \right)$$

$$- .275 \left(H_i + \frac{h}{2} k_{2i} \right) \left(M_i + \frac{h}{2} j_{2i} \right)$$

$$j_{3i} = 4.00 \left(H_i + \frac{h}{2} k_{2i} \right) - 3.296 \left(M_i + \frac{h}{2} j_{2i} \right)$$

$$- .206 \left(H_i + \frac{h}{2} k_{2i} \right) \left(M_i + \frac{h}{2} j_{2i} \right)$$

$$k_{4i} = .275 (M_i + h j_{3i}) - .231 (H_i + h k_{3i}) - .275 (H_i + h k_{3i})(M_i + h j_{3i})$$

$$j_{4i} = 4.00 (H_i + h k_{3i}) - 3.296 (M_i + h j_{3i}) - .206 (H_i + h k_{3i})(M_i + h j_{3i})$$

Suppose that initially no humans and ten percent of the mosquitos are affected with malaria. Then $H(0)=0$ and $M(0)=1.9418$. With $h=.5$ (two weeks), some typical values given by the Runge-Kutta method are

i	H_i	M_i
0	.000	1.9418
1	.112	.688
2	.142	.327
3	.153	.227
4	.159	.201

i	H_i	M_i
5	.163	.197
6	.166	.198
12	.186	.221
18	.204	.243
24	.222	.263
30	.236	.280
36	.248	.295
42	.258	.307
48	.267	.317
54	.273	.325
60	.278	.331
72	.285	.340
84	.289	.345
96	.292	.348
108	.293	.349
120	.294	.350

Notice that H_i and M_i are the approximate affected human and mosquito populations after $2i$ weeks. It appears from the data that H_i and M_i are approaching limits as $i \to \infty$ of about .294 and .352, respectively. In fact, it can be shown (see Exercise 8) that $H(t) \to .295$ and $M(t) \to .352$ as $t \to \infty$, where the limiting values have been rounded to three decimal places.

Also notice that the affected human population seems to increase monotonically, while the affected mosquito population initially decreases, reaches a minimum, and then increases monotonically. Thus after about ten weeks both the affected human and mosquito populations increase to their equilibrium values of .295 and .352, respectively.   ◀

Exercises   SECTION 5.10

In Exercises 1–4 use the Runge-Kutta method to approximate the solution to the given initial-value problem on the interval $[0, 1]$. Take $n = 10$ and $h = 1/10$. Compare the approximations with the true solution.

1. $\dfrac{dx}{dt} = -y, \quad x(0) = 1$

   $\dfrac{dy}{dt} = x, \quad y(0) = 1$

2. $\dfrac{dx}{dt}=x+y,\quad x(0)=0$

$\dfrac{dy}{dt}=x-y,\quad y(0)=1$

3. $\dfrac{dx}{dt}=y+t,\quad x(0)=0$

$\dfrac{dy}{dt}=x-t,\quad y(0)=0$

4. $\dfrac{dx}{dt}=x+y+1,\quad x(0)=-1$

$\dfrac{dy}{dt}=x-y-t,\quad y(0)=1$

5. When a charged surface is in contact with a liquid medium, it attracts ions of the opposite charge and repels ions of the same charge. In the study of such a situation in one dimension, the one-dimensional Poisson-Boltzman equation arises. This equation has the form

$$\frac{d^2z}{dt^2}=A\sinh(Bz)$$

where A and B are positive constants. A complete discussion of this problem may be found in Section 6.1 of *Biological Interfaces* by M. Jones (Elsevier Publishing Co., Amsterdam, 1975). For the special case $A=B=1$, $z(0)=1$, and $z'(0)=0$, use the Runge-Kutta method to approximate the solution on $[0,2]$ with $n=20$ and $h=.1$.

6. Enzymes have the ability to catalyze numerous chemical reactions in living organisms. One model for the action of enzymes is

$$\frac{d^2z}{dt^2}+\frac{az}{b+z}=0$$

where z is the concentration of the substance acted upon by the enzyme, and a,b are positive constants. [L. B. Wingard Jr., *Enzyme Engineering*, Advances in Biological Engineering 2, edited by T. K. Ghose, A. Fiechter, and N. Blakebrough, Springer-Verlag, New York, 1972, page 38.] Use the Runge-Kutta method with $n=10$ and $h=.1$ to approximate the solution to the equation for the special case $a=1$, $b=10$, $z(0)=.5$, and $z'(0)=0$.

7. Van der Pol's equation

$$\frac{d^2x}{dt^2}+c(x^2-1)\frac{dx}{dt}+x=0$$

arises in the study of electrical circuits containing vacuum tubes. Use the Runge-Kutta method to approximate the solution of van der Pol's equation that satisfies the initial condition $x(0)=1$, $x'(0)=0$ for the special case $c=.1$. Use $h=.1$ and $n=100$.

8. The limiting values of M and H in Example 3 turn out to be constants M_0 and H_0 such that the functions $M(t)\equiv M_0$ and $H(t)\equiv H_0$ are solutions of the system of differential equations found there. Find H_0 and M_0.

6 Series Solutions of Differential Equations

6.1 Power Series

Suppose that we have a sequence $\{a_n\}_{n=1}^{\infty}$ of numbers and a number a. Then for each number x we can form a power series in $(x-a)$:

$$\sum_{n=0}^{\infty} a_n(x-a)^n = a_0 + a_1(x-a) + a_2(x-a)^2 + \cdots \qquad \text{(6.1)}$$

The terms of the sequence $\{a_n\}_{n=1}^{\infty}$ are called the **coefficients** of the power series. The series

$$\sum_{n=0}^{\infty} \frac{1}{n!}(x-2)^n \quad \text{and} \quad \sum_{n=0}^{\infty} \frac{1}{n+1}x^n$$

are examples of power series. In the first of these, $a=2$ and $a_n=1/n!$, while in the second, $a=0$ and $a_n=1/(n+1)$.

The power series in (6.1) converges, at least for $x=a$. In fact, it can be shown that only one of the following statements is true:

(i) The series converges only for $x=a$.

(ii) The series converges for every number x.

(iii) There is a positive number R, called the **radius of convergence,** such that the series converges for every number x in the open interval $(a-R, a+R)$ and diverges for every number x not in the closed interval $[a-R, a+R]$.

For ease of notation we will refer to the radius of convergence as ∞ if the series converges, for all numbers x, and as zero if the series converges only for $x=a$. Thus the set of all numbers at which the power series in (6.1) converges, called the **interval of convergence,** is either a single point or an interval. In practice this set is rarely a single point, so it is aptly named.

It is easy to compute the derivative of a power series by differentiating term by term:

$$\frac{d}{dx}\left(\sum_{n=0}^{\infty} a_n(x-a)^n\right)=\sum_{n=0}^{\infty} na_n(x-a)^{n-1}$$

$$=\sum_{n=1}^{\infty} na_n(x-a)^{n-1}$$

It can be proved that the radius of convergence of the derivative is the same as the radius of convergence of the power series. It follows that if the power series has a nonzero radius of convergence R, then it has derivatives of all orders on the interval $(a-R, a+R)$. Thus, if we denote the power series in (6.1) by $f(x)$, then for every number x in $(a-R, a+R)$ we have

$$f'(x)=\sum_{n=1}^{\infty} na_n(x-a)^{n-1}$$

$$f''(x)=\sum_{n=2}^{\infty} n(n-1)a_n(x-a)^{n-2} \tag{6.2}$$

$$\cdots\cdots\cdots\cdots\cdots\cdots\cdots\cdots\cdots$$

$$f^{(k)}(x)=\sum_{n=k}^{\infty} n(n-1)\ldots(n-k+1)(x-a)^{n-k}$$

Using equation (6.1) and those in (6.2), we have

$$f(a)=a_0$$
$$f'(a)=a_1$$
$$f''(a)=2a_2$$
$$\cdots\cdots\cdots\cdots$$
$$f^{(k)}(a)=k!a_k$$

Thus the coefficients of the power series are determined uniquely by the values of f and its derivatives at $x=a$:

$$a_k = \frac{f^{(k)}(a)}{k!}$$

This enables us to rewrite the power series as

$$f(x) = \sum_{n=0}^{\infty} \frac{f^{(n)}(a)}{n!}(x-a)^n \qquad \textbf{(6.3)}$$

The series in (6.3) is called the **Taylor series** for $f(x)$ at a. Since the coefficients of the power series are uniquely determined, we have the following theorem:

THEOREM 6.1 Let $f(x) = \sum_{n=0}^{\infty} a_n(x-a)^n$ and $g(x) = \sum_{n=0}^{\infty} b_n(x-a)^n$ for every number x in an open interval I containing a. If $f(x) = g(x)$ for every x in I, then $a_n = b_n$ for every n. In particular, if $f(x) = 0$ for every x in I, then $a_n = 0$ for every n.

In the following sections this theorem will serve as the basis for methods of finding solutions of linear differential equations with constant coefficients.

The most commonly used method of computing the radius of convergence of a power series $\sum_{n=0}^{\infty} a_n(x-a)^n$ is the **ratio test**: If $\lim_{n\to\infty} |a_{n+1}/a_n|$ exists and equals a number r, then

1. The radius of convergence is $1/r$ if $r \neq 0$.

2. The radius of convergence is ∞ if $r=0$.

The following examples illustrate how the preceding theory works in practice.

▶ **Example 1** Consider the power series $\sum_{n=0}^{\infty}(x-2)^n/n!$ For this series $a=2$ and $a_n = 1/n!$, so

$$r = \lim_{n\to\infty} \frac{1/(n+1)!}{1/n!} = \lim_{n\to\infty} \frac{n!}{(n+1)!} = \lim_{n\to\infty} \frac{1}{n+1} = 0$$

Thus the radius of convergence is infinite and the series converges for all numbers x. ◀

▶ **Example 2** Consider the power series $\sum_{n=0}^{\infty} x^n/(n+1)$. For this series $a=0$ and $a_n = 1/(n+1)$, so

$$r = \lim_{n\to\infty} \frac{1/(n+1)}{1/n} = \lim_{n\to\infty} \frac{n}{n+1} = 1$$

Thus the radius of convergence is 1. When $x=1$ and $x=-1$ the series becomes

$$\sum_{n=0}^{\infty} \frac{1}{n+1} \quad \text{and} \quad \sum_{n=0}^{\infty} \frac{(-1)^n}{n+1}$$

respectively. Since the former diverges and the latter converges, the interval of convergence of the power series is $[-1,1)$.

Now consider the derivative of the power series

$$\frac{d}{dx}\left(\sum_{n=0}^{\infty} \frac{1}{n+1} x^n \right) = \sum_{n=1}^{\infty} \frac{n}{n+1} x^{n-1} = \sum_{n=2}^{\infty} \frac{n+1}{n+2} x^n$$

The radius of convergence of this series is $1/r$, where

$$r= \lim_{n\to\infty} \frac{n+2}{n+3} \Big/ \frac{n+1}{n+2} = \lim_{n\to\infty} \frac{(n+2)^2}{(n+1)(n+3)} = 1$$

Thus the radius of convergence of the derivative of the series coincides with that of the power series itself. When $x=1$ and $x=-1$ are inserted in the derivative, we obtain the series

$$\sum_{n=2}^{\infty} \frac{n+1}{n+2} \quad \text{and} \quad \sum_{n=2}^{\infty} (-1)^n \frac{n+1}{n+2}$$

Since the nth term of each series does not tend to zero as $n\to\infty$, recall that each series diverges. Therefore the interval of convergence of the derivative is $(-1,1)$, which is not the interval of convergence of the original series.

This is an example of a general principle. The radius of convergence of a power series and its derivative are identical. Nonetheless, the intervals of convergence of the two series need not be identical.   ◀

In order to simplify the discussions in the following sections we introduce the following definition.

DEFINITION 6.1   A function f is said to be **analytic** at x_0 if its Taylor series about x_0 exists and converges to $f(x)$ for every x in some open interval I containing x_0. That is,

$$f(x)= \sum_{n=0}^{\infty} \frac{f^{(n)}(x_0)}{n!} (x-x_0)^n$$

for every x in I.

Note that all constant functions and all polynomials are analytic functions at every x_0. Moreover, the "standard" functions of calculus, such as $\sin x$, $\cos x$, $\tan x$, $\ln x$, e^x, $\arcsin x$, $\arccos x$, and $\arctan x$, are analytic at every x_0 inside their domains. It can also be shown that the set of all functions that are analytic at x_0 possesses important algebraic properties. If f and g are analytic at x_0, then $f+g$ and fg are also analytic at x_0. Moreover, if $g(x_0) \neq 0$, then f/g is also analytic at x_0. Hence quotients of polynomials are analytic at each x_0 that is not a root of the denominator.

When using power series to find solutions of a differential equation, it is usually necessary to combine several power series into one. Frequently these power series have the form

$$\sum_{n=0}^{\infty} f(n)x^{n+k}$$

where f is a function defined on the nonnegative integers and k is an integer. Setting $m=n+k$, so that $n=m-k$, we have

$$\sum_{n=0}^{\infty} f(n)x^{n+k} = \sum_{m=k}^{\infty} f(m-k)x^m$$

Since m in the second summation is merely an index of summation, or "dummy variable," we may replace it by any symbol that does not already appear in the summation. In particular, if we replace m by n, we obtain the identity

$$\sum_{n=0}^{\infty} f(n)x^{n+k} = \sum_{n=k}^{\infty} f(n-k)x^n$$

Notice that the right summation is obtained by simply replacing n by $n-k$ in the left summation.

▶ **Example 3** Let $\{a_n\}_{n=1}^{\infty}$ be a sequence of numbers and let k be an integer. The identities

$$\sum_{n=0}^{\infty} a_n x^{n+k} = \sum_{n=k}^{\infty} a_{n-k} x^n$$

$$\sum_{n=0}^{\infty} n a_n x^{n+k} = \sum_{n=k}^{\infty} (n-k) a_{n-k} x^n$$

$$\sum_{n=0}^{\infty} n(n-1) a_n x^{n+k} = \sum_{n=k}^{\infty} (n-k)(n-k-1) a_{n-k} x^n$$

can be obtained by replacing n by $n-k$ in each of the left summations. It is important that the reader understand how these identities are obtained because the technique used to obtain them will be used repeatedly in the remainder of this chapter. ◀

Exercises   SECTION 6.1

In Exercises 1–6 determine the Taylor series for each of the given functions at the indicated number.

1. $\sin x$, $a=0$

2. $\sin x$, $a=\pi/2$

3. e^x, $a=0$

4. e^x, $a=1$

5. $\dfrac{1}{1+x}$, $a=0$

6. $(1+x)^{1/2}$, $a=0$

7. Prove Theorem 6.1.

In Exercises 8–15 determine the radius and interval of convergence of each of the given power series.

8. $\displaystyle\sum_{n=0}^{\infty} \frac{1}{2^n}(x-1)^n$

9. $\displaystyle\sum_{n=0}^{\infty} \frac{n}{3^n}(x+1)^n$

10. $\displaystyle\sum_{n=0}^{\infty} \frac{2^n}{n!}(x+3)^n$

11. $\displaystyle\sum_{n=0}^{\infty} \frac{n^2+1}{5n-3}x^n$

12. $\displaystyle\sum_{n=0}^{\infty} (x-2)^n$

13. $\displaystyle\sum_{n=0}^{\infty} \frac{\sin n}{n!}(x+4)^n$

14. $\displaystyle\sum_{n=0}^{\infty} n!x^n$

15. $\displaystyle\sum_{n=0}^{\infty} \frac{1}{n2^n}x^n$

In Exercises 16–21 determine all numbers x_0 at which the given functions are **not** analytic.

16. $\dfrac{x^2+3x+7}{x^2-5x+4}$

17. $\dfrac{\cos x}{x}+\dfrac{e^x}{x^2-9}$

18. $\tan x$

19. $\dfrac{\sec x}{x^2-3x}$

20. $\dfrac{1}{1-\sin x}$

21. $\begin{cases} \dfrac{\sin x}{x} & \text{if } x\neq 0 \\ 1 & \text{if } x=0 \end{cases}$

6.2 Series Solutions about Ordinary Points

In this section we will consider differential equations of the form

$$\frac{d^2y}{dx^2}+b_1(x)\frac{dy}{dx}+b_0(x)y=0$$

where the functions b_0 and b_1 are analytic at a number x_0. In practice the functions b_0 and b_1 are frequently quotients of polynomials. With this in mind, we write this equation in a slightly different, but equivalent, form and make the following definition.

DEFINITION 6.2   The number x_0 is called an **ordinary point** of the differential equation

$$a_2(x)\frac{d^2y}{dx^2}+a_1(x)\frac{dy}{dx}+a_0(x)y=0 \qquad\qquad \textbf{(6.4)}$$

if both a_1/a_2 and a_0/a_2 are analytic functions at x_0. Otherwise, x_0 is called a **singular point** of the differential equation.

We will consider only the case $x_0=0$. In Exercise 14 the reader is asked to show that this results in no loss of generality.

If we rewrite equation (6.4) as

$$\frac{d^2y}{dx^2}+\frac{a_1(x)}{a_2(x)}\frac{dy}{dx}+\frac{a_0(x)}{a_2(x)}y=0$$

we see that Theorem 2.1 assures the existence of solutions on an open interval containing x_0. In fact, equation (6.4) has analytic solutions that may be found by the following procedure.

The coefficients $b_0, b_1, \ldots$ of an analytic solution $y(x) = \sum_{n=0}^{\infty} b_n x^n$ of equation (6.4) can be computed as follows:

(1) Insert $y(x) = \sum_{n=0}^{\infty} b_n x^n$ into the differential equation.

(2) Rearrange the left-hand side of the equation so that the equation has the form $\sum_{n=0}^{\infty} c_n x^n = 0$.

(3) By Theorem 6.1, $c_n = 0$ for every n. Solve these equations to determine $b_0, b_1, \ldots$.

Before considering an equation with variable coefficients we will consider an equation with constant coefficients. This will permit us to illustrate steps (2) and (3) in an elementary setting and to compare the resulting answer with the answer computed using the technique described in Section 2.3.

▶ **Example 1**   Consider the differential equation

$$\frac{d^2 y}{dx^2} - y = 0 \tag{6.5}$$

Evidently $x_0 = 0$ is an ordinary point of this equation since its coefficients are constants. Inserting $\sum_{n=0}^{\infty} a_n x^n$ into equation (6.5) yields

$$\sum_{n=0}^{\infty} n(n-1) a_n x^{n-2} - \sum_{n=0}^{\infty} a_n x^n = 0 \tag{6.6}$$

We will now change the index in the first summation so that the two summations may be combined into one. Replacing n by $n+2$ in the first summation, we have

$$\sum_{n=0}^{\infty} n(n-1) a_n x^{n-2} = \sum_{n=-2}^{\infty} (n+2)(n+1) a_{n+2} x^n \tag{6.7}$$

Noting that $(n+2)(n+1) = 0$ when $n = -2$ and $n = -1$, we have

$$\sum_{n=-2}^{\infty} (n+2)(n+1) a_{n+2} x^n = \sum_{n=0}^{\infty} (n+2)(n+1) a_{n+2} x^n$$

so that

$$\sum_{n=0}^{\infty} n(n-1) a_n x^{n-2} = \sum_{n=0}^{\infty} (n+2)(n+1) a_{n+2} x^n$$

With this identity, equation (6.6) can be rewritten as

$$\sum_{n=0}^{\infty} (n+2)(n+1)a_{n+2}x^n - \sum_{n=0}^{\infty} a_n x^n = 0$$

or, equivalently,

$$\sum_{n=0}^{\infty} \left[(n+2)(n+1)a_{n+2} - a_n\right]x^n = 0$$

By Theorem 6.1 we have

$$(n+2)(n+1)a_{n+2} - a_n = 0$$

for $n = 0, 1, 2, \ldots$, so that

$$a_{n+2} = \frac{1}{(n+1)(n+2)} a_n$$

for $n = 0, 1, 2, \ldots$. The coefficients a_n may now be calculated in terms of the two coefficients a_0 and a_1:

$$a_2 = \frac{1}{2}a_0 = \frac{1}{2!}a_0 \qquad\qquad a_3 = \frac{1}{2 \cdot 3}a_1 = \frac{1}{3!}a_1$$

$$a_4 = \frac{1}{3 \cdot 4}a_2 = \frac{1}{3 \cdot 4}\frac{1}{2!}a_0 \qquad a_5 = \frac{1}{4 \cdot 5}a_3 = \frac{1}{4 \cdot 5}\frac{1}{3!}a_1$$

$$= \frac{1}{4!}a_0 \qquad\qquad\qquad\quad = \frac{1}{5!}a_1$$

$$a_6 = \frac{1}{5 \cdot 6}a_4 = \frac{1}{5 \cdot 6}\frac{1}{4!}a_0 \qquad a_7 = \frac{1}{6 \cdot 7}a_5 = \frac{1}{6 \cdot 7}\frac{1}{5!}a_1$$

$$= \frac{1}{6!}a_0 \qquad\qquad\qquad\quad = \frac{1}{7!}a_1$$

$$\cdots\cdots\cdots\cdots\cdots\cdots\cdots \qquad \cdots\cdots\cdots\cdots\cdots\cdots\cdots$$

Proceeding in this manner, we find that

$$a_n = \frac{1}{n!}\begin{cases} a_0 & \text{if } n \text{ is even} \\ a_1 & \text{if } n \text{ is odd} \end{cases}$$

Thus

$$y(x) = a_0 + a_1 x + a_2 x^2 + a_3 x^3 + a_4 x^4 + \cdots$$

$$= a_0 + a_1 x + \frac{1}{2!} a_0 x^2 + \frac{1}{3!} a_1 x^3 + \frac{1}{4!} a_0 x^4 + \cdots$$

$$= a_0 \left(1 + \frac{1}{2!} x^2 + \frac{1}{4!} x^4 + \cdots \right) + a_1 \left(x + \frac{1}{3!} x^3 + \frac{1}{5!} x^5 + \cdots \right)$$

$$= a_0 \sum_{n=0}^{\infty} \frac{x^{2n}}{(2n)!} + a_1 \sum_{n=0}^{\infty} \frac{x^{2n+1}}{(2n+1)!}$$

is a solution of equation (6.5) for any choice of the coefficients a_0 and a_1. In particular, when $a_0 = 1$ and $a_1 = 1$, we have that

$$y_1(x) = \left(1 + \frac{1}{2!} x^2 + \frac{1}{4!} x^4 + \cdots \right) + \left(x + \frac{1}{3!} x^3 + \frac{1}{5!} x^5 + \cdots \right)$$

$$= \sum_{n=0}^{\infty} \frac{1}{n!} x^n$$

$$= e^x$$

is a solution of equation (6.5). When $a_0 = 1$ and $a_1 = -1$, we have that

$$y_2(x) = \left(1 + \frac{1}{2!} x^2 + \frac{1}{4!} x^4 + \cdots \right) - \left(x + \frac{1}{3!} x^3 + \frac{1}{5!} x^5 + \cdots \right)$$

$$= \sum_{n=0}^{\infty} \frac{1}{n!} (-1)^n x^n$$

$$= e^{-x}$$

is also a solution of equation (6.5).

Since the auxiliary equation $r^2 - 1 = 0$ for equation (6.5) has 1 and -1 as its roots, a general solution of equation (6.5) is

$$y(x) = c_1 e^x + c_2 e^{-x}$$

Thus the procedure outlined in steps (1), (2), and (3) above generates series representations for functions that can be used to form a general solution of equation (6.5).   ◀

We will now apply the method outlined in steps (1), (2), and (3) to differential equations with variable coefficients.

▶ **Example 2** Consider the equation

$$\frac{d^2y}{dx^2} + x\frac{dy}{dx} + y = 0 \tag{6.8}$$

Evidently $x_0 = 0$ is an ordinary point of this equation. Insertion of $y(x) = \sum_{n=0}^{\infty} a_n x^n$ into equation (6.8) yields

$$\sum_{n=0}^{\infty} n(n-1)a_n x^{n-2} + x\sum_{n=0}^{\infty} na_n x^{n-1} + \sum_{n=0}^{\infty} a_n x^n = 0$$

Since x is independent of the index of summation n, this equation can be rewritten as

$$\sum_{n=0}^{\infty} n(n-1)a_n x^{n-2} + \sum_{n=0}^{\infty} na_n x^n + \sum_{n=0}^{\infty} a_n x^n = 0$$

Replacing n by $n+2$ in the first summand, we have

$$\sum_{n=-2}^{\infty} (n+2)(n+1)a_{n+2} x^n + \sum_{n=0}^{\infty} na_n x^n + \sum_{n=0}^{\infty} a_n x^n = 0 \tag{6.9}$$

Noting that $(n+2)(n+1)=0$ when $n=-2$ and $n=-1$, we have $\sum_{n=-2}^{\infty}(n+2)(n+1)a_{n+2}x^n = \sum_{n=0}^{\infty}(n+2)(n+1)a_{n+2}x^n$. The summations in (6.9) can now be condensed into the single summation, and equation (6.9) becomes

$$\sum_{n=0}^{\infty} \left[(n+2)(n+1)a_{n+2} + (n+1)a_n\right]x^n = 0$$

By Theorem 6.1

$$(n+2)(n+1)a_{n+2} + (n+1)a_n = 0$$

for each n or, equivalently,

$$(n+2)a_{n+2} + a_n = 0$$

Hence for each n we have

$$a_{n+2} = -\frac{1}{n+2}a_n$$

so that

$$a_2 = -\frac{1}{2}a_0 \qquad\qquad a_3 = -\frac{1}{3}a_1$$

$$a_4 = -\frac{1}{4}a_2 = -\frac{1}{4}\left(-\frac{1}{2}a_0\right) \qquad a_5 = -\frac{1}{5}a_3 = -\frac{1}{5}\left(-\frac{1}{3}a_1\right)$$

$$= \frac{1}{2\cdot4}a_0 \qquad\qquad = \frac{1}{3\cdot5}a_1$$

$$a_6 = -\frac{1}{6}a_4 = -\frac{1}{6}\left(\frac{1}{2\cdot4}a_0\right) \qquad a_7 = -\frac{1}{7}a_5 = -\frac{1}{7}\left(\frac{1}{3\cdot5}a_1\right)$$

$$= -\frac{1}{2\cdot4\cdot6}a_0 \qquad\qquad = -\frac{1}{3\cdot5\cdot7}a_1$$

. . . .

Thus

$$y(x) = a_0 + a_1 x + a_2 x^2 + a_3 x^3 + a_4 x^4 + a_5 x^5 + a_6 x^6 + a_7 x^7 + \cdots$$

$$= a_0\left(1 - \frac{1}{2}x^2 + \frac{1}{2\cdot4}x^4 - \frac{1}{2\cdot4\cdot6}x^6 + \cdots\right)$$

$$+ a_1\left(x - \frac{1}{3}x^3 + \frac{1}{3\cdot5}x^5 - \frac{1}{3\cdot5\cdot7}x^7 + \cdots\right)$$

is a solution of equation (6.8) for every choice of the numbers a_0 and a_1. We will now simplify the form of this solution. We begin by noting that the even integers and odd integers can be written as $2n$ and $2n+1$, respectively, for $n=0,1,2,\ldots$. With this notation we observe that for $n=0,1,2\ldots$, we have

$$a_{2n} = (-1)^n \frac{1}{2\cdot4\cdot6\cdots(2n)}a_0$$

$$= (-1)^n \frac{1}{2^n} \frac{1}{1\cdot2\cdot3\cdots n}a_0$$

$$= (-1)^n \frac{1}{2^n n!}a_0$$

and

$$a_{2n+1} = (-1)^n \frac{1}{1\cdot3\cdot5\cdot7\cdots(2n+1)} a_1$$

$$= (-1)^n \frac{2\cdot4\cdots(2n)}{1\cdot2\cdot3\cdot4\cdots(2n+1)} a_1$$

$$= (-1)^n \frac{2^n(1\cdot2\cdots n)}{(2n+1)!} a_1$$

$$= (-1)^n \frac{2^n n!}{(2n+1)!} a_1$$

Thus

$$y(x) = \sum_{n=0}^{\infty} a_n x^n$$

$$= \left(a_0 + a_2 x^2 + a_4 x^4 + \cdots \right) + \left(a_3 x + a_5 x^5 + a_7 x^7 + \cdots \right)$$

$$= \sum_{n=0}^{\infty} a_{2n} x^{2n} + \sum_{n=0}^{\infty} a_{2n+1} x^{2n+1}$$

$$= a_0 \sum_{n=0}^{\infty} (-1)^n \frac{1}{2^n n!} x^{2n} + a_1 \sum_{n=0}^{\infty} (-1)^n \frac{2^n n!}{(2n+1)!} x^{2n+1}$$

When $a_0 = 1$, $a_1 = 0$ and $a_0 = 0$, $a_1 = 1$, we obtain the two solutions

$$y_1(x) = \sum_{n=0}^{\infty} (-1)^n \frac{1}{2^n n!} x^{2n}, \quad y_2(x) = \sum_{n=0}^{\infty} (-1)^n \frac{2^n n!}{(2n+1)!} x^{2n+1}$$

respectively. We will show that y_1 and y_2 are linearly independent. Notice that $y_1(0) = 1$ and $y_2(0) = 0$. It follows that y_1 is not a constant multiple of y_2 since $y_1(0) \neq k y_2(0)$ for any number k. If y_2 is a constant multiple c of y_1, then c must be zero since $0 = y_2(0) = c y_1(0) = c$. In such a case we have $y_2(x) = 0$ for every x in I. By Theorem 6.1 the solution $y_2(x)$ cannot be zero for every x in I. Therefore, y_2 cannot be a constant multiple of y_1. Since neither y_1 nor y_2 is a constant multiple of the other, the two functions are linearly independent.

Notice that the solution y_1 can be written as

$$y_1(x) = \sum_{n=0}^{\infty} \frac{1}{n!} \left(-\frac{x^2}{2} \right)^n$$

If we now recall that

$$e^t = \sum_{n=0}^{\infty} \frac{1}{n!} t^n$$

we conclude that

$$y_1(x) = e^{-x^2/2}$$

The solution y_2 can be written in terms of a series in the variable x^2:

$$y_2(x) = x \sum_{n=0}^{\infty} (-1)^n \frac{2^n n!}{(2n+1)!} (x^2)^n \qquad \text{(6.10)}$$

Applying the ratio test to the series

$$\sum_{n=0}^{\infty} (-1)^n \frac{2^n n!}{(2n+1)!} (t)^n \qquad \text{(6.11)}$$

we have

$$\lim_{n \to \infty} \left| \frac{a_{n+1}}{a_n} \right| = \lim_{n \to \infty} \left| \frac{\dfrac{(-1)^{n+1} 2^{n+1}(n+1)!}{(2(n+1)+1)!}}{\dfrac{(-1)^n 2^n n!}{(2n+1)!}} \right|$$

$$= \lim_{n \to \infty} \frac{2^{n+1}(n+1)!(2n+1)!}{2^n n!(2n+3)!}$$

$$= \lim_{n \to \infty} \frac{1}{2n+3} = 0$$

so that the series in (6.11) converges for all values of t. Consequently, the series in (6.10) converges for all values of x. Thus y_1 and y_2 are linearly independent solutions of equation (6.8) on the interval $(-\infty, \infty)$.

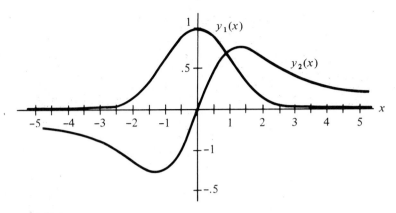

FIGURE 6.1

Therefore

$$y(x)=c_1 y_1(x)+c_2 y_2(x)$$

is a general solution of equation (6.8) on $(-\infty, \infty)$. The graphs of y_1 and y_2 are shown in Figure 6.1. ◀

▶ **Example 3** Consider the equation

$$(x+1)\frac{d^2 y}{dx^2}+\frac{dy}{dx}+xy=0 \qquad \textbf{(6.12)}$$

Evidently $x_0=0$ is an ordinary point of this equation. Insertion of $y(x)=\Sigma_{n=0}^{\infty} b_n x^n$ into equation (6.12) yields

$$(x+1)\sum_{n=0}^{\infty} n(n-1)b_n x^{n-2} + \sum_{n=0}^{\infty} nb_n x^{n-1} + x\sum_{n=0}^{\infty} b_n x^n =0$$

Since x is independent of the index of summation n, this equation can be rewritten as

$$\sum_{n=0}^{\infty} n(n-1)b_n x^{n-1} + \sum_{n=0}^{\infty} n(n-1)b_n x^{n-2}$$

$$+ \sum_{n=0}^{\infty} nb_n x^{n-1} + \sum_{n=0}^{\infty} b_n x^{n+1} =0$$

If we replace n by $n+1$ in the first and third summations, n by $n+2$ in the second summation, and n by $n-1$ in the fourth summation, we obtain

$$\sum_{n=-1}^{\infty} (n+1)nb_{n+1}x^n + \sum_{n=-2}^{\infty} (n+2)(n+1)b_{n+2}x^n$$

$$+ \sum_{n=-1}^{\infty} (n+1)b_{n+1}x^n + \sum_{n=1}^{\infty} b_{n-1}x^n = 0$$

which can be simplified by noting that $n+2=0$, $n+1=0$, and $n=0$ when $n=-2$, $n=-1$, and $n=0$, respectively:

$$\sum_{n=1}^{\infty} (n+1)nb_{n+1}x^n + \sum_{n=0}^{\infty} (n+2)(n+1)b_{n+2}x^n$$

$$+ \sum_{n=0}^{\infty} (n+1)b_{n+1}x^n + \sum_{n=1}^{\infty} b_{n-1}x^n = 0$$

In order to combine these four summations into one summation, we need to have each index of summation begin at the same value. This can be done by writing the second and third summations in an equivalent form:

$$\sum_{n=1}^{\infty} (n+1)nb_{n+1}x^n + 2b_2 + \sum_{n=1}^{\infty} (n+2)(n+1)b_{n+2}x^n + b_1$$

$$+ \sum_{n=1}^{\infty} (n+1)b_{n+1}x^n + \sum_{n=1}^{\infty} b_{n-1}x^n = 0$$

The four terms containing the summation signs can now be combined to yield the equation

$$2b_2 + b_1 + \sum_{n=1}^{\infty} \left[(n+1)nb_{n+1} + (n+2)(n+1)b_{n+2} \right.$$

$$\left. + (n+1)b_{n+1} + b_{n-1} \right] x^n = 0$$

Notice that this equation has the form $\sum_{n=0}^{\infty} c_n x^n = 0$, where

$$c_0 = 2b_2 + b_1$$

and

$$c_n = (n+1)nb_{n+1} + (n+2)(n+1)b_{n+2} + (n+1)b_{n+1} + b_{n-1}$$

$$= (n+1)^2 b_{n+1} + (n+2)(n+1)b_{n+2} + b_{n-1} \qquad (n \geqslant 1)$$

By Theorem 6.1 we have

$$0 = c_0 = 2b_2 + b_1$$

$$0 = c_n = (n+1)^2 b_{n+1} + (n+2)(n+1)b_{n+2} + b_{n-1} \qquad (n \geq 1)$$

Hence

$$b_2 = -\tfrac{1}{2}b_1$$

$$b_{n+2} = -\frac{n+1}{n+2}b_{n+1} - \frac{1}{(n+1)(n+2)}b_{n-1} \qquad (n \geq 1)$$

Notice that b_0 and b_1 are arbitrary since they are not determined by these equations. Thus

$$b_2 = -\tfrac{1}{2}b_1$$

$$b_3 = -\tfrac{2}{3}b_2 - \tfrac{1}{6}b_0$$

$$= \tfrac{1}{3}b_1 - \tfrac{1}{6}b_0$$

$$b_4 = -\tfrac{3}{4}b_3 - \tfrac{1}{12}b_1$$

$$= -\tfrac{3}{4}\left(\tfrac{1}{3}b_1 - \tfrac{1}{6}b_0\right) - \tfrac{1}{12}b_1$$

$$= -\tfrac{1}{3}b_1 + \tfrac{1}{8}b_0$$

$$b_5 = -\tfrac{4}{5}b_4 - \tfrac{1}{20}b_2$$

$$= -\tfrac{4}{5}\left(-\tfrac{1}{3}b_1 + \tfrac{1}{8}b_0\right) - \tfrac{1}{20}\left(-\tfrac{1}{2}\right)b_1$$

$$= \tfrac{7}{24}b_1 + \tfrac{1}{15}b_0$$

$$\cdots \cdots \cdots \cdots \cdots \cdots$$

A solution of the differential equation has the form

$$y(x) = \sum_{n=0}^{\infty} b_n x^n$$

$$= b_0 + b_1 x + b_2 x^2 + b_3 x^3 + b_4 x^4 + b_5 x^5 + \cdots$$

$$= b_0 + b_1 x - \tfrac{1}{2}b_1 x^2 + \left(\tfrac{1}{3}b_1 - \tfrac{1}{6}b_0\right)x^3 + \left(-\tfrac{1}{3}b_1 + \tfrac{1}{8}b_0\right)x^4$$

$$+ \left(\tfrac{7}{24}b_1 + \tfrac{1}{15}b_0\right)x^5 + \cdots$$

$$= b_0\left(1 - \tfrac{1}{6}x^3 + \tfrac{1}{8}x^4 + \tfrac{1}{15}x^5 + \cdots\right)$$

$$+ b_1\left(x - \tfrac{1}{2}x^2 + \tfrac{1}{3}x^4 + \tfrac{7}{24}x^5 + \cdots\right) \qquad \textbf{(6.13)}$$

In the previous example we were able to find a simple relationship between b_n and the quantities b_0 and b_1. In this example this does not appear to be possible. Hence we are not able to write solutions as a sum of infinite series whose coefficients are easily calculated. Nonetheless, as demonstrated above, we are able to calculate the coefficients sequentially. When $b_0 = 1$, $b_1 = 0$ and $b_0 = 0$, $b_1 = 1$ we obtain the solutions

$$y_1(x) = 1 - \tfrac{1}{6}x^3 + \tfrac{1}{8}x^4 + \tfrac{1}{15}x^5 + \cdots$$

$$y_2(x) = x - \tfrac{1}{2}x^2 + \tfrac{1}{3}x^4 + \tfrac{7}{24}x^5 + \cdots$$

Since $y_1(0) = 1$ and $y_2(0) = 0$, an argument analogous to that in the previous example shows that y_1 and y_2 are linearly independent. Hence the solution in (6.13) is a general solution of the differential equation in (6.12). ◀

As illustrated in Example 3, it is frequently not possible to find an explicit expression for b_n in terms of b_0 and b_1. In such a case the first few terms of the series often will provide an approximation that is sufficiently accurate for the application that gave rise to the differential equation. If a few terms will not suffice, it is possible to use a computer to compute several of the coefficients for specific choices of b_0 and b_1.

Exercises　SECTION 6.2

Find series solutions of the form $\sum_{n=0}^{\infty} a_n x^n$ for each of the following equations.

1. $\dfrac{d^2y}{dx^2} + y = 0$

2. $\dfrac{d^2y}{dx^2} + x\dfrac{dy}{dx} + 2y = 0$

3. $\dfrac{d^2y}{dx^2} + x^2\dfrac{dy}{dx} + xy = 0$

4. $(1+x^2)\dfrac{d^2y}{dx^2} + 2\dfrac{dy}{dx} - 2y = 0$

5. $\dfrac{d^2y}{dx^2} + (1+x)\dfrac{dy}{dx} + y = 0$

6. $\dfrac{d^2y}{dx^2} - x\dfrac{dy}{dx} + x^2y = 0$

7. $(1+x)\dfrac{d^2y}{dx^2} + 2x\dfrac{dy}{dx} - 4y = 0$

8. $\dfrac{d^2y}{dx^2}+2(1+x)\dfrac{dy}{dx}+xy=0$

9. $(1-x)\dfrac{d^2y}{dx^2}-\dfrac{dy}{dx}+y=0$

10. $\dfrac{d^2y}{dx^2}+\dfrac{dy}{dx}+(1+x^2)y=0$

11. The differential equation

$$(1-x^2)\dfrac{d^2y}{dx^2}-2x\dfrac{dy}{dx}+P(P+1)y=0$$

called Legendre's equation, arises in applications such as the temperature distribution in a heated sphere.

(a) Find two linearly independent solutions of the form $\sum_{n=0}^{\infty}a_nx^n$.

(b) Show that if P is a nonnegative integer, then one of the solutions found in part (a) is a polynomial of degree P.

12. The differential equation called Hermite's equation,

$$\dfrac{d^2y}{dx^2}-2x\dfrac{dy}{dx}+ky=0$$

arises in many areas of mathematics and physics.

(a) Find two linearly independent solutions of the form $\sum_{n=0}^{\infty}a_nx^n$.

(b) Show that if k is an even integer $2n$, then one of the solutions in part (a) is a polynomial of degree n.

13. The differential equation, called Tchebycheff's equation,

$$(1-x^2)\dfrac{d^2y}{dx^2}-x\dfrac{dy}{dx}+k^2y=0$$

arises in many areas of mathematics and physics.

(a) Find two linearly independent solutions of the form $\sum_{n=0}^{\infty}a_nx^n$.

(b) Show that if k is an integer n, then one of the solutions in part (a) is a polynomial of degree n. These polynomials are of particular interest in numerical analysis.

14. Suppose that x_0 is an ordinary point of equation (6.4). Show that the change of variables $t=x-x_0$ transforms equation (6.4) into an equation having zero as an ordinary point.

6.3 Solutions about Regular Singular Points, I

In 1824, while studying the motion of the planets, Friedrich Bessel discovered functions that satisfy the differential equation

$$x^2 \frac{d^2 y}{dx^2} + x \frac{dy}{dx} + (x^2 - p^2) y = 0 \qquad (6.14)$$

where $0 \leqslant p$. Notice that Theorem 2.1, which assured the existence of solutions to equations considered earlier, does not assure the existence of solutions to the initial-value problem involving equation (6.14) when the initial data is given at $x=0$. Thus there is no guarantee that there is a solution of equation (6.14) on any open interval containing zero. In this section we will define a class of differential equations that includes equation (6.14), and describe a procedure to determine solutions of differential equations in this class. As in the case of equation (6.14), this procedure will frequently yield a solution on an open interval containing a singular point of the differential equation in question. At the least, this procedure yields a solution defined on a set of the form $(x_0 - A, x_0) \cup (x_0, x_0 + A)$, where x_0 is a singular point of the differential equation and A is some positive number. In Exercise 2, the reader is asked to show that

1. If p is not an integer, then there is no analytic solution at $x=0$, other than $y(x)=0$, of equation (6.14).

2. If p is an integer, then there is an analytic solution at $x=0$ of equation (6.14), but there are not two linearly independent analytic solutions at $x=0$.

Thus the solutions we will determine are not necessarily analytic at $x=0$.

Since $x=0$ is the only singular point of equation (6.14), the reader may be wondering why we do not choose some $x_0 \neq 0$ and form a general solution on an interval about x_0. This could certainly be done using the method described in the previous section. Such a solution is usually not satisfactory because:

(i) Equation (6.14) has solutions that are unbounded at $x=0$. A general solution consisting of power series does not allow the nature of solutions near $x=0$ to be analyzed easily.

(ii) In practice, initial values of the solution and its derivative are frequently given at $x=0$. The use of power series solutions at $x=x_0 \neq 0$ is not an effective means of handling such initial-value problems.

Fortunately, differential equations having a singular point x_0 that arise in applications frequently have a special form permitting the existence of solutions at x_0 similar to power series. We will first define a class of differential equations and then describe a procedure for determining solutions of equations in this class. In Section 6.5 we will study equation (6.14) and its solutions in greater detail.

DEFINITION 6.3   Let $a_0(x)$, $a_1(x)$, and $a_2(x)$ be analytic functions at a number x_0. If:

(a) at least one of $a_1(x)/a_2(x)$ and $a_0(x)/a_2(x)$ is not analytic at x_0, and
(b) both $(x-x_0)a_1(x)/a_2(x)$ and $(x-x_0)^2 a_0(x)/a_2(x)$ are analytic at x_0,

then x_0 is called a **regular singular point** of the differential equation

$$a_2(x)\frac{d^2y}{dx^2} + a_1(x)\frac{dy}{dx} + a_0(x)y = 0$$

Before proceeding, it should be noted that if x_0 is a regular singular point of this equation, then $a_2(x_0)=0$.

► **Example 1**   Consider the equation

$$x^2\frac{d^2y}{dx^2} + 2x\frac{dy}{dx} + (x^2-1)y = 0$$

Evidently the functions $2x/x^2$ and $(x^2-1)/x^2$ are not analytic at $x_0=0$. However, $x(2x/x^2)$ and $x^2((x^2-1)/x^2)$ are analytic at $x_0=0$ since the former is a constant function and the latter is a polynomial. Therefore $x_0=0$ is a regular singular point of the differential equation. ◄

There is a theorem assuring the existence about a regular singular point x_0 of at least one solution of the form $(x-x_0)^r\sum_{n=0}^{\infty}b_n(x-x_0)^n$. This result is analogous to the existence of power series solutions at an ordinary point of a second-order linear differential equation.

THEOREM 6.2   If x_0 is a regular singular point of

$$a_2(x)\frac{d^2y}{dx^2} + a_1(x)\frac{dy}{dx} + a_0(x)y = 0 \qquad\qquad \textbf{(6.15)}$$

then there is *at least one* solution of the form

$$(x-x_0)^r \sum_{n=0}^{\infty} b_n(x-x_0)^n \qquad\qquad \textbf{(6.16)}$$

where r is a number (which may be a non-integer), and the series

$$\sum_{n=0}^{\infty} b_n(x-x_0)^n$$

converges in an interval $|x-x_0|<R$ for some $R>0$.

Note that this theorem assures the existence of only one solution of the form in (6.16). If there are two such solutions that are linearly independent, then a general solution can be formed from them. If there are not two linearly independent solutions of the form in (6.16), it is still possible to form a general solution. This will be discussed in the next section.

We will now describe a procedure, called the **method of Frobenius,** to find a solution of equation (6.15) with the form of (6.16). This procedure is essentially the same as the one for computing series solutions about an ordinary point. As in the previous section, we will assume that x_0 is zero. If this is not the case, the change of variable $t=x-x_0$ transforms equation (6.15) into an equation which does have $t=0$ as a regular singular point. The exponent r and the coefficients $b_0, b_1, \dots$ of a solution $y(x)=x^r\sum_{n=0}^{\infty}b_n x^n$ may be computed as follows.

(1) Insert $y(x)=x^r\sum_{n=0}^{\infty}b_n x^n$ into the differential equation.

(2) Rearrange and simplify the left-hand side of the equation so that the equation has the form $\sum_{n=0}^{\infty}c_n x^n=0$ with $c_0\neq 0$. By Theorem 6.1 we must have $c_n=0$ for every n.

(3) The equation $c_0=0$ determines a quadratic equation for r, called the **indicial equation** of the differential equation (6.15). This equation determines the only possible values for the exponent r. Let r_1 and r_2, $r_2\leq r_1$ denote these two numbers.

(4) The larger value r_1 for r is now used in the remaining equations $c_1=0, c_2=0, \dots$ to determine the coefficients $b_0, b_1, \dots$.

(5) If $r_1\neq r_2$, we may repeat step (4) with r replaced by r_2. In this way a second solution is obtained. However, the two solutions may not be linearly independent. A method of handling this situation is discussed in the next section: *It can be shown that if r_1-r_2 is not an integer, then the two solutions are linearly independent.*

In performing step (1), it is convenient to write $y(x)$ in the form $y(x)=\sum_{n=0}^{\infty}a_n x^{n+r}$ so that $y'(x)=\sum_{n=0}^{\infty}(n+r)a_n x^{n+r-1}$ and $y''(x)=\sum_{n=0}^{\infty}(n+r)(n+r-1)a_n x^{n+r-2}$.

▶ **Example 2**   Consider the differential equation

$$9x^2\frac{d^2y}{dx^2}+9x\frac{dy}{dx}+(9x^2-1)y=0 \tag{6.17}$$

Since $x(9x/9x^2)$ and $x^2((9x^2-1)/9x^2)$ are polynomials, they are analytic at $x=0$. Hence $x=0$ is a regular singular point of the differential equation in (6.17). Inserting $y(x)=\sum_{n=0}^{\infty}a_nx^{n+r}$ in the differential equation yields

$$9x^2\sum_{n=0}^{\infty}(n+r)(n+r-1)a_nx^{n+r-2}+9x\sum_{n=0}^{\infty}(n+r)a_nx^{n+r-1}$$

$$+(9x^2-1)\sum_{n=0}^{\infty}a_nx^{n+r}=0$$

which can be rewritten as

$$x^r\left[\sum_{n=0}^{\infty}9(n+r)(n+r-1)a_nx^n+\sum_{n=0}^{\infty}9(n+r)a_nx^n+\sum_{n=0}^{\infty}9a_nx^{n+2}\right.$$

$$\left.-\sum_{n=0}^{\infty}a_nx^n\right]=0$$

If we replace n by $n-2$ in the third summation, this equation can be rewritten as

$$x^r\left[9r(r-1)a_0+9(r+1)ra_1x+\sum_{n=2}^{\infty}9(n+r)(n+r-1)a_nx^n+9ra_0\right.$$

$$+9(r+1)a_1x+\sum_{n=2}^{\infty}9(n+r)a_nx^n+\sum_{n=2}^{\infty}9a_{n-2}x^n-a_0-a_1x$$

$$\left.-\sum_{n=2}^{\infty}a_nx^n\right]=0$$

which simplifies to

$$x^r\left[(9r^2-1)a_0+(9r^2+18r+8)a_1x\right.$$

$$\left.+\sum_{n=2}^{\infty}\left\{(9(n+r)^2-1)a_n+9a_{n-2}\right\}x^n\right]=0$$

This equation can be written in the form

$$x^r \sum_{n=0}^{\infty} c_n x^n = 0 \tag{6.18}$$

where

$$c_0 = (9r^2 - 1)a_0$$

$$c_1 = (9r^2 + 18r + 8)a_1 \tag{6.19}$$

$$c_n = \left(9(n+r)^2 - 1\right)a_n + 9a_{n-2} \quad (n \geqslant 2)$$

Since we want equation (6.18) to hold for all x, $x \neq 0$, in some open interval I containing zero, we must have

$$\sum_{n=0}^{\infty} c_n x^n = 0$$

for all x, $x \neq 0$, in I. By Theorem 6.1, $c_n = 0$ for each n. We will use the equation $c_0 = 0$ to determine the exponent r. Clearly $c_0 = 0$ when

$$9r^2 - 1 = 0$$

This is the indicial equation for the differential equation (6.17) and has $r_2 = -\frac{1}{3}$ and $r_1 = \frac{1}{3}$ as its roots. With $r = \frac{1}{3}$, we have from (6.19)

$$0 = c_1 = 15a_1$$

$$0 = c_n = \left(9\left(n + \tfrac{1}{3}\right)^2 - 1\right)a_n + 9a_{n-2}$$

$$= (9n^2 + 6n)a_n + 9a_{n-2} \quad (n \geqslant 2)$$

Since a_0 is not determined by the equations, it is arbitrary, and

$$a_1 = 0$$

$$a_n = -\frac{3}{n(3n+2)}a_{n-2} \quad (n \geqslant 2)$$

Hence

$$a_2 = -\frac{3}{2\cdot 8}a_0 \qquad\qquad a_3 = -\frac{3}{3\cdot 11}a_1 = 0$$

$$a_4 = -\frac{3}{4\cdot 14}a_2 \qquad\qquad a_5 = -\frac{3}{5\cdot 17}a_3 = 0$$

$$= -\frac{3}{4\cdot 14}\left(-\frac{3}{2\cdot 8}\right)a_0$$

$$= \frac{3^2}{2\cdot 4\cdot 8\cdot 14}a_0$$

$$a_6 = -\frac{3}{6\cdot 20}a_4 \qquad\qquad a_7 = -\frac{3}{7\cdot 23}a_5 = 0$$

$$= -\frac{3}{6\cdot 20}\left(\frac{3^2}{2\cdot 4\cdot 8\cdot 14}\right)a_0$$

$$= -\frac{3^3}{2\cdot 4\cdot 6\cdot 8\cdot 14\cdot 20}$$

. . . .

Evidently $a_n = 0$ whenever n is an odd integer. Moreover,

$$a_{2n} = \frac{(-1)^n 3^n}{2\cdot 4\cdot 6\cdots(2n)\cdot 8\cdot 14\cdot 20\cdots(6n+2)}a_0 \qquad (n\geqslant 1)$$

so that upon choosing $a_0 = 1$, we have that

$$y_1(x) = x^{1/3}\left[1 + \sum_{n=1}^{\infty}\frac{(-1)^n 3^n x^{2n}}{2\cdot 4\cdot 6\cdots(2n)\cdot 8\cdot 14\cdot 20\cdots(6n+2)}\right] \qquad \textbf{(6.20)}$$

is a solution of the differential equation in (6.17). If we now use the other root, $r = -\frac{1}{3}$, of the indicial equation, we can find a second solution of the differential equation. With this choice for r, the second and third equations in (6.19) become

$$0 = c_1 = 3a_1$$

$$0 = c_n = \left(9\left(n-\tfrac{1}{3}\right)^2 - 1\right)a_n + 9a_{n-2}$$

$$= (9n^2 - 6n)a_n + 9a_{n-2} \qquad (n\geqslant 2)$$

Since a_0 is not determined by the equations, it is arbitrary, and

$$a_1 = 0$$

$$a_n = -\frac{3}{n(3n-2)} a_{n-2} \qquad (n \geqslant 2)$$

Hence

$$a_2 = -\frac{3}{2 \cdot 4} a_0 \qquad\qquad a_3 = -\frac{3}{3 \cdot 7} a_1 = 0$$

$$a_4 = -\frac{3}{4 \cdot 10} a_2 \qquad\qquad a_5 = -\frac{3}{5 \cdot 13} a_3 = 0$$

$$= -\frac{3}{4 \cdot 10} \left(-\frac{3}{2 \cdot 4} a_0 \right)$$

$$= \frac{3^2}{2 \cdot 4 \cdot 4 \cdot 10} a_0$$

$$a_6 = -\frac{3}{6 \cdot 16} a_4 \qquad\qquad a_7 = -\frac{3}{7 \cdot 19} a_5 = 0$$

$$= -\frac{3}{6 \cdot 16} \left(\frac{3^2}{2 \cdot 4 \cdot 4 \cdot 10} a_0 \right)$$

$$= -\frac{3^3}{2 \cdot 4 \cdot 6 \cdot 4 \cdot 10 \cdot 16} a_0$$

$\cdots\cdots\cdots\cdots\cdots\cdots\cdots\cdots\cdots\cdots\cdots\cdots$

Evidently $a_n = 0$ whenever n is an odd integer. Moreover,

$$a_{2n} = \frac{(-1)^n 3^n}{2 \cdot 4 \cdot 6 \cdots (2n) \cdot 4 \cdot 10 \cdot 16 \cdots (6n-2)} a_0 \qquad (n \geqslant 1)$$

so that upon choosing $a_0 = 1$, we have that

$$y_2(x) = x^{-1/3} \left[1 + \sum_{n=1}^{\infty} \frac{(-1)^n 3^n x^{2n}}{2 \cdot 4 \cdot 6 \cdots (2n) \cdot 4 \cdot 10 \cdot 16 \cdots (6n-2)} \right] \qquad \text{(6.21)}$$

is also a solution of the differential equation in (6.17). We will now show that y_1 and y_2 are linearly independent. We begin by noting that the functions in the brackets in the representations of y_1 and y_2 are analytic functions, and as such they are not identically zero on any interval (Theorem 6.1). Hence neither y_1 nor y_2 is identically zero on any interval. Notice that $y_1(0) = 0$, while $\lim_{x \to 0^+} y_2(x) = +\infty$. It follows that neither y_1

nor y_2 is a constant multiple of the other. Therefore y_1 and y_2 are linearly independent.

This example illustrates the importance of the method of Frobenius. It can be shown that the infinite series in (6.20) and (6.21) converge for all x. Therefore y_1 is defined for all x, while y_2 is defined for all nonzero x. Thus, if y is any solution of equation (6.17) on an interval I not containing zero, then there are constants C_1 and C_2 such that

$$y(x) = C_1 y_1(x) + C_2 y_2(x)$$

for every x in I.

Evidently any nonzero number x_0 is an ordinary point of equation (6.17), so we may apply the method of Section 6.2 to find two linearly independent solutions:

$$z_1(x) = \sum_{n=0}^{\infty} c_n (x - x_0)^n$$

$$z_2(x) = \sum_{n=0}^{\infty} d_n (x - x_0)^n$$

which converge for $|x - x_0| < R_1$ and $|x - x_0| < R_2$, respectively. Since the solutions are linearly independent on the interval determined by $|x - x_0| < \min\{R_1, R_2\}$, there are constants D_1 and D_2 such that

$$y_2(x) = D_1 z_1(x) + D_2 z_2(x)$$

for $|x - x_0| < \min\{R_1, R_2\}$. Since y_2 is undefined at zero, we must have $\min\{R_1, R_2\} \leq |x_0|$. Therefore the radius of convergence of either z_1 or z_2 is at most $|x_0|$. It follows that y_2 is not a linear combination of z_1 and z_2 on the interval $(0, \infty)$. Thus there are solutions of equation (6.17) that cannot be written in the form

$$Az_1(x) + Bz_2(x)$$

on the interval $(0, \infty)$. If we are interested in solutions on the interval $(0, \infty)$, the method of Frobenius enables us to compute a general solution of equation (6.17), whereas the method presented in Section 6.2 does not. The method of Frobenius frequently determines solutions of differential equations with regular singular points which converge on larger intervals than the solutions determined by the method of Section 6.2. In addition, obtaining the solutions by the method of Frobenius allows the behavior of solutions near $x = 0$ to be easily investigated. ◀

▶ **Example 3**   Consider the differential equation

$$x^2 \frac{d^2y}{dx^2} - 2y = 0 \tag{6.22}$$

Since $x^2(-2/x^2)$ is a polynomial, it is analytic at $x=0$. Hence $x=0$ is a regular singular point of equation (6.22). Insertion of $y(x) = \sum_{n=0}^{\infty} a_n x^{n+r}$ into (6.22) yields

$$x^2 \sum_{n=0}^{\infty} (n+r)(n+r-1)a_n x^{n+r-2} - 2 \sum_{n=0}^{\infty} a_n x^{n+r} = 0$$

which can be rewritten as

$$x^r \left[\sum_{n=0}^{\infty} (n+r)(n+r-1)a_n x^n - 2 \sum_{n=0}^{\infty} a_n x^n \right] = 0$$

This equation simplifies to

$$x^r \sum_{n=0}^{\infty} \left[(n+r)(n+r-1)-2 \right] a_n = 0$$

which can be rewritten as

$$x^r \sum_{n=0}^{\infty} c_n x^n = 0 \tag{6.23}$$

where

$$c_n = \left[(n+r)(n+r-1)-2 \right] a_n \qquad (n \geq 0) \tag{6.24}$$

Since we want equation (6.23) to hold for all x, $x \neq 0$, in some open interval I containing zero, we must have

$$\sum_{n=0}^{\infty} c_n x^n = 0$$

for all x, $x \neq 0$, in I. By Theorem 6.1, $c_n = 0$ for each n. Clearly $c_0 = 0$ whenever

$$r^2 - r - 2 = 0$$

This is the indicial equation for the differential equation in (6.22), and it has $r_2 = -1$ and $r_1 = 2$ as its roots. With $r=2$ we have from (6.24)

$$0 = c_n = (n^2 + 3n)a_n \qquad (n \geq 1)$$

so that a_0 is arbitrary and $a_n = 0$ for $n \geq 1$. Hence

$$y(x) = x^2 \sum_{n=0}^{\infty} a_n x^n$$

$$= a_0 x^2$$

Setting $a_0 = 1$, we have that $y_1(x) = x^2$ is a solution of equation (6.22). We will now use the other root, $r = -1$, of the indicial equation to compute another solution. With $r = -1$ we have from (6.24)

$$0 = c_n = (n^2 - 3n)a_n \qquad (n \geq 1)$$

Notice that $n^2 - 3n = 0$ whenever $n = 0$ or $n = 3$. Therefore a_0 and a_3 are arbitrary, while $a_n = 0$ for $n \neq 0$ and $n \neq 3$. Hence

$$y(x) = x^{-1}(a_0 + a_3 x^3)$$

Setting $a_0 = 1$ and $a_3 = 0$, we have that $y_2(x) = x^{-1}$ is a solution of (6.22). Evidently y_1 and y_2 are linearly independent. Thus, if the roots of the indicial equation differ by an integer, there may be two linearly independent solutions of the form $x^r \sum_{n=0}^{\infty} a_n x^n$. However, this is not always the case as we will see in the next example. ◀

▶ **Example 4** Consider the differential equation

$$x^2 \frac{d^2 y}{dx^2} + x \frac{dy}{dx} + (x^2 - 1)y = 0 \qquad \text{(6.25)}$$

Evidently $x = 0$ is a regular singular point of the equation. Inserting $y(x) = \sum_{n=0}^{\infty} a_n x^{n+r}$ into (6.25) yields

$$\sum_{n=0}^{\infty} (n+r)(n+r-1)a_n x^{n+r} + \sum_{n=0}^{\infty} (n+r)a_n x^{n+r} + \sum_{n=0}^{\infty} a_n x^{n+2}$$

$$- \sum_{n=0}^{\infty} a_n x^n = 0$$

Replacing n by $n-2$ in the third summation enables to write this equation as

$$x^r \left[\sum_{n=0}^{\infty} \left((n+r)^2 - 1 \right) a_n x^n + \sum_{n=2}^{\infty} a_{n-2} x^n \right] = 0$$

or, equivalently, as

$$x^r \left[(r^2 - 1) a_0 + \left((1+r)^2 - 1 \right) a_1 + \sum_{n=2}^{\infty} \left(\left((n+r)^2 - 1 \right) a_n + a_{n-2} \right) \right] x^n = 0$$

This equation may now be written in the form

$$x^r \sum_{n=0}^{\infty} c_n x^n$$

where

$$c_0 = (r^2 - 1) a_0$$

$$c_1 = \left[(1+r)^2 - 1 \right] a_1 \qquad \text{(6.26)}$$

$$c_n = \left[(n+r)^2 - 1 \right] a_n + a_{n-2} \qquad (n \geqslant 2)$$

As in the previous examples, we must have $c_n = 0$ for each n. The equation $c_0 = 0$ determines r. Thus

$$r^2 - 1 = 0$$

is the indicial equation for the differential equation (6.25), and it has $r_2 = -1$ and $r_1 = 1$ as its roots. With $r = 1$ we have from (6.26)

$$0 = c_1 = 3a_1$$

$$0 = c_n = n(n+2) a_n + a_{n-2}$$

so that a_0 is arbitrary, $a_1 = 0$, and

$$a_n = -\frac{a_{n-2}}{n(n+2)} \qquad (n \geqslant 2)$$

Hence

$$a_2 = -\frac{a_0}{2\cdot4} \qquad\qquad a_3 = -\frac{a_1}{3\cdot5} = 0$$

$$a_4 = -\frac{a_2}{4\cdot6} = \frac{a_0}{2\cdot4\cdot4\cdot6} \qquad\qquad a_5 = -\frac{a_3}{5\cdot7} = 0$$

$$a_6 = -\frac{a_4}{6\cdot8} = -\frac{a_0}{2\cdot4\cdot6\cdot4\cdot6\cdot8} \qquad\qquad a_7 = -\frac{a_5}{7\cdot9} = 0$$

. . . .

Evidently $a_n = 0$ whenever n is an odd integer. Moreover,

$$a_{2n} = \frac{(-1)^n}{2\cdot4\cdots(2n)\cdot4\cdot6\cdots(2n+2)}a_0$$

$$= \frac{2(-1)^n}{2\cdot4\cdots(2n)\cdot2\cdot4\cdot6\cdots(2n+2)}a_0$$

$$= \frac{2(-1)^n}{2^n(1\cdot2\cdots n)2^{n+1}(1\cdot2\cdot3\cdots(n+1))}a_0$$

$$= \frac{(-1)^n}{2^{2n}n!(n+1)!}a_0$$

so that

$$y_1(x) = a_0 x \left[1 + \sum_{n=1}^{\infty} \frac{(-1)^n}{2^{2n}n!(n+1)!}x^{2n} \right]$$

$$= a_0 \sum_{n=0}^{\infty} \frac{(-1)^n}{2^{2n}n!(n+1)!}x^{2n+1}$$

is a solution of equation (6.25). We will now use the other root, $r = -1$, of the indicial equation to compute another solution. With $r = -1$ we have from (6.26)

$$0 = c_1 = -a_1$$

$$0 = n(n-2)a_n + a_{n-2} \qquad (n \geqslant 2)$$

When $n=2$ we have

$$0=2(2-2)a_2+a_0=a_0$$

Thus $a_0=0$, $a_1=0$, and

$$a_n=-\frac{a_{n-2}}{n(n-2)} \qquad (n\geq 3)$$

Hence

$$a_3=-\frac{a_1}{1\cdot 3}=0 \qquad a_4=-\frac{a_2}{2\cdot 4}$$

$$a_5=-\frac{a_3}{3\cdot 5}=0 \qquad a_6=-\frac{a_4}{4\cdot 6}=\frac{a_2}{2\cdot 4\cdot 4\cdot 6}$$

$$a_7=-\frac{a_5}{5\cdot 7}=0 \qquad a_8=-\frac{a_6}{6\cdot 8}=-\frac{a_2}{2\cdot 4\cdot 6\cdot 4\cdot 6\cdot 8}$$

. . . .

Evidently $a_n=0$ whenever n is an odd integer. Moreover,

$$a_{2n}=\frac{(-1)^{n+1}}{2\cdot 4\cdots(2n-2)\cdot 4\cdot 6\cdots(2n)}a_2$$

$$=\frac{2(-1)^{n+1}}{2\cdot 4\cdots(2n-2)\cdot 2\cdot 4\cdot 6\cdots(2n)}a_2$$

$$=\frac{2(-1)^{n+1}}{2^{n-1}(1\cdot 2\cdots(n-1))2^n(1\cdot 2\cdot 3\cdots n)}a_2$$

$$=\frac{(-1)^{n+1}}{2^{2n-2}(n-1)!n!}a_2$$

whenever $n\geq 2$ so that

$$y_2(x)=a_2x^{-1}\left[x^2+\sum_{n=2}^{\infty}\frac{(-1)^{n+1}}{2^{2n-2}(n-1)!n!}x^{2n}\right]$$

is a solution of equation (6.25). If we now replace n by $n+1$ and notice that $(-1)^{n+2}=(-1)^n$, we have

$$y_2(x)=a_2x^{-1}\left[x^2+\sum_{n=1}^{\infty}\frac{(-1)^n}{2^n n!(n+1)!}x^{2n+2}\right]$$

$$=a_2\sum_{n=0}^{\infty}\frac{(-1)^n}{2^n n!(n+1)!}x^{2n+1}$$

Clearly y_2 is a constant multiple of y_1. Thus, if the roots of the indicial equation differ by an integer, there may only be one solution of the form $x^r\sum_{n=0}^{\infty}a_nx^n$. ◀

If the difference r_1-r_2 between the roots of the indicial equation is not an integer, then it can be shown that the method of Frobenius yields two linearly independent solutions, one for each root of the indicial equation. If the difference is an integer, then, as the previous two examples show, the method of Frobenius may or may not yield two linearly independent solutions. Notice that in these examples the smaller root of the indicial equation determines a solution which for appropriate choices of arbitrary constants is the solution determined by the larger root. In fact, it can be shown that the smaller root determines either the solution $y(x)\equiv0$ or all solutions of the form in (6.16). If two linearly independent solutions are not determined by the smaller root, then some other method must be used to obtain a second linearly independent solution. A method for determining such a second solution is presented in the following section.

Exercises   SECTION 6.3

1. Zero is not an ordinary point of

$$x^2\frac{d^2y}{dx^2}+3x\frac{dy}{dx}+2y=0$$

Show that this equation has no solution of the form $y(x)=\sum_{n=0}^{\infty}a_nx^n$, except $y(x)\equiv0$. (Hence the x^r term introduced by Frobenius is important.)

2. (a) If p is not an integer, show that there is no analytic solution at $x=0$, other than $y(x)\equiv0$, of equation (6.14).
   (b) If p is an integer, show that there is an analytic solution at $x=0$ of equation (6.14), but there are not two linearly independent analytic

solutions. Hint: Assume there is an analytic solution $\sum_{n=0}^{\infty} b_n x^n$, insert this solution into equation (6.14), and evaluate the coefficients b_n.

In Exercises 3–18 find a solution guaranteed by Theorem 6.2. If possible, find a general solution consisting of solutions having the form of (6.16).

3. $2x^2 \dfrac{d^2y}{dx^2} + 5x \dfrac{dy}{dx} + (1+x)y = 0$

4. $9x^2 \dfrac{d^2y}{dx^2} + 9x \dfrac{dy}{dx} - (1-x^2)y = 0$

5. $x^2 \dfrac{d^2y}{dx^2} + x \dfrac{dy}{dx} + (x^2-4)y = 0$

6. $x \dfrac{d^2y}{dx^2} - 3 \dfrac{dy}{dx} + xy = 0$

7. $x^2 \dfrac{d^2y}{dx^2} + x(1+x) \dfrac{dy}{dx} - y = 0$

8. $4x^2 \dfrac{d^2y}{dx^2} + 4x \dfrac{dy}{dx} - y = 0$

9. $4x^2 \dfrac{d^2y}{dx^2} + 4x \dfrac{dy}{dx} - (1+x^2)y = 0$

10. $x^2 \dfrac{d^2y}{dx^2} + x^2 \dfrac{dy}{dx} - 2y = 0$

11. $2x^2 \dfrac{d^2y}{dx^2} + 5x \dfrac{dy}{dx} + y = 0$

12. $2x \dfrac{d^2y}{dx^2} + 3 \dfrac{dy}{dx} + y = 0$

13. $3x \dfrac{d^2y}{dx^2} + 4 \dfrac{dy}{dx} + 2y = 0$

14. $2x \dfrac{d^2y}{dx^2} + (x+1) \dfrac{dy}{dx} + 3y = 0$

15. $x^2 \dfrac{d^2y}{dx^2} + 5x \dfrac{dy}{dx} + 3y = 0$

16. $x^2 \dfrac{d^2y}{dx^2} + x \dfrac{dy}{dx} + (x^2-9)y = 0$

17. $x \dfrac{d^2y}{dx^2} + \dfrac{dy}{dx} + y = 0$

18. $x \dfrac{d^2y}{dx^2} + \dfrac{dy}{dx} + 2y = 0$

19. A supply of hot air can be obtained by passing the air through a heated cylindrical tube. It can be argued that the temperature T of the air in the tube satisfies the differential equation

$$\frac{d^2T}{dx^2} - \frac{upC}{kA}\frac{dT}{dx} + \frac{2\pi rh}{kA}(T_w - T) = 0$$

where

x = distance from intake end of tube
u = flow rate of air
p = density of air
C = heat capacity of air
k = thermal conductivity
A = cross-sectional area of tube
r = radius of tube
h = heat transfer coefficient of air (nonconstant)
T_w = temperatures of tube

(See page 100 of *Mathematical Methods in Chemical Engineering* [Academic Press, London, 1977] by V. G. Jenson and G. V. Jefferys.) In the reference, the parameters are given values such that the differential equation becomes

$$\frac{d^2T}{dx^2} - 26{,}200\frac{dT}{dx} - 11{,}430x^{-1/2}(T_w - T) = 0$$

(a) Show that the change of variables $y = T_w - T$ and $x = z^2$ transforms the equation into

$$z\frac{d^2y}{dz} - (1 + 52{,}400z^2)\frac{dy}{dz} - 45{,}720z^2 y = 0$$

(b) Show that 0 and 2 are the roots of the indicial equation.
(c) Using the method of Frobenius, compute the first five terms of a series solution.

By Theorem 6.2 this series converges. However, note the size of the fifth coefficient. This example shows that even though the series converges, the coefficients a_n may not be small.

6.4 Solutions about Regular Singular Points, II

In the previous section we found that the method of Frobenius always yields one solution of a differential equation having a regular singular point, but it may not yield two linearly independent solutions. If the method of Frobenius does not yield two linearly independent solutions, it is still possible to obtain a second solution that is linearly independent of the first. Unfortunately, the calculations in such a case are complicated, but there is no other way to proceed. The following theorem gives the form of this second solution.

THEOREM 6.3   Let x_0 be a regular singular point of

$$a_2(x)\frac{d^2y}{dx^2}+a_1(x)\frac{dy}{dx}+a_0(x)y=0$$

and let r_1, r_2 $(r_2 \leqslant r_1)$ be the roots of the indicial equation. If the method of Frobenius yields only one linearly independent solution y_1, then there is a solution y_2 that is linearly independent of y_1 having the form

$$y_2(x)=(x-x_0)^{r_2}\sum_{n=0}^{\infty} b_n(x-x_0)^n+cy_1(x)\ln|x-x_0| \tag{6.27}$$

where $b_0 \neq 0$, $c \neq 0$, and the series $\sum_{n=0}^{\infty}b_n(x-x_0)^n$ converges in an interval $|x-x_0|<R$, for some $R>0$.

The constant c and the coefficients $b_0, b_1,\dots$ in the representation of y_2 in (6.27) can be determined by inserting y_2 into the differential equation and proceeding in a manner analogous to that in the method of Frobenius. We will illustrate this procedure by computing a second solution for the differential equation considered in Example 4 of Section 6.3.

▶ **Example 1**   We found previously that the only solutions of

$$x^2\frac{d^2y}{dx^2}+x\frac{dy}{dx}+(x^2-1)y=0 \tag{6.28}$$

having the form

$$y(x)=x^r\sum_{n=0}^{\infty} a_nx^n$$

are constant multiples of

$$y_1(x) = \sum_{n=0}^{\infty} \frac{(-1)^n}{2^{2n} n!(n+1)!} x^{2n+1}$$

Recall that the roots of the indicial equation are $r_1 = 1$ and $r_2 = -1$. Inserting

$$y_2(x) = x^{-1} \sum_{n=0}^{\infty} b_n x^n + c y_1(x) \ln|x|$$

$$= \sum_{n=0}^{\infty} b_n x^{n-1} + c y_1(x) \ln|x|$$

into the differential equation, we obtain

$$x^2 \left[\sum_{n=0}^{\infty} (n-1)(n-2) b_n x^{n-3} + c y_1''(x) \ln|x| + 2cx^{-1} y_1'(x) - cx^{-2} y_1(x) \right]$$

$$+ x \left[\sum_{n=0}^{\infty} (n-1) b_n x^{n-2} + c y_1'(x) \ln|x| + cx^{-1} y_1(x) \right]$$

$$+ (x^2 - 1) \left[\sum_{n=0}^{\infty} b_n x^{n-1} + c y_1(x) \ln|x| \right] = 0$$

which can be rearranged to

$$\sum_{n=0}^{\infty} \left[(n-1)(n-2) + (n-1) - 1 \right] b_n x^{n-1} + \sum_{n=0}^{\infty} b_n x^{n+1}$$

$$+ \left[x^2 y_1'' + x y_1' + (x^2 - 1) y_1 \right] c \ln|x| + 2cx y_1'(x) = 0$$

Since y_1 is a solution, $x^2 y_1'' + x y_1' + (x^2 - 1) y_1 = 0$. Thus the equation reduces further to

$$\sum_{n=0}^{\infty} n(n-2) b_n x^{n-1} + \sum_{n=0}^{\infty} b_n x^{n+1} + 2c \sum_{n=0}^{\infty} a_{2n+1} x^{2n+1} = 0$$

where $a_{2n+1} = \dfrac{(2n+1)(-1)^n}{2^{2n} n!(n+1)!}$. In the first summation we replace n by $n+1$, while in the second and third summations we replace n by $n-1$, to obtain

$$\sum_{n=-1}^{\infty}(n+1)(n-1)b_{n+1}x^n+\sum_{n=1}^{\infty}b_{n-1}x^n+2c\sum_{n=1}^{\infty}a_{2n-1}x^{2n-1}=0$$

or

$$-b_1+\sum_{n=1}^{\infty}\left[(n+1)(n-1)b_{n+1}+b_{n-1}\right]x^n+2c\sum_{n=1}^{\infty}a_{2n-1}x^{2n-1}=0$$

We now break the first summation into two summations: one with only even powers of x, and the other with only odd powers of x. Noting that for $n=1,2,\dots$, we have $2n-1=1,3,5,\dots$. Hence

$$-b_1+\sum_{n=1}^{\infty}\left[((2n-1)+1)((2n-1)-1)b_{2n}+b_{2n-2}+2ca_{2n-1}\right]x^{2n-1}$$

$$+\sum_{n=1}^{\infty}\left[(2n+1)(2n-1)b_{2n+1}+b_{2n-1}\right]x^{2n}=0$$

or

$$-b_1+\sum_{n=1}^{\infty}\left[4n(n-1)b_{2n}+b_{2n-2}+2ca_{2n-1}\right]x^{2n-1}$$

$$+\sum_{n=1}^{\infty}\left[(4n^2-1)b_{2n+1}+b_{2n-1}\right]x^{2n}=0$$

This equation has the form $\sum_{n=0}^{\infty}c_nx^n=0$, where

$$c_0=-b_1$$
$$c_{2n}=(4n^2-1)b_{2n+1}+b_{2n-1}\qquad(n\geqslant1)$$
$$c_{2n-1}=4n(n-1)b_{2n}+b_{2n-2}+2ca_{2n-1}\qquad(n\geqslant1)$$

In order for $\sum_{n=0}^{\infty}c_nx^n=0$ to hold for all x, $x\neq0$ in an open interval containing zero, we must have $c_n=0$ for every n. Hence

$$0=c_0=-b_1$$
$$0=c_1=b_0+2ca_1$$
$$=b_0-c$$
$$b_{2n}=\frac{-1}{4n(n-1)}\left[b_{2n-2}+2ca_{2n-1}\right]\qquad(n\geqslant2)$$
$$b_{2n+1}=\frac{-1}{4n^2-1}b_{2n-1}\qquad(n\geqslant1)$$

From the first two equations we have $b_1 = 0$ and $c = b_0$. Note that b_2 is not determined by these equations and therefore is arbitrary. Then

$$b_4 = -\frac{1}{4 \cdot 2 \cdot 1}[b_2 + 2ca_3] \qquad\qquad b_3 = -\tfrac{1}{3}b_1 = 0$$

$$b_6 = -\frac{1}{4 \cdot 3 \cdot 2}[b_4 + 2ca_5] \qquad\qquad b_5 = -\tfrac{1}{15}b_3 = 0$$

$$= \frac{1}{4^2 \cdot 3 \cdot 2 \cdot 2 \cdot 1}[b_2 + 2ca_3] - \frac{2c}{4 \cdot 3 \cdot 2 \cdot 2 \cdot 1}a_5$$

$$b_8 = -\frac{1}{4 \cdot 4 \cdot 3}[b_6 + 2ca_7] \qquad\qquad b_7 = -\tfrac{1}{35}b_5 = 0$$

$$= \frac{-1}{4^3 \cdot 4 \cdot 3 \cdot 2 \cdot 3 \cdot 2}[b_2 + 2ca_3] - \frac{2c}{4 \cdot 4 \cdot 3}a_7 + \frac{2c}{4^3 \cdot 3^2 \cdot 2^2}a_5$$

· ·

Evidently $b_n = 0$ whenever n is an odd integer. Moreover,

$$b_{2n} = \frac{(-1)^{n+1}}{4^{n-1}n!(n-1)!}(b_2 + 2ca_3) - c[\text{terms involving } a_3, a_5, \ldots, a_{2n-1}]$$

$$= \frac{(-1)^{n+1}}{4^{n-1}n!(n-1)!}(b_2 - \tfrac{3}{8}b_0) - c[\text{terms involving } a_3, a_5, \ldots, a_{2n-1}]$$

(6.29)

With these values for the b_n we have

$$y_2(x) = x^{-1}\sum_{n=0}^{\infty} b_n x^n + cy_1(x)\ln|x|$$

$$= \sum_{n=0}^{\infty} b_{2n}x^{2n-1} + cy_1(x)\ln|x|$$

is a solution of equation (6.28). Recall that $b_0 = c$. If we should set $b_0 = 0$, then, using (6.29), y_2 becomes

$$x^{-1}\sum_{n=1}^{\infty}\frac{(-1)^{n+1}}{4^{n-1}n!(n-1)!}b_2 x^{2n} = b_2 x^{-1}\sum_{n=0}^{\infty}\frac{(-1)^{n+2}}{4^n n!(n+1)!}x^{2(n+1)}$$

$$= b_2\sum_{n=0}^{\infty}\frac{(-1)^n}{2^{2n}n!(n+1)!}x^{2n+1}$$

$$= b_2 y_1(x)$$

Thus when $b_0 = 0$ we have the solution obtained using the method of Frobenius. This is why $b_0 \neq 0$ is required in Theorem 6.3. However, there is no condition which b_2 must satisfy. If we choose b_2 to be zero, then we obtain a slightly less complicated solution that is linearly independent of y_1. With $b_2 = 0$, $b_0 = c = 1$, and $a_n = \dfrac{(-1)^n}{2^{2n} n!(n+1)!}$ the solution becomes

$$y_2(x) = x^{-1} + \frac{1}{2^5}x^3 - \frac{1}{2^5(3!)^2}x^5 + \frac{5}{2^6(3!)(4!)^3}x^7 + \cdots + y_1(x)\ln|x|$$

The graphs of y_1 and y_2 are shown in Figure 6.2. ◀

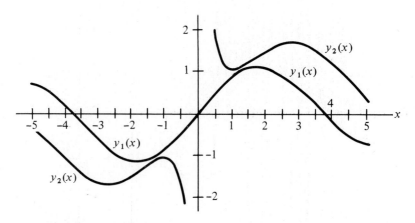

FIGURE 6.2

Exercises SECTION 6.4

In exercises 1–5 find a general solution of the given differential equation.

1. $x^2\dfrac{d^2y}{dx^2} + x(x-1)\dfrac{dy}{dx} + (1-x)y = 0$

2. $x\dfrac{d^2y}{dx^2} - x\dfrac{dy}{dx} + y = 0$

3. $x^2\dfrac{d^2y}{dx^2} + (x^2 - x)\dfrac{dy}{dx} + y = 0$

4. $x\dfrac{d^2y}{dx^2} + (1+x)\dfrac{dy}{dx} + y = 0$

5. $x^2\dfrac{d^2y}{dx^2} - x\dfrac{dy}{dx} + (x^2+1)y=0$

6. In 1732 Daniel Bernoulli, a Swiss mathematician, was studying the oscillations of a heavy flexible chain suspended with its lower end free. He obtained a differential equation of the form

$$x\frac{d^2y}{dx^2} + \frac{dy}{dx} + k^2y=0$$

where k is a positive number. Find a general solution for this equation for the special case $k=1$.

6.5 Bessel Functions

At the beginning of Section 6.3, the equation

$$x^2\frac{d^2y}{dx^2} + x\frac{dy}{dx} + (x^2-p^2)y=0 \qquad \textbf{(6.30)}$$

called **Bessel's equation of order p**, was introduced. This equation arises when determining the motion of a vibrating circular membrane, the electrostatic potential inside a cylinder, the temperature distribution in a disk, and in many other problems involving circles and cylinders. In applications, p is often a nonnegative integer, and that is the case we will consider here.

One of the most direct methods for finding solutions of equation (6.30) is the method of Frobenius. By Theorem 6.2 there is at least one solution of the form

$$y(x)= \sum_{n=0}^{\infty} a_n x^{n+r}$$

where r is some number. Inserting this function into equation (6.30) yields

$$\sum_{n=0}^{\infty} (n+r)(n+r-1)a_n x^{n+r} + \sum_{n=0}^{\infty} (n+r)a_n x^{n+r} - p^2 \sum_{n=0}^{\infty} a_n x^{n+r}$$

$$+ \sum_{n=0}^{\infty} a_n x^{n+2+r}=0$$

which simplifies to

$$\sum_{n=0}^{\infty} \left((n+r)^2-p^2\right)a_n x^{n+r} + \sum_{n=2}^{\infty} a_{n-2} x^{n+r} = 0$$

Hence

$$x^r\left[\left(r^2-p^2\right)a_0 + \left((r+1)^2-p^2\right)a_1 x + \sum_{n=2}^{\infty}\left[\left((n+r)^2-p^2\right)a_n + a_{n-2}\right]x^n\right] = 0$$

This equation is of the form $x^r\sum_{n=0}^{\infty}c_n x^n = 0$, where

$$c_0 = \left(r^2-p^2\right)a_0$$

$$c_1 = \left[(r+1)^2-p^2\right]a_1$$

$$c_n = \left[(n+r)^2-p^2\right]a_n + a_{n-2} \qquad (n \geqslant 2)$$

As before, we must have $c_n = 0$ for every n. The equation $c_0 = 0$ determines the indicial equation

$$r^2 - p^2 = 0$$

for Bessel's equation. Clearly $r_1 = -p$ and $r_2 = p$ are the roots of the indicial equation. With $r = p$, the equations $c_n = 0$ become

$$0 = c_1 = \left[(p+1)^2-p^2\right]a_1$$

$$= (2p+1)a_1$$

$$0 = c_n = \left[(n+p)^2-p^2\right]a_n + a_{n-2}$$

$$= n(n+2p)a_n + a_{n-2} \qquad (n \geqslant 2)$$

so that

$$a_1 = 0$$

$$a_n = -\frac{1}{n(n+2p)}a_{n-2}$$

Hence

$$a_2 = -\frac{1}{2(2+2p)}a_0 \qquad\qquad a_3 = -\frac{1}{3(3+2p)}a_1 = 0$$

$$a_4 = -\frac{1}{4(4+2p)}a_2 \qquad\qquad a_5 = -\frac{1}{5(5+2p)}a_3 = 0$$

$$= \frac{1}{2\cdot4(2+2p)(4+2p)}a_0$$

$$a_6 = -\frac{1}{6(6+2p)}a_4 \qquad\qquad a_7 = -\frac{1}{7(7+2p)}a_5 = 0$$

$$= \frac{1}{2\cdot4\cdot6(2+2p)(4+2p)(6+2p)}a_0$$

. . . .

Evidently $a_n = 0$ whenever n is an odd integer and

$$a_{2n} = \frac{(-1)^n}{2\cdot4\cdot6\cdots(2n)(2+2p)(4+2p)(6+2p)\cdots(2n+2p)}a_0$$

$$= \frac{(-1)^n}{2^n(1\cdot2\cdot3\cdots n)2^n(1+p)(2+p)(3+p)\cdots(n+p)}a_0$$

$$= \frac{(-1)^n}{2^{2n}n!(1+p)(2+p)(3+p)\cdots(n+p)}a_0$$

Thus a solution of Bessel's equation is

$$y(x) = a_0\left[1 + \sum_{n=1}^{\infty}\frac{(-1)^n}{2^{2n}n!(1+p)(2+p)\cdots(n+p)}x^{2n}\right]x^p \qquad \textbf{(6.31)}$$

Since p is a nonnegative integer, this solution may be written in the more compact form:

$$y(x) = a_0\left[1 + \sum_{n=1}^{\infty}\frac{(-1)^n p!}{2^{2n}n!(n+p)!}x^{2n}\right]x^p$$

$$= a_0 2^p p!\sum_{n=0}^{\infty}\frac{(-1)^n}{n!(n+p)!}\left(\frac{x}{2}\right)^{2n+p} \qquad \textbf{(6.32)}$$

The infinite series in (6.32), which converges for all values of x, is called the **Bessel function of the first kind of order p** and is denoted by J_p:

$$J_p(x) = \sum_{n=0}^{\infty} \frac{(-1)^n}{n!(n+p)!} \left(\frac{x}{2} \right)^{2n+p}$$

Before finding another solution of Bessel's equation that is linearly independent of J_p, we will discuss the graph of J_p. It is convenient to change variables according to $y = x^{-1/2}z$, so that

$$\frac{dy}{dx} = -\tfrac{1}{2}x^{-3/2}z + x^{-1/2}\frac{dz}{dx}$$

$$\frac{d^2y}{dx^2} = \tfrac{3}{4}x^{-5/2}z - x^{-3/2}\frac{dz}{dx} + x^{-1/2}\frac{d^2z}{dx^2}$$

Using these identities, Bessel's equation (6.30) can be written as

$$x^{3/2}\frac{d^2z}{dx^2} + \left(x^{3/2} + \left(\tfrac{1}{4}-p^2\right)x^{-1/2} \right)z = 0$$

or, equivalently,

$$\frac{d^2z}{dx^2} + \left[1 + \left(\frac{\tfrac{1}{4}-p^2}{x^2} \right) \right]z = 0$$

For large values of x, this equation is approximately

$$\frac{d^2z}{dx^2} + z = 0$$

which has $\sin x$ and $\cos x$ as linearly independent solutions. Therefore we might expect Bessel's equation to have solutions that resemble $x^{-1/2}\cos x$ and $x^{-1/2}\sin x$ for large values of x. In fact, it can be shown that all solutions of Bessel's equation tend to zero and oscillate between positive and negative values as $x \to \infty$. The graphs of J_0, J_1, and J_2 are shown in Figure 6.3.

A solution $J_{-p}(x)$ for equation (6.30) corresponding to $r = -p$ can be computed in a manner analogous to that used to obtain J_p. Unfortunately, J_p and J_{-p} are linearly dependent. From Theorem 6.3 we know that there is a solution linearly independent of J_p having the form

$$\sum_{n=0}^{\infty} b_n x^{n-p} + cJ_p(x)\ln|x|$$

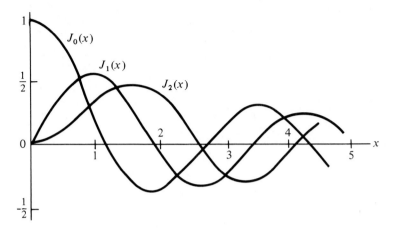

FIGURE 6.3

It is traditional to choose the "second" solution of Bessel's equation to be the sum of a solution of this form and $\gamma J_p(x)$, where γ is Euler's constant and defined by

$$\gamma = \lim_{k \to \infty} \left(1 + \tfrac{1}{2} + \tfrac{1}{3} + \cdots + \frac{1}{k} - \ln k \right)$$

$$= .5772157 \cdots$$

For ease of writing this solution, we introduce the function defined by

$$\varphi(0) = 0$$

$$\varphi(k) = \sum_{j=0}^{k} \frac{1}{j}, \quad \text{if } k \text{ is a positive integer}$$

A solution of Bessel's equation, called a **Bessel function of the second kind of order p**, that is linearly independent of $J_p(x)$ is

$$Y_p(x) = \frac{2}{\pi} \left[\left(\gamma + \ln \frac{x}{2} \right) J_p(x) - \frac{1}{2} \sum_{n=0}^{p-1} \frac{(p-n-1)!}{n!} \left(\frac{x}{2} \right)^{2n-p} \right.$$

$$\left. + \frac{1}{2} \sum_{n=0}^{\infty} (-1)^{n+1} \frac{\varphi(n) + \varphi(p+n)}{n!(n+p)!} \left(\frac{x}{2} \right)^{2n+p} \right]$$

The choice of Y_p over other "second" solutions is due to the exceptional computational properties possessed by Y_p. A discussion of these properties is beyond the scope of this text. As for $J_p(x)$, the solution $Y_p(x)$ tends to zero and oscillates between positive and negative values as $x \to \infty$. The graphs of Y_0 and Y_1 are shown in Figure 6.4.

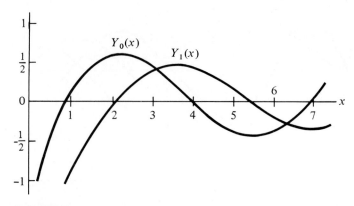

FIGURE 6.4

We are now able to write a general solution of equation (6.30), Bessel's equation of order p, as

$$y(x) = c_1 J_p(x) + c_2 Y_p(x)$$

We have seen that Bessel functions are solutions of specific linear differential equations with variable coefficients. In general, differential equations with variable coefficients are difficult, if not impossible, to solve in closed form. Bessel functions provide us with one of the few techniques available to solve such equations. A somewhat general form for the types of equations that can be solved by Bessel functions can be developed as follows. Let $Z_p(X)$ denote a general solution of the Bessel's equation

$$X^2 \frac{d^2 Y}{dX^2} + X \frac{dY}{dX} + (X^2 - p^2) Y = 0 \qquad \textbf{(6.33)}$$

If we make use of the substitutions

$$Y = \frac{y}{g(x)} \qquad X = f(x)$$

and notice that

$$\frac{d}{dX} = \frac{dx}{dX} \frac{d}{dx} = \frac{1}{f'} \frac{d}{dx}$$

equation (6.33) can be rewritten as

$$\frac{f^2}{f'}\frac{d}{dx}\left[\frac{1}{f'}\frac{d}{dx}\left(\frac{y}{g}\right)\right]+f\left[\frac{1}{f'}\frac{d}{dx}\left(\frac{y}{g}\right)\right]+(f^2-p^2)\frac{y}{g}=0 \qquad \text{(6.34)}$$

The general solution of (6.34) is given by

$$y(x)=g(x)Z_p[f(x)]$$

In particular, if

$$g(x)=x^a e^{-bx^r} \quad \text{and} \quad f(x)=cx^s$$

equation (6.34) can be written as

$$x^2\frac{d^2y}{dx^2}+x[(1-2a)+2rbx^r]\frac{dy}{dx}$$

$$+\left(a^2-p^2s^2+s^2c^2x^{2s}-rb(2a-r)x^r+r^2b^2x^{2r}\right)y=0 \qquad \text{(6.35)}$$

with the general solution

$$y(x)=x^a e^{-bx^r}Z_p(cx^s)$$

In Exercises 9–12 the reader is asked to find general solutions of various second-order linear differential equations with variable coefficients. One of the most common variants of Bessel's equation is

$$x^2\frac{d^2y}{dx^2}+x\frac{dy}{dx}+x\frac{dy}{dx}+(k^2x^2-p^2)y=0 \qquad \text{(6.36)}$$

If in equation (6.35) we choose $a=r=b=0$, $c=k$, and $s=1$, we obtain equation (6.36). Hence a general solution of equation (6.36) is $y(x)=Z_p(kx)$.

Before considering some applications, we need to make the following observations, which follow directly from the series expansions for J_p and Y_p:

$$J_0(0)=1$$

$$J_p(0)=0 \quad \text{if } p>0$$

$$\lim_{x\to 0}|Y_p(x)|=\infty \quad \text{if } p\geq 0$$

► **Example 1** Probably the most common example illustrating the use of Bessel functions is that of determining the motion of a vibrating circular membrane that is clamped along its edge. We begin by assuming that the center of the membrane is at the origin of a polar coordinate system and that the edge of the membrane lies on the circle $r=a$. We will denote the displacement of a point (r, θ) of the membrane at time t by $u(r, \theta, t)$. We will further assume that the membrane is thin, homogeneous, perfectly flexible, maintained in a state of uniform tension, and subject to no external forces. Under these assumptions the equation of motion of the membrane is

$$\frac{\partial^2 u}{\partial t^2} = c^2 \left(\frac{\partial^2 u}{\partial r^2} + \frac{1}{r} \frac{\partial u}{\partial r} + \frac{1}{r^2} \frac{\partial^2 u}{\partial \theta^2} \right) \tag{6.37}$$

where c is a positive constant depending on the tension and physical properties of the membrane. Since we are assuming that the membrane is clamped along its edge, we have

$$u(a, \theta, t) = 0 \tag{6.38}$$

for all θ and positive t.

We assume that the membrane is set into motion by being displaced from its equilibrium position and then released. The complete determination of the motion of the membrane is beyond the scope of this text and will not be attempted. However, we will determine some of the possible ways in which the membrane can vibrate.

Since we are assuming that there are no external forces, we hypothesize that there are possible modes of vibration in which the motion of each point is periodic. A **normal mode of vibration** is one in which all points of the membrane vibrate with the same period and pass through their equilibrium positions at the same time. We look for normal modes of vibration by considering possible displacement functions of the form

$$u(r, \theta, t) = v(r, \theta) \cos(wt + d)$$

for some constants w and d.

Recalling that the membrane is circular, we must also have that the function v is periodic in θ with period 2π. One possible choice for v is

$$v(r, \theta) = R(r) \cos(n\theta)$$

where n is a nonnegative integer. We will find normal modes of vibration having the form

$$u(r,\theta,t)=R(r)\cos(n\theta)\cos(wt+d)$$

Inserting this choice of u into equation (6.37) and into the boundary condition (6.38) yields

$$r^2\frac{d^2R}{dr^2}+r\frac{dR}{dr}+\left[\left(\frac{w}{c}\right)^2r^2-n^2\right]R=0,\quad R(a)=0$$

This differential equation is equation (6.36) with x, y, k, and p replaced by r, R, w/c, and n, respectively. Therefore

$$R(r)=c_1J_n\left(\frac{w}{c}r\right)+c_2Y_n\left(\frac{w}{c}r\right)$$

for some choice of the constants c_1 and c_2. In order to evaluate one of these constants, we make the physically realistic assumption that the displacement at the origin is bounded. Since $|Y_n(wr/c)|\to\infty$ as $r\to0$, we must have $c_2=0$. In order that $R(a)=0$, we must also have

$$0=R(a)=c_1J_n\left(\frac{w}{c}a\right)$$

Thus the constant w must be chosen so that

$$w=cj/a$$

where j is a root of J_n. From our earlier discussion we know that there are infinitely many choices for j. We have shown that any function of the form

$$u(r,\theta,t)=c_1J_n\left(\frac{j}{a}r\right)\cos(n\theta)\cos\left(\frac{cj}{a}t+d\right)$$

where n is a nonnegative integer, j a zero of J_n, and c_1 and d are arbitrary, gives a normal mode of vibration for the circular membrane. The following table contains the values of the pth positive zero $j_{n,p}$ of the Bessel function J_n.

p	1	2	3	4	5
$j_{0,p}$	2.4048	5.5201	8.6537	11.7915	14.9309
$j_{1,p}$	3.8317	7.0156	10.1735	13.3237	16.4706
$j_{2,p}$	5.1356	8.4172	11.6198	14.7960	17.9598
$j_{3,p}$	6.3802	9.7610	13.0152	16.2235	19.4094
$j_{4,p}$	7.5883	11.0647	14.3725	17.6160	20.8269
$j_{5,p}$	8.7714	12.3386	15.7002	18.9801	22.2178

◄

▶ **Example 2**   Consider a wedge-shaped canal of uniform depth h that empties into the open sea (Figure 6.5). We will assume that the water

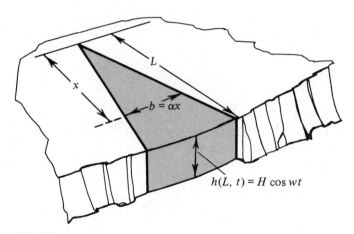

$$h(L, t) = H \cos wt$$

FIGURE 6.5

level at the mouth of the canal varies harmonically with time. That is, the depth at the mouth of the river is given by $H \cos wt$, where H and w are positive constants. The intent of this assumption is to simulate the motion of the tides. It can be argued that the function $h(x, t)$, which gives the depth at a distance x from the inland end of the canal at time t, has the form

$$h(x, t) = y(x) \cos wt$$

where y satisfies the differential equation

$$x^2 \frac{d^2 y}{dx^2} + x \frac{dy}{dx} + k^2 x^2 y = 0$$

and $k = W/(gh)^{1/2}$ (g being the force due to gravity). This is equation (6.36) with $p = 0$, and it has a general solution of the form

$$y(x) = c_1 J_0(kx) + c_2 Y_0(kx)$$

At $x = 0$ the depth of the water must be finite at all times. This means that $\lim_{x \to 0} y(x)$ has to be finite. Since $\lim_{x \to 0} |Y_0(kx)| = \infty$, we must have $c_2 = 0$ so that $y(x) = c_1 J_0(kx)$. The depth $h(x, t)$ can now be written as $h(x, t) = c_1 J_0(kx) \cos wt$. We assume that the depth at the mouth of the river is given by $H \cos wt$. Hence

$$H \cos wt = h(L, t) = c_1 J_0(kL) \cos wt$$

so that

$$c_1 = \frac{H}{J_0(kL)}$$

whenever $J_0(kL) \neq 0$. The depth h is now given by

$$h(x,t) = H\frac{J_0(kx)}{J_0(kL)}\cos wt$$

Recalling that $J_0(0) = 1$, we notice that $J_0(x) \neq 0$ for all x sufficiently small. Thus $J_0(kL) \neq 0$ whenever $kL = WL/(gh)^{1/2}$ is sufficiently small. A complete derivation and discussion of this and related problems may be found in *Hydrodynamics* (6th ed., Dover, New York, 1945) by H. Lamb, beginning on page 275.   ◀

Exercises   SECTION 6.5

In Exercises 1–4 verify the given identity.

1. $\dfrac{d}{dx}[x^p J_p(x)] = x^p J_{p-1}(x)$

2. $\dfrac{d}{dx}[x^{-p} J_p(x)] = -x^{-p} J_{p+1}(x)$

3. $\dfrac{d}{dx}[J_p(x)] = J_{p-1}(x) - \dfrac{p}{x}J_p(x)$ (Use Exercise 1.)

4. $J_{p-1}(x) + J_{p+1}(x) = \dfrac{2p}{x}J_p(x)$ (Use Exercises 1 and 2.)

5. Show that the initial-value problem

$$x^2\frac{d^2y}{dx^2} + x\frac{dy}{dx} + (x^2 - p^2)y = 0, \quad y(0) = 0, \quad y^{(p)}(0) = 2^{-p}$$

has $J_p(x)$ as its unique solution for every positive integer p.

6. Let n be a positive integer, and define a function f_n by

$$f_n(x) = \int_0^\pi \cos(nt - x\sin t)\, dt$$

The purpose of this exercise is to show that there is a constant A_n such that $J_n(x) = A_n f_n(x)$.

(a) Show that

$$f_n'(x) = \int_0^\pi \sin(nt - x\sin t)\sin t \, dt$$

$$f_n''(x) = -\int_0^\pi \cos(nt - x\sin t)\sin^2 t \, dt$$

(b) Use integration by parts to show that

$$f_n'(x) = \int_0^\pi \cos(nt - x\sin t)(n - x\cos t)\cos t \, dt$$

(c) Show that

$$x\int_0^\pi \cos(nt - x\sin t)\cos t \, dt$$

$$= -\int_0^\pi \cos(nt - x\sin t)(n - x\cos t)\, dt + n\int_0^\pi \cos(nt - x\sin t)\, dt$$

$$= n\int_0^\pi \cos(nt - x\sin t)\, dt$$

(d) Use parts (a), (b), and (c) to show that f_n is a solution of Bessel's equation of order n.

(e) Show that there is a constant A_n such that $J_n(x) = A_n f_n(x)$.

(f) Use Exercise 5 to show that $A_1 = 1/\pi$ and $A_2 = 1/\pi$. In fact, Exercise 5 can be used to show that $J_n(x) = \pi^{-1} f_n(x)$ for every positive integer n.

(g) Assuming that $J_n(x) = \pi^{-1} f_n(x)$, show that $|J_n(x)| \leq 1$ for every positive integer n and every real number x.

In Exercises 7 and 8 evaluate the given limit.

7. $\lim_{x\to 0} x^{-p} J_p(x)$

8. $\lim_{x\to 0+} x Y_1(x)$

In Exercises 9–12 find a general solution of the given differential equation.

9. $x^2\dfrac{d^2y}{dx^2} + x\dfrac{dy}{dx} - (16 - 4x^2)y = 0$

10. $\dfrac{d^2y}{dx^2} + 4x^{-5/3}y = 0$

11. $x\dfrac{d^2y}{dx^2} - 3\dfrac{dy}{dx} + xy = 0$

12. $x^2\dfrac{d^2y}{dx^2} + x\dfrac{dy}{dx} + (7x^2 - 25) = 0$

13. Show that the initial-value problem

$$x^2\dfrac{d^2y}{dx^2} + x\dfrac{dy}{dx} + 16x^4 y = 0, \quad y(0) = 1$$

has a unique solution.

14. Show that the initial-value problem

$$x^2\dfrac{d^2y}{dx^2} - 7x\dfrac{dy}{dx} + (6 + 16x^4)y = 0, \quad y(0) = 1$$

has no solution.

15. The following is one of the classical problems of applied mathematics. It was first studied in 1732 by Daniel Bernoulli and later in 1781 by Leonhard Euler. Consider a uniform heavy flexible chain fixed at the upper end and free at the lower end. If the lower end is slightly displaced by a distance s from its equilibrium position, then it can be shown that the horizontal displacement y, at time t after the free end is released, of an element whose distance from the fixed end is x, satisfies the equation

$$\dfrac{\partial^2 y}{\partial t^2} = g(L - x)\dfrac{\partial^2 y}{\partial x^2} - g\dfrac{\partial y}{\partial x}$$

where L is the length of the chain and g is the constant acceleration due to the force of gravity. See Figure 6.6.

(a) Set $z = L - x$ and show that a mode of vibration of the form $y = u(z)e^{kti}$ satisfies the equation

$$z\dfrac{d^2u}{dz^2} + \dfrac{du}{dz} + \dfrac{k^2}{g}u = 0$$

(b) Find a general solution of the equation found in part (a).
(c) At time $t = 0$ the chain is fixed at the end $x = 0$ and displaced by an amount s at the end $x = L$. Thus we want to find solutions of the equation in part (a) that satisfy the boundary conditions (remember that u is a function of z and that $z = L - x$) $u(0) = s$ and $u(L) = 0$. Find

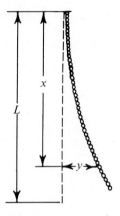

FIGURE 6.6

all values of k for which there are solutions satisfying these boundary conditions.

16. Consider a vertical column of length L, such as a telephone pole, as shown in Figure 6.7. In certain cases it can be shown that

$$\frac{d^2y}{dx^2} + \frac{P}{EI_0}e^{kx/L}(y-a)=0 \qquad\qquad (6.39)$$

where E is the modulus of elasticity, I_0 the moment of inertia at the base of the column about an axis perpendicular to the plane of bending, and both P and k are positive constants.

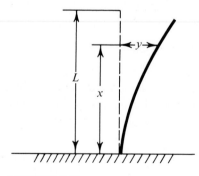

FIGURE 6.7

(a) Use the changes of variable $z=y-a$, $2b=k/L$, and $n=\sqrt{P/EI_0}$ to write the equation as

$$\frac{d^2z}{dx^2} + n^2e^{2bx}z=0$$

(b) Find a general solution of the equation in part (a). Hint: Use the change of variable $t=ne^{bx}$.

(c) Find a general solution of equation (6.39).

This and similar problems are discussed in N. J. Durant, Struts of variable flexibility, Philosophical Magazine and Journal of Science **36** (1945), 572–577.

17. Consider the canal emptying into the sea in Example 2. Now we assume that the depth l is not uniform, but varies with x according to $l(x)=\beta x$.

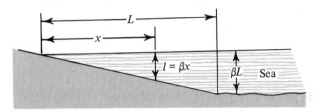

FIGURE 6.8

Figure 6.8 shows a lengthwise cross section of the canal. As before, we assume that the water level at the mouth of the canal varies harmonically. It can be argued that the function describing the depth, $h(x,t)$, has the form

$$h(x,t)=y(x)\cos wt$$

where y satisfies the differential equation

$$x^2\frac{d^2y}{dx^2}+2x\frac{dy}{dx}+k^2xy=0$$

where $k=\sqrt{w^2/\beta g}$. Find $h(x,t)$.

18. Consider a pendulum undergoing small oscillations. Instead of having a fixed length, the length $L(t)$ of the pendulum at time t increases according to $L(t)=at+b$, where a and b are positive numbers. The angle θ that the pendulum makes with a vertical line satisfies the differential equation

$$(at+b)\frac{d^2\theta}{dt^2}+2b\frac{d\theta}{dt}+g\sin\theta=0$$

where g is the acceleration due to gravity. For small values of θ, $\sin\theta\approx\theta$. Thus the above equation is approximately

$$(at+b)\frac{d^2\theta}{dt^2}+2b\frac{d\theta}{dt}+g\theta=0 \tag{6.40}$$

for small values of θ. Let the initial displacement of the pendulum be θ_0, i.e., $\theta(0)=\theta_0$. For the special case $a=2b$, find the approximate motion of the pendulum by solving the differential equation in (6.40) subject to the initial condition $\theta(0)=\theta_0$. Hint: Use the change of variable $x=at+b$.

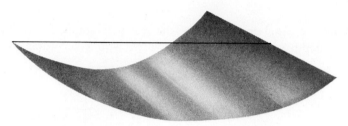

FIGURE 6.9

19. Consider a long, flat triangular piece of metal whose ends are joined by an inextensible piece of string of length slightly less than that of the piece of metal (see Figure 6.9). The line of the string is taken as the x-axis, with the left end as the origin. The deflection $y(x)$ of the piece of metal from horizontal at x satisfies the differential equation

$$EI\frac{d^2y}{dx^2}+Ty=0$$

where T is the tension in the string, E is Young's modulus, and I is the moment of inertia of a cross section of the piece of metal. Since the metal is triangular, $I=ax$ for some number a. The above equation can now be written as

$$x\frac{d^2y}{dx^2}+k^2y=0$$

where $k^2=T/Ea$. Find the general form of a solution of this equation that is bounded at $x=0$.

6.6 Modified Bessel Functions (OPTIONAL)

Variants of the Bessel functions discussed in the previous section occasionally arise in applications. Below is a hint of their applications; further details may be found in advanced books.

In Example 1 of Section 6.5 we considered a vibrating circular membrane which had no external forces acting upon it. We now assume that the membrane does have external forces acting upon it that cause the vibrations to tend to zero exponentially with time. In particular, we assume that the displacement $u(r, \theta, t)$ of a point (r, θ) of the membrane at time t has the form

$$u(r, \theta, t) = e^{-st} R(r) \cos n\theta \tag{6.41}$$

where s is a positive number and n is a nonnegative integer. As before, the equation of motion is

$$\frac{\partial^2 u}{\partial t^2} = c^2 \left(\frac{\partial^2 u}{\partial r^2} + \frac{1}{r} \frac{\partial u}{\partial r} + \frac{1}{r^2} \frac{\partial^2 u}{\partial \theta^2} \right) \tag{6.42}$$

If we now insert our choice for the displacement u, equation (6.41), into equation (6.42), we obtain

$$r^2 \frac{d^2 R}{dr^2} + r \frac{dR}{dr} - \left[\left(\frac{s}{c} \right)^2 r^2 + n^2 \right] R = 0 \tag{6.43}$$

This differential equation is equation (6.36) with x, y, k, and p replaced by r, R, is/c, and n, respectively. Hence a general solution of equation (6.43) is $Z_n(isr/c)$, where $Z_n(x)$ is a general solution of Bessel's equation of order n. We will determine linearly independent solutions of equation (6.43) that are real-valued. For simplicity of notation we set $x = sr/c$. We begin by considering the solution $J_n(ix)$ of equation (6.43):

$$J_n(ix) = \sum_{k=0}^{\infty} \frac{(-1)^k}{k!(n+k)!} \left(\frac{ix}{2} \right)^{2k+n}$$

$$= \sum_{k=0}^{\infty} \frac{(-1)^k}{k!(n+k)!} \left(\frac{x}{2} \right)^{2k+n} (i)^{2k+n}$$

Since $i^{2k} = (i^2)^k = (-1)^k$, we have

$$J_n(ix) = i^n \sum_{k=0}^{\infty} \frac{1}{k!(n+k)!} \left(\frac{x}{2} \right)^{2k+n} \tag{6.44}$$

The number i^n is real if and only if n is an even integer. Recalling that any constant multiple, including a complex constant multiple, of a solution of a homogeneous linear differential equation is also a solution, we define a real-valued solution of equation (6.43) by

$$I_n(x)=i^{-n}J_n(ix)$$

By equation (6.44)

$$I_n(x)=\sum_{k=0}^{\infty}\frac{1}{k!(n+k)!}\left(\frac{x}{2}\right)^{2k+n}$$

The function $I_n(x)$ is called the **modified Bessel function of the first kind.**
A "second" solution is usually defined by

$$K_n(x)=\frac{\pi}{2}i^{n+1}\left[J_n(ix)+iY_n(ix)\right]$$

The function $K_n(x)$, called the **modified Bessel function of the second kind**, is a solution of equation (6.43), with $x=sr/c$, since each summand is a solution. It is not apparent that $K_n(x)$ is a real-valued function. Nonetheless it is. In Exercise 1 the reader is asked to show that

$$K_n(x)=(-1)^{n+1}\left(\gamma+\ln\frac{x}{2}\right)I_n(x)+\frac{1}{2}\sum_{k=0}^{n-1}\frac{(n-k-1)!}{k!}(-1)^k\left(\frac{x}{2}\right)^{2k+n}$$

$$+(-1)^n\frac{1}{2}\sum_{k=0}^{\infty}\frac{\varphi(k)+\varphi(n+k)}{k!(n+k)!}\left(\frac{x}{2}\right)^{2k+n} \qquad \textbf{(6.45)}$$

From the series expansions for $J_n(x)$ and $K_n(x)$, we find that

$$I_n(0)=J_n(0)=\begin{cases}1 & \text{if } n=0 \\ 0 & \text{if } n\geqslant 1\end{cases}$$

$$\lim_{x\to 0}|K_n(x)|=\infty$$

It can also be shown that

$$\lim_{x\to\infty}\sqrt{2\pi x}\,e^{-x}I_p(x)=1 \qquad \text{and} \qquad \lim_{x\to\infty}\frac{2x}{\pi}e^x K_p(x)=1$$

so that

$$\lim_{x\to\infty}I_p(x)=\infty \qquad \text{and} \qquad \lim_{x\to\infty}K_p(x)=0$$

The graphs of I_0, I_1, K_0, and K_1 are shown in Figure 6.10.

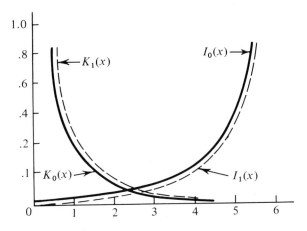

FIGURE 6.10

If $Z_n(ix)$ is a general solution of a homogeneous linear differential equation, then

$$y(x)=c_1 I_n(x)+c_2 K_n(x)$$

is also a general solution. Thus a general solution of equation (6.43) is

$$R(r)=c_1 I_n\left(\frac{s}{c}r\right)+c_2 K_n\left(\frac{s}{c}r\right)$$

If we again require that the displacement at the origin is bounded, then

$$R(r)=c_1 I_n\left(\frac{s}{c}r\right)$$

We have shown that functions of the form

$$u(r,\theta,t)=c_1 e^{-st} I_n\left(\frac{s}{c}r\right)\cos n\theta$$

where c_1 is a arbitrary number, s a positive number, and n a nonnegative integer, give modes of vibration which decay exponentially as $t\to\infty$.

▶ **Example 1**   In studying the flow of current through electrical transmission lines, one encounters the differential equation

$$x^2\frac{d^2E}{dx^2}-Ax\frac{dE}{dx}-k^2 x^B E=0 \qquad \text{(6.46)}$$

where A, B are positive numbers and E represents the potential difference (with respect to one end of the line) at a point a distance x from that end of the line. A complete discussion of this problem may be found on pages 126–128 of *Bessel Functions for Engineers* (2nd ed., Oxford University Press, London, 1955) by N. W. McLachlan. Comparing the differential equation in (6.46) to the differential equation in (6.36), we see that upon choosing $r=0$, $b=0$, $1-2a=-A$, $a^2-p^2s^2=0$, $s^2c^2=-k^2$, and $2s=B$, the two equations are identical. Thus, a general solution of equation (6.46) is given by

$$E(x)=x^aZ_p(cx^s)$$

where Z_p is a general solution of Bessel's equation of order p, $a=\frac{1}{2}(1+A)$, $s=\frac{1}{2}B$, $c=2ki/B$, and $p=(1+A)/B$. Hence a general solution of (6.46) is

$$E(x)=x^{1/2(1+A)}\left[c_1 I_{(1+A)/B}\left(\frac{2k}{B}x\right)+c_2 K_{(1+A)/B}\left(\frac{2k}{B}x\right)\right] \quad \blacktriangleleft$$

Exercises   SECTION 6.6

1. Given that $\ln\frac{ix}{2}=\ln x+i\frac{\pi}{2}$ whenever $x>0$, verify the identity in (6.45).

In Exercises 2–5 find a general solution of the given differential equation.

2. $x^2\dfrac{d^2y}{dx^2}+5x\dfrac{dy}{dx}-(12+36x^4)y=0$

3. $4x^2\dfrac{d^2y}{dx^2}+(1-x)y=0$

4. $x^2\dfrac{d^2y}{dx^2}-x\dfrac{dy}{dx}-(8+x^2)y=0$

5. $x\dfrac{d^2y}{dx^2}+\dfrac{dy}{dx}-x^2y=0$

6. Consider the wedge-shaped cooling fin depicted in Figure 6.11. It is possible to describe the heat flow through and from this wedge by the differential equation

$$x^2\frac{d^2y}{dx^2}+x\frac{dy}{dx}-Axy=0$$

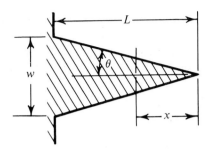

FIGURE 6.11

where

> $x=$ distance from tip of fin
> $T=$ temperature of fin at x
> $T_0=$ temperature of surrounding air (assumed constant)
> $y=T-T_0$
> $h=$ heat-transfer coefficient from outside surface of fin to the surrounding air
> $k=$ thermal conductivity of fin material
> $L=$ length of fin
> $w=$ thickness of fin at its base
> $\theta=$ one-half the wedge angle
> $$A=\frac{2h\sec(L\theta)}{kw}$$

(A derivation of this equation may be found on page 117 of *Applied Mathematics in Chemical Engineering* [McGraw-Hill, 1957] by H. S. Mickley, T. S. Sherwood, and C. E. Reed.) Assuming that the temperature is finite at each point of the fin, find the temperature distribution T if:

$T_0=120°F$ $T(L)=190°F$
$L=2$ ft $\theta=\pi/12$ radians
$h=2$ Btu/(hr)(ft)^2(°F) $w=1$ in $=\frac{1}{12}$ ft
$k=200$ Btu/(hr)(ft)(°F)

7. Suppose that an object is repelled from the origin with a force per unit mass that is numerically equal to $\frac{1}{4}$ the object's distance x, measured in meters, from the origin. Moreover, suppose that the object gains mass so that its mass at time t, measured in seconds, is $3+t$ grams.

(a) Use Newton's second law of motion (force is equal to the rate of change of momentum) to find a differential equation for the object's distance from the origin.

(b) Find a general solution for the equation found in part (a). Hint: Set $s = 3 + t$.

(c) Assuming that the object is one meter from the origin and has a velocity of $\frac{1}{4}$ meter/sec when $t = 1$, find the object's distance from the origin at time t.

8. Consider a horizontal beam of length $2L$ supported at both ends and laterally fixed at its left end. The beam carries a uniform load w per unit length and is subjected to a constant "end-pull" P from the right. (See Figure 6.12.) Suppose that the moment of inertia of a cross section of the

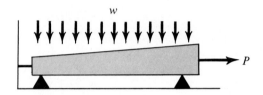

FIGURE 6.12

beam at a distance s from its left end is $I = 2(s + 1)$. For convenience we take the origin to be the middle of the bar. The equation for the vertical deflection $y(x)$ at x is

$$2E(x+1+L)\frac{d^2y}{dx^2} - Py = \frac{1}{2}w(x+L)^2 - w(x+L)$$

where E is Young's modulus. Find a general solution. A derivation of this equation and a complete discussion of this problem may be found in Section 7.12 of *Applied Bessel Functions* (Blackie & Son Limited, London and Glasgow, 1946) by F. E. Relton. Hint: Find a particular solution of the form $A(x+L)^2 + B(x+L) + C$.

9. Consider a thin tapered rod of constant density ρ lying along the x-axis with cross-sectional area Ax^2 and moment of inertia Bx for some constants A and B. Suppose that the rod is undergoing oscillatory motion. The displacement $y(x, t)$ of a point x at time t satisfies the partial differential equation

$$\rho A x^2 \frac{\partial^2 y}{\partial t^2} = -\frac{\partial^2}{\partial x^2}\left\{ EBx^4 \frac{\partial^2 y}{\partial x^2} \right\} \tag{6.47}$$

where E is Young's modulus. Suppose that this equation has solutions of the form $y(x, t) = X(x) \sin wt$.

(a) Insert this form for y into equation (6.47) and find an ordinary differential equation for X. For convenience set $k^4 = A\rho w^2 / BE$.

(b) Show that the differential equation found in part (a) can be written as either (where $D=d/dx$)

$$(xD^2+3D+k^2)(xD^2+3D-k^2)X=0$$

or

$$(xD^2+3D-k^2)(xD^2+3D+k^2)X=0$$

(c) Use part (b) to show that if X_1 and X_2 are solutions of

$$(xD^2+3D-k^2)Y=0$$

and

$$(xD^2+3D+k^2)Y=0$$

respectively, then X_1 and X_2 are solutions of the differential equation found in part (a).

(d) Find a general solution of the equation found in part (a).

7 Boundary-Value Problems

7.1 Fourier Sine and Cosine Series

Consider a hot metal rod of length L that is insulated on its sides and cooled at its ends as indicated in Figure 7.1, which shows a lengthwise section of the arrangement.

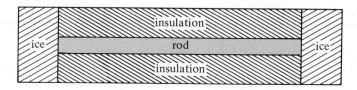

FIGURE 7.1

Around 1800 the nature of heat was a topic of much study. To aid in this study, exacting experiments were performed on heat flow in a cooling rod. The results were puzzling. If the rod initially had hot and cold patches determined by a sine wave (the temperature at time $t=0$ at a distance x from one end of the rod is $\sin(n\pi x/L)$), then the temperature throughout the entire rod reached equilibrium uniformly with time. Figure 7.2 shows the initial temperature distribution in the rod, T vs. x, as well as temperature distributions at

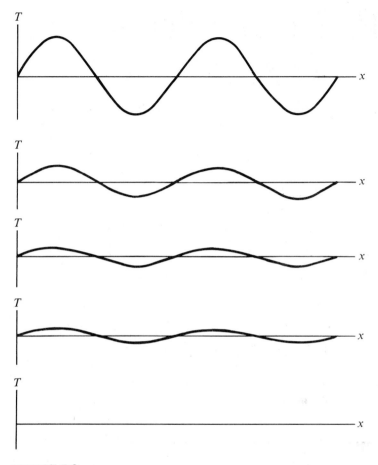

FIGURE 7.2

progressively later times. However, if the hot and cold patches were not uniformly distributed, they tended to move around as the rod cooled. Some portions of the rod actually became hotter. Figure 7.3 shows the initial temperature distribution and temperature distributions at progressively later times for such a case.

If a heated rod is insulated on its sides and cooled on its ends as shown in Figure 7.1, then the temperature at any point in the rod depends only on the time t and the distance x of the point from one end of the rod. Using Newton's law of cooling (the rate of change of temperature with respect to time is proportional to the difference of temperatures in the object and the surrounding material), it is possible to show that the temperature $T(x, t)$ at a point a distance x from one end of the rod at time t satisfies the partial differential

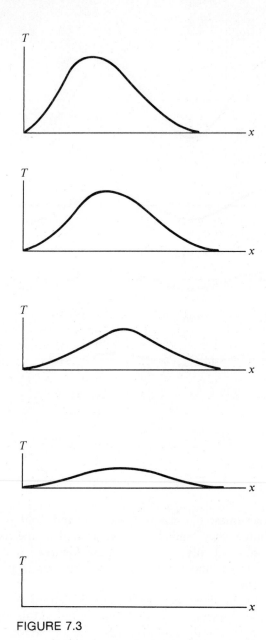

FIGURE 7.3

equation

$$\frac{\partial T}{\partial t} = c^2 \frac{\partial^2 T}{\partial x^2} \tag{7.1}$$

We assume that the temperature distribution is known at $t=0$. That is, we assume that there is a known function $f(x)$ such that

$$T(x,0) = f(x) \tag{7.2}$$

for $0 \le x \le L$. We also assume that beginning at $t=0$ the ends of the rod are held at a temperature of zero on some unspecified scale:

$$T(0,t) = 0 \quad \text{and} \quad T(L,t) = 0 \tag{7.3}$$

for $0 < t$.

   We want to find a function that satisfies equation (7.1) along with the initial and boundary conditions in (7.2) and (7.3). One of the simplest forms such a solution could have is that of the product of a function $X(x)$ and a function $S(t)$. We will show that it is possible to choose the functions $X(x)$ and $S(t)$ so that $X(x)S(t)$ is a solution of equation (7.1) that satisfies the boundary condition in (7.3). We then show that the temperature distribution $T(x,t)$ can be written as the sum of such solutions.

   If the function $X(x)S(t)$ is inserted into equation (7.1), we obtain

$$X(x)\frac{dS}{dt}(t) = c^2 \frac{d^2X}{dx^2}(x)S(t)$$

Rearranging the terms of this equation yields

$$\frac{1}{S(t)}\frac{dS}{dt}(t) = c^2 \frac{1}{X(x)}\frac{d^2X}{dx^2}(x) \tag{7.4}$$

Notice that the left-hand side of (7.4) is a function of the variable t alone and hence is independent of the variable x. Likewise, the right-hand side is independent of the variable t. Therefore each side of (7.4) is independent of both variables and hence must be a constant, which we will denote by $-k$ for notational convenience.

$$\frac{1}{S(t)}\frac{dS}{dt}(t) = c^2 \frac{1}{X(x)}\frac{d^2X}{dx^2}(x) = -k$$

This leads to two differential equations:

$$\frac{dS}{dt} + kS = 0 \tag{7.5}$$

and

$$\frac{d^2 X}{dt^2} + \frac{k}{c^2} X = 0$$

From the boundary conditions in (7.3) we obtain

$$X(0)S(t) = T(0,t) = 0 \quad \text{and} \quad X(L)S(t) = T(L,t) = 0 \tag{7.6}$$

for $0 < t$. If $f(x) \neq 0$ for some x between 0 and L, then the temperature distribution $T(x,t)$ must be nonzero for some values of x and t. Therefore $S(t) \neq 0$ for some value of t. From the equations in (7.6) we conclude that $X(0) = 0$ and $X(L) = 0$. The constant k must be chosen so that

$$\frac{d^2 X}{dx^2} + \frac{k}{c^2} X = 0 \quad (X(x) \neq 0) \tag{7.7}$$

$$X(0) = 0, \quad X(L) = 0$$

There are three cases to consider: $k < 0$, $k = 0$, and $0 < k$. In Exercise 1 the reader is asked to show that if $k \leq 0$, then there is no solution to the problem in (7.7). If $0 < k$, then a general solution is

$$X(x) = c_1 \cos\left(\frac{\sqrt{k}}{c} x\right) + c_2 \sin\left(\frac{\sqrt{k}}{c} x\right)$$

The boundary conditions $X(0) = 0$, $X(L) = 0$ dictate that

$$0 = X(0) = c_1$$

$$0 = X(L) = c_2 \sin\left(\frac{\sqrt{k}}{c} L\right)$$

Since we do not want $X(x) \equiv 0$, we must have $c_2 \neq 0$ and $\sin(\sqrt{k}\,L/c) = 0$. Thus $\sqrt{k}\,L/c = n\pi$, where n is a positive integer, and we have

$$k = \frac{c^2 n^2 \pi^2}{L^2}, \quad n = 1, 2, 3, \cdots$$

For each positive integer n, the function $X_n(x) = b_n \sin(n\pi x/L)$, $b_n \neq 0$, is a solution of the problem in (7.7). Having determined appropriate values of k, we are now able to solve equation (7.5) for $S(t)$:

$$S(t)=c_n e^{-(c^2 n^2 \pi^2/L^2)t}$$

where $c_n \neq 0$. Thus

$$T_n(x,t)=a_n e^{-(c^2 n^2 \pi^2/L^2)t}\sin\left(\frac{n\pi}{L}x\right) \tag{7.8}$$

is a solution of equation (7.1) that satisfies the boundary conditions in (7.3) for every $a_n \neq 0$. This confirms the remark made earlier in this section: If a rod initially has alternating hot and cold patches determined by a sine wave, then the temperature throughout the entire rod tends uniformly to zero. Moreover, according to the solution in (7.8), this cooling occurs more rapidly as the number of hot and cold patches increases. Notice that this result is intuitive. If there are a large number of hot and cold patches, the heat need not travel far to cancel out the cold spots.

Unless $f(x)$ is a constant multiple of $\sin(n\pi x/L)$ for some positive integer n, we have not found a solution satisfying the initial condition in (7.2). We find such a solution by considering a sum of solutions of the form in (7.8). That is, we assume that there is a solution of the original problem having the form

$$T(x,t)= \sum_{n=1}^{\infty} a_n e^{-(c^2 n^2 \pi^2/L^2)t}\sin\left(\frac{n\pi}{L}x\right) \tag{7.9}$$

Evidently this choice of $T(x,t)$ satisfies the boundary conditions in (7.3). If we assume that the order of summation and differentiation can be interchanged, then this function also satisfies equation (7.1) since each summand satisfies this equation. In order for this solution to satisfy the initial condition in (7.2), we must have

$$f(x)=T(x,0)= \sum_{n=1}^{\infty} a_n \sin\left(\frac{n\pi}{L}x\right)$$

Thus the function in (7.9) is a solution to equation (7.1) and satisfies the boundary and initial conditions in (7.2) and (7.3), provided it is possible to choose the numbers a_n so that ·

$$f(x)= \sum_{n=1}^{\infty} a_n \sin\left(\frac{n\pi}{L}x\right) \tag{7.10}$$

Most functions arising in applications can be written in this form. Assuming that $f(x)$ is such a function, we will compute the coefficients a_n. If we multiply each side of equation (7.10) by $\sin(m\pi x/L)$, assume that summation and integration can be interchanged, and integrate from 0 to L, we obtain

$$\int_0^L f(x)\sin\left(\frac{m\pi}{L}x\right)dx = \sum_{n=1}^{\infty} a_n \int_0^L \sin\left(\frac{n\pi}{L}x\right)\sin\left(\frac{m\pi}{L}x\right)dx$$

Using integration by parts, it is possible to show that

$$\int_0^L \sin\left(\frac{n\pi}{L}x\right)\sin\left(\frac{m\pi}{L}x\right)dx = \begin{cases} \dfrac{L}{2} & \text{if } m=n \\ 0 & \text{if } m\neq n \end{cases}$$

Thus

$$a_m = \frac{2}{L}\int_0^L f(x)\sin\left(\frac{m\pi}{L}x\right)dx \qquad (7.11)$$

▶ **Example 1**   Suppose that $L=\pi$ and

$$f(x) = \begin{cases} x\left(\dfrac{\pi}{2}-x\right) & \text{if } 0\leqslant x\leqslant \pi/2 \\ 0 & \text{if } \pi/2 < x \end{cases}$$

Then the temperature distribution $T(x,t)$ in the rod is given by (7.9), and the coefficients a_n in that equation are given by (7.11). The graphs of the resulting temperature distributions at various times are shown in Figure 7.4.   ◀

When the coefficients a_n are chosen as in (7.11), the series in (7.10) is called the **Fourier sine series** for f on the interval $(0, L)$. The following theorem gives conditions which guarantee that a function can be represented by its Fourier sine series. Before proceeding, the reader should review the concept of a piecewise-continuous function introduced in Section 4.1.

THEOREM 7.1   Let f be a piecewise-continuous function on the interval $(0, L)$ whose derivative f' is also piecewise continuous on $(0, L)$. Then at each point x_0 of $(0, L)$ the Fourier sine series has the value $\frac{1}{2}[\lim_{x\to x_0^+} f(x) + \lim_{x\to x_0^-} f(x)]$.

Immediate consequences of Theorem 7.1 are:

1. If f is continuous at x_0, then the value of the Fourier sine series for f at x_0 is $f(x_0)$.

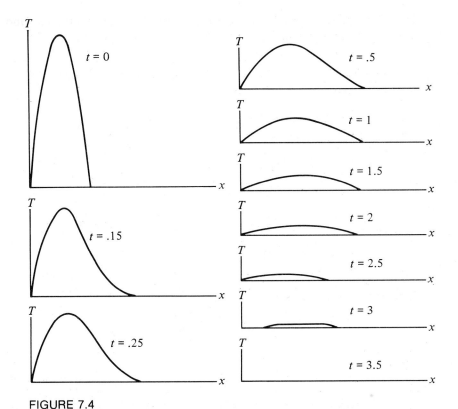

FIGURE 7.4

2. If f has a jump at x_0, then the value of the Fourier sine series for f at x_0 is the average value of the jump.

Although it is an abuse of notation, it is customary to write

$$f(x) = \sum_{n=1}^{\infty} a_n \sin\left(\frac{n\pi}{L}x\right)$$

where

$$a_n = \frac{2}{L}\int_0^L f(x)\sin\left(\frac{n\pi}{L}x\right)dx$$

with the understanding that equality holds only at points of continuity of f.

Even though the Fourier sine series coincides on the interval $(0, L)$ with a continuous function satisfying the hypotheses of Theorem 7.1, it may happen that the values of the series and of the function do not coincide at the end points of the interval. This phenomenon is illustrated in part of the following example.

▶ **Example 2**   Let $f(x)=\pi/2$ for $0\leqslant x\leqslant\pi$. The Fourier sine series for f on $(0,\pi)$ is

$$\sum_{n=1}^{\infty} a_n\sin nx \qquad\qquad (7.12)$$

where

$$a_n=\frac{2}{\pi}\int_0^\pi \frac{\pi}{2}\sin nx\,dx=-\frac{1}{n}(\cos n\pi-1)=\begin{cases}\dfrac{2}{n} & \text{if } n \text{ is odd}\\[2mm] 0 & \text{if } n \text{ is even}\end{cases}$$

Thus

$$\frac{\pi}{2}=\sum_{n=1}^{\infty} a_n\sin nx=\sum_{m=1}^{\infty}\frac{2}{2m-1}\sin\big[(2m-1)x\big]$$

for every x in the interval $(0,\pi)$. In particular, when $x=\pi/2$ we have

$$\frac{\pi}{4}=\sum_{m=1}^{\infty}\frac{1}{2m-1}\sin\Big[(2m-1)\frac{\pi}{2}\Big]$$

$$=1-\frac{1}{3}+\frac{1}{5}-\frac{1}{7}+\cdots$$

It should be noted that at $x=0$ and $x=\pi$ the value of the Fourier series is zero, while $f(0)=f(\pi)=\pi/2$. Thus the continuity of a function on the closed interval $[0, L]$ does not assure that it coincides with its Fourier sine series on that interval.

Let $f_n(x)$ denote the sum of the first n terms of the series in (7.12), so that

$$f_1(x)=2\sin x$$

$$f_3(x)=2\Big(\sin x+\frac{\sin 3x}{3}\Big)$$

· · · · · · · · · · · · · · · · ·

The graphs of f, f_1, and f_3 are shown in Figure 7.5, while the graphs of f and f_{13} appear in Figure 7.6.

For large values of n, the graph of f_n rises sharply from the origin, oscillates in a narrow strip about the line $y=\pi/2$, and then decreases sharply to the point $(\pi,0)$ on the x-axis. As n increases, the strip in which the graph of f_n oscillates becomes narrower and the portion of the graph

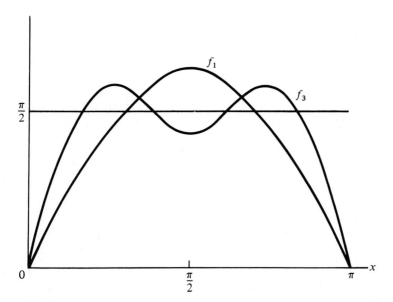

FIGURE 7.5

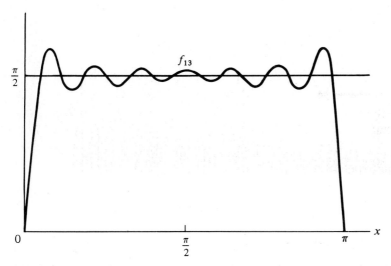

FIGURE 7.6

outside of this strip approaches the two line segments from $(0,0)$ to $(0, \pi/2)$ and from $(\pi,0)$ to $(\pi, \pi/2)$. Letting $n \to \infty$, we obtain the graph of the Fourier sine series for f on $(0, \pi)$, which coincides with the graph of f on the interval $(0, \pi)$.   ◀

In a manner similar to that for a Fourier sine series, it is possible to represent a function on an interval $(0, L)$ by a series of cosine terms:

$$f(x) = \sum_{n=0}^{\infty} b_n \cos\left(\frac{n\pi}{L}x\right) \qquad (7.13)$$

where

$$b_0 = \frac{1}{L}\int_0^L f(x)\,dx$$

$$b_n = \frac{2}{L}\int_0^L f(x)\cos\left(\frac{n\pi}{L}x\right)dx \qquad (n \neq 0)$$

With this choice of coefficients b_n, the series in (7.13) is called the **Fourier cosine series** for f on the interval $(0, L)$. As might be expected, Theorem 7.1 remains true if "Fourier sine series" is replaced by "Fourier cosine series."

▶ **Example 3**   Let $g(x) = x$ for $0 < x < 1$. The Fourier cosine series for g on $(0, 1)$ is

$$\sum_{n=0}^{\infty} b_n \cos n\pi x \qquad (7.14)$$

where

$$b_0 = \int_0^1 x\,dx = \frac{1}{2}$$

$$b_n = 2\int_0^1 x \cos n\pi x\,dx = \frac{2}{(n\pi)^2}(\cos n\pi - 1) = \begin{cases} 0 & \text{if } n \text{ is even} \\[2mm] \dfrac{-4}{(n\pi)^2} & \text{if } n \text{ is odd} \end{cases}$$

Thus

$$x = \frac{1}{2} - \frac{4}{\pi^2}\sum_{m=1}^{\infty} \frac{1}{(2m-1)^2}\cos\big[(2m-1)\pi x\big]$$

for every x in the interval $(0, 1)$.

Let $g_n(x)$ denote the sum of the first n terms of the series in (7.14), so that

$$g_1(x)=\frac{1}{2}$$

$$g_3(x)=\frac{1}{2}-\frac{4}{\pi^2}\cos\pi x$$

. . . .

The graphs of g, g_1, and g_3 are shown in Figure 7.7. The graph of g_n oscillates in a narrow strip about the graph of g. As n increases, the strip becomes narrower, so that in the limit the graph of the Fourier cosine series coincides with the graph of g on the interval $(0, 1)$. ◄

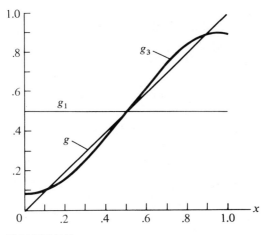

FIGURE 7.7

Exercises SECTION 7.1

1. Let $0<L$ and consider the problem of finding a solution of

$$\frac{d^2X}{dx^2}+kX=0$$

which satisfies the boundary conditions

$$X(0)=0 \qquad X(L)=0$$

(a) If $k\leqslant0$, show that $X(x)\equiv0$ is the only such solution.
(b) If $0<k$, show that there is a solution other than $X(x)\equiv0$ if and only if $k=(n\pi/L)^2$, where n is a positive integer.

2. Show that if

$$f(x) = \sum_{n=0}^{\infty} b_n \cos\left(\frac{n\pi}{L} x\right)$$

on an interval $(0, L)$, then

$$b_0 = \frac{1}{L} \int_0^L f(x)\, dx$$

$$b_n = \frac{2}{L} \int_0^L f(x) \cos\left(\frac{n\pi}{L} x\right) dx \qquad (n \neq 0)$$

You may assume that summation and integration can be interchanged.

In Exercises 3–10 find the Fourier sine series for each function on the interval $[0, 1]$. What is the value of the series when $x = 0, \frac{1}{4}, \frac{1}{2}, \frac{3}{4}, 1$?

3. x

4. 1

5. e^x

6. $1 + x$

7. $\cos \pi x$

8. $\begin{cases} 0 & \text{if } 0 < x < \frac{1}{2} \\ 1 & \text{if } \frac{1}{2} \leqslant x < 1 \end{cases}$

9. $\begin{cases} x & \text{if } 0 \leqslant x < \frac{1}{2} \\ x - 1 & \text{if } \frac{1}{2} \leqslant x \leqslant 1 \end{cases}$

10. $\begin{cases} 0 & \text{if } 0 \leqslant x \leqslant \frac{1}{3} \\ 1 & \text{if } \frac{1}{3} < x < \frac{2}{3} \\ 0 & \text{if } \frac{2}{3} \leqslant x \leqslant 1 \end{cases}$

In Exercises 11–18 find the Fourier cosine series for each of the given functions on the interval $[0, 1]$. What is the value of the series when $x = \frac{1}{4}, \frac{1}{2}, \frac{3}{4}$?

11. x

12. 1

13. e^x

14. $1 + x$

15. $\cos \pi x$

16. $\begin{cases} 0 & \text{if } 0 \leqslant x < \frac{1}{2} \\ 1 & \text{if } \frac{1}{2} \leqslant x \leqslant 1 \end{cases}$

17. $\begin{cases} x & \text{if } 0 \leqslant x < \frac{1}{2} \\ x-1 & \text{if } \frac{1}{2} \leqslant x \leqslant 1 \end{cases}$

18. $\begin{cases} 0 & \text{if } 0 \leqslant x \leqslant \frac{1}{3} \\ 1 & \text{if } \frac{1}{3} < x < \frac{2}{3} \\ 0 & \text{if } \frac{2}{3} \leqslant x \leqslant 1 \end{cases}$

19. Sketch the graph of the sum of the first two terms of the Fourier sine series for $f(x) = e^x$ computed in Exercise 5.

20. Sketch the graph of the sum of the first two terms of the Fourier cosine series for $f(x) = e^x$ computed in Exercise 13.

21. A cycloid is the path of a point P on a circle as the circle rolls without slipping along the x-axis. If C denotes the center of the circle, Q the point of contact between the circle and x-axis, and θ the angle $\angle PCQ$, then the cycloid can be described by the parametric equations (where a is the radius of the circle)

$$x = a(\theta - \sin\theta) \qquad y = a(1 - \cos\theta)$$

for $0 \leqslant \theta \leqslant 2\pi$. See Figure 7.8.

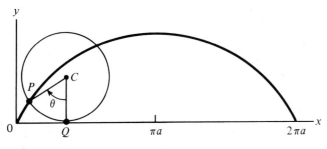

FIGURE 7.8

For simplicity we will take the radius of the circle to be one, i.e., $a = 1$.

We assume that for each positive integer n the Bessel function J_n can be written in the form

$$J_n(x) = \frac{1}{\pi} \int_0^\pi \cos(n\theta - x\sin\theta)\, d\theta \qquad (7.15)$$

(See Exercise 6 of Section 6.5.) Using this identity, we will determine y as a function of x. Let b_n denote the coefficients of the Fourier cosine series for y (with respect to x) on $(0, \pi)$.

(a) Show that

$$b_0 = \frac{1}{\pi} \int_0^\pi y\, dx = \frac{1}{\pi} \int_0^\pi (1 - \cos\theta)^2\, d\theta = \frac{3}{2}$$

(b) Show that for $n \neq 0$

$$b_n = \frac{2}{\pi} \int_0^\pi y \cos nx\, dx$$

$$= \frac{2}{\pi} \int_0^\pi \cos(n\theta - n\sin\theta)(1 - \cos\theta)^2\, d\theta$$

$$= \frac{2}{\pi} \int_0^\pi \cos(n\theta - n\sin\theta)(2 - 2\cos\theta - \sin^2\theta)\, d\theta$$

(c) Show that

$$2 \int_0^\pi \cos(n\theta - n\sin\theta)\cos\theta\, d\theta$$

$$= \int_0^\pi \cos[(n+1)\theta - n\sin\theta]\, d\theta + \int_0^\pi \cos[(n-1)\theta - n\sin\theta]$$

$$= \pi J_{n+1}(n) + \pi J_{n-1}(n)$$

$$= 2\pi J_n(n)$$

(See Exercise 4 of Section 6.5.)

(d) Use the representation for $J_n(x)$ in (7.14) to show that

$$\pi J_n''(x) = -\int_0^\pi \cos(n\theta - x\sin\theta)\sin^2\theta\, d\theta$$

(e) Show that $b_n = 2J''(n)$ for $n \neq 0$.

(f) Use Bessel's equation to show that $J_n''(n) = -\frac{1}{n} J'(n)$.

(g) Show that

$$y = \frac{3}{2} - 2 \sum_{n=1}^\infty n^{-1} J_n'(n) \cos nx$$

on the interval $(0, \pi)$.

(h) Noting that the cycloid is symmetric about the line $x=\pi$, find y as a function of x on the interval $(0, 2\pi)$.

22. In applications (for example, see Exercise 8 in the next section) it is sometimes desirable to represent a function f on an interval $(-L, L)$ instead of on the interval $(0, L)$. In many such cases it is advantageous to represent f as

$$f(x)=a_0+ \sum_{n=1}^{\infty} a_n\cos\left(\frac{n\pi}{L}x\right)+b_n\sin\left(\frac{n\pi}{L}x\right) \qquad \text{(7.16)}$$

Assuming that the order of summation and integration can be interchanged, determine the coefficients a_0, a_n, and b_n. The series in (7.16) is called the **Fourier series** for the function f on the interval $(-L, L)$.

7.2 Separation of Variables

In the previous section we found the solution of the partial differential equation

$$\frac{\partial T}{\partial t}=c^2\frac{\partial^2 T}{\partial t^2}$$

for $0<x<L$ and $0<t$ that satisfies the initial condition

$$T(x,0)=f(x)$$

and the boundary conditions

$$T(0, t)=0 \qquad T(L, t)=0$$

The technique we used to obtain this solution is a particular application of a method called the **method of separation of variables**.

For simplicity of exposition we will assume that the desired solution^T of a partial differential equation is a function of two independent variables x and y. The method of separation of variables may be summarized as follows:

1. Assume that the partial differential equation has solutions of the form $X(x)Y(y)$.

2. Insert this form for a solution into the partial differential equation, rearrange terms, and obtain two ordinary differential equations containing a parameter k for X and Y.

3. Use boundary conditions given for T to obtain boundary conditions for either X or Y. For the purposes of this discussion we assume that boundary conditions for X are determined. The partial differential equations we will consider lead to the differential equation

$$\frac{d^2 X}{dx^2} + kX = 0$$

and boundary conditions of one of the following forms:

(i) $X(0)=0$, $X(L)=0$ (ii) $X'(0)=0$, $X'(L)=0$
(iii) $X'(0)=0$, $X(L)=0$ (iv) $X(0)=0$, $X'(L)=0$

for some $0 < L$.

4. Determine values k_n, $n = 1, 2, \cdots$, for k that allow the ordinary differential equation for X to have solutions X_n other than $X(x) \equiv 0$ that satisfy the boundary conditions.

5. The boundary or initial conditions for T may determine at least one initial condition for Y of the form $Y(0)=0$ or $Y'(0)=0$. Using the values of k found above, find solutions Y_n of the ordinary differential equation for Y that satisfy any initial conditions.

6. Form a series solution

$$T(x, y) = \sum_{n=1}^{\infty} a_n X_n(x) Y_n(y)$$

Assuming that the order of differentiation and summation can be interchanged, this function is a solution of the partial differential equation that satisfies all but one of the boundary and initial conditions.

7. The remaining boundary or initial condition allows the coefficients a_n to be determined as the Fourier sine or cosine coefficients of a given function of y.

We will now illustrate the method of undetermined coefficients with two examples.

▶ **Example 1**   Consider a vibrating string of constant linear density ρ stretched between the points $x=0$ and $x=L$ on the x-axis. The transverse displacement $u(x, t)$ at time t of a point x satisfies the partial differential equation

$$\frac{\partial^2 u}{\partial t^2} = \frac{T}{\rho} \frac{\partial^2 u}{\partial x^2} \tag{7.17}$$

where T is the tension in the string. Since the ends of the string remain fixed, we have the boundary conditions

$$u(0, t) = 0 \qquad u(L, t) = 0 \tag{7.18}$$

for all $0 \leq t$. Recalling that velocity is the rate of change of position with respect to time, we note that $\partial u / \partial t(x, 0)$ is the initial velocity of the point x on the string. If the string is set in motion by displacing it from its equilibrium position along the x-axis, holding it steady, and then releasing it, we obtain the following initial conditions:

$$u(x, 0) = f(x) \qquad \frac{\partial u}{\partial t}(x, 0) = 0 \tag{7.19}$$

for $0 \leq x \leq L$ and for some function f.

We will use the method of separation of variables to find a solution of equation (7.17) that satisfies the boundary and initial conditions in (7.18) and (7.19). We begin by finding solutions of equation (7.17) having the form $X(x)S(t)$. If such a function is inserted into equation (7.17) and the terms rearranged, we obtain

$$\frac{1}{X} \frac{d^2 X}{dx^2} = \frac{1}{c^2 S} \frac{d^2 S}{dt^2} \tag{7.20}$$

where we have set $c = \sqrt{T/\rho}$ for algebraic convenience. The left side of equation (7.20) is independent of the variable t, while the right side is independent of the variable x. Therefore there is a constant k such that

$$\frac{1}{X} \frac{d^2 X}{dx^2} = \frac{1}{c^2 S} \frac{d^2 S}{dt^2} = -k$$

or, equivalently,

$$\frac{d^2 X}{dx^2} + kX = 0 \tag{7.21}$$

$$\frac{d^2 S}{dt^2} + kc^2 S = 0 \tag{7.22}$$

From the boundary conditions in (7.18) we have

$$X(0)S(t) = 0 \qquad X(L)S(t) = 0$$

for all $0 \leq t$. Since we want a solution other than $X(x)S(t) \equiv 0$, we must have

$$X(0)=0 \qquad X(L)=0 \tag{7.23}$$

From the second initial condition in (7.19) we have

$$X(x)\frac{dS}{dt}(0)=0$$

The requirement that $X(x)S(t)\neq0$ dictates that we have

$$\frac{dS}{dt}(0)=0 \tag{7.24}$$

In Exercise 1 of Section 7.1 we found that equation (7.21) has solutions that satisfy the boundary conditions in (7.23) if and only if $k=(n\pi/L)^2$ for some positive integer n. In such cases solutions are constant multiples of

$$X_n(x)=\sin\left(\frac{n\pi}{L}x\right)$$

With $k=(n\pi/L)^2$, the solutions of equation (7.22) that satisfy the initial condition in (7.24) are constant multiples of

$$S_n(t)=\cos\left(\frac{n\pi c}{L}t\right)$$

Thus, for each positive integer n, the function

$$u_n(x,t)=\sin\left(\frac{n\pi}{L}x\right)\cos\left(\frac{n\pi c}{L}t\right)$$

called the **nth normal mode of vibration**, is a solution of equation (7.17) that satisfies the boundary conditions in (7.18) and the initial condition $\partial u(x,0)/\partial t=0$. The **fundamental mode**, which determines the pitch of the note sounded by the vibrating string, corresponds to $n=1$. The frequency F of the fundamental mode is given by

$$F=\frac{c}{2L}=\frac{1}{2L}\sqrt{\frac{T}{\rho}}$$

Notice that the frequency varies directly as the square root of the tension and inversely as the length and the square root of the linear density of the string. In 1636, long before the development of the mathematical theory, the French philosopher and mathematician Marin Mersenne formulated these observations (now called Mersenne's laws) based upon experimental evidence.

A piano illustrates the use of Mersenne's laws. The 7.25 octave range of a modern piano contains notes whose frequencies range from 27 to 4096 cycles per second (Hz). If each string had the same linear density and the same tension, the longest string would be approximately 152 times as long as the shortest. To avoid this great discrepancy in length, the linear densities and tensions of various strings are modified. Unduly long bass strings are avoided by increasing their linear densities, usually by wrapping them with soft wire. Very short treble strings are avoided by using longer strings with increased tensions.

We also notice that the frequency, $(n/2L)\sqrt{T/\rho}$, of the nth normal mode is n times the frequency of the fundamental mode. When the frequency of a vibration is increased n-fold, the same note is sounded except that it is raised by n harmonics. Thus the notes sounded by the normal modes are identical except for their pitch. If the note sounded by the vibrations corresponding to the fundamental mode is C, then the following normal modes sound the notes C, G, C, E, B^b, C, ...

In order to find a solution that satisfies the remaining initial condition, $u(x,0)=f(x)$, we set

$$u(x,t)= \sum_{n=1}^{\infty} a_n \sin\left(\frac{n\pi}{L}x\right)\cos\left(\frac{n\pi c}{L}t\right) \tag{7.25}$$

Assuming that summation and integration can be interchanged, this function is a solution of equation (7.17) that satisfies the boundary conditions in (7.18) and the right-hand initial condition in (7.19). We now choose the coefficients a_n so that the remaining condition is satisfied. That is, so that

$$f(x)=u(x,0)= \sum_{n=1}^{\infty} a_n \sin\left(\frac{n\pi}{L}x\right)$$

Thus the a_n are the coefficients of the Fourier sine series for the function f on the interval $(0, L)$:

$$a_n = \frac{2}{L}\int_0^L f(x)\sin\left(\frac{n\pi}{L}x\right)dx$$

With this choice of coefficients, the function in (7.25) satisfies the partial differential equation in (7.17) along with the boundary and initial conditions in (7.18) and (7.19).

We have shown that the vibration of the string is the "sum" of vibrations having frequencies that are integer multiples of the frequency of the fundamental mode. That is, the sound of the vibrating string is the "sum" of the notes sounded by the fundamental mode and notes that are an integral number of harmonics higher in pitch. This accounts for the fuller, brighter, and richer sound of a plucked harp string than the sound of a tuning fork vibrating only at the frequency of the fundamental mode of the string.   ◄

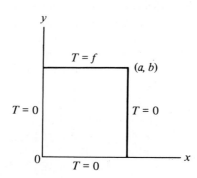

FIGURE 7.9

► **Example 2**   Consider a homogeneous rectangular plate with insulated faces, as shown in Figure 7.9. The edges $x=0, x=a$, and $y=0$ are maintained at zero degrees, while the fourth edge $y=b$ is maintained at a temperature distribution $f(x)$ that does not change with time. After a sufficiently long period of time, the temperature at any point in the plate no longer varies with time; that is, a steady state has been attained. The steady-state temperature $T(x, y)$ at a point (x, y) in the plate can be shown to satisfy the partial differential equation

$$\frac{\partial^2 T}{\partial x^2} + \frac{\partial^2 T}{\partial y^2} = 0 \tag{7.26}$$

for $0<x<a$ and $0<y<b$. The prescribed temperatures along the edges of the plate dictate that T also satisfy the boundary conditions

$$T(0, y)=0 \qquad T(a, y)=0 \qquad T(x,0)=0 \tag{7.27}$$

$$T(x, b)=f(x) \tag{7.28}$$

for $0<x<a$ and $0<y<b$.

We begin by finding solutions of equation (7.26) having the form $X(x)Y(y)$ and satisfying the boundary conditions in (7.27). If we insert such a function into equation (7.26) and rearrange terms, we obtain

$$-\frac{1}{X}\frac{d^2X}{dx^2} = \frac{1}{Y}\frac{d^2Y}{dy^2}.$$

The left side of this equation is independent of the variable y, while the right side is independent of the variable x. Therefore each side must be a constant k. Thus we are led to consider the pair of ordinary differential equations

$$\frac{d^2X}{dx^2} + kX = 0 \qquad\qquad\qquad \textbf{(7.29)}$$

$$\frac{d^2Y}{dy^2} - kY = 0 \qquad\qquad\qquad \textbf{(7.30)}$$

From the boundary conditions in (7.27) we have

$$X(0)Y(y) = 0 \qquad X(a)Y(y) = 0 \qquad X(x)Y(0) = 0$$

for all $0 < x < a$ and $0 < y < b$. Since we want a solution other than $X(x)Y(y) \equiv 0$, we must have

$$X(0) = 0 \qquad X(a) = 0 \qquad\qquad \textbf{(7.31)}$$

$$Y(0) = 0 \qquad\qquad\qquad\qquad\qquad \textbf{(7.32)}$$

In Exercise 1 of Section 7.1 we found that equation (7.29) has a solution other than $X(x) \equiv 0$ that satisfies the boundary conditions in (7.31) if and only if $k = (n\pi/a)^2$, where n is a positive integer. When k assumes such a value, the corresponding solutions of equation (7.29) are constant multiples of

$$X_n(x) = \sin\left(\frac{n\pi}{a}x\right).$$

With $k = (n\pi/a)^2$, a general solution of equation (7.30) is

$$Y(y) = c_1 e^{(n\pi/a)y} + c_2 e^{-(n\pi/a)y}.$$

We now choose the coefficients c_1 and c_2 so that the initial condition in (7.32) is satisfied. A short calculation shows that we must have $c_2 = -c_1$,

so that

$$Y(y)=2c_1\frac{e^{(n\pi/a)y}-e^{-(n\pi/a)y}}{2}$$

$$=2c_1\sinh\left(\frac{n\pi}{a}y\right)$$

Thus solutions of equation (7.30) that satisfy the initial condition in (7.32) are constant multiples of

$$Y_n(y)=\sinh\left(\frac{n\pi}{a}y\right)$$

We now set

$$T(x,y)=\sum_{n=1}^{\infty}a_n\sin\left(\frac{n\pi}{a}x\right)\sinh\left(\frac{n\pi}{a}y\right)\tag{7.33}$$

Assuming that summation and differentiation can be interchanged, this function is a solution of equation (7.26) that satisfies the boundary conditions in (7.27). We now choose the coefficients a_n so that the remaining boundary condition is satisfied. That is, so that

$$f(x)=T(x,b)=\sum_{n=1}^{\infty}a_n\sin\left(\frac{n\pi}{a}x\right)\sinh\left(\frac{n\pi}{a}b\right)$$

We observe that $a_n\sinh(n\pi b/a)$ must be the nth coefficient of the Fourier sine series for f on the interval $(0,a)$. Thus

$$a_n=\left(\sinh\left(\frac{n\pi}{a}b\right)\right)^{-1}\frac{2}{a}\int_0^a f(x)\sin\left(\frac{n\pi}{a}x\right)dx$$

With this choice of coefficients, the function in (7.33) satisfies the partial differential equation in (7.26) along with the boundary conditions in (7.27) and (7.28).

For the special case $a=b=\pi$ and $f(x)=\sin 3x$, a short calculation shows that

$$a_n=\begin{cases}0 & \text{if } n\neq 3\\ (\sinh 3\pi)^{-1} & \text{if } n=3\end{cases}$$

so that

$$T(x,y)=\sin 3x\sinh 3y(\sinh 3\pi)^{-1}$$

The graph of the temperature distribution T for this case is given in Figure 7.10. ◀

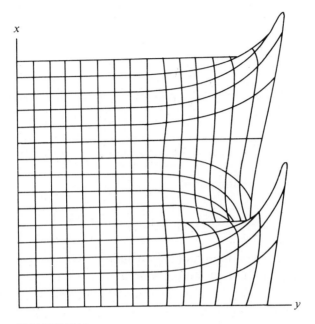

x

y

FIGURE 7.10

Exercises   SECTION 7.2

1. Consider the vibrating string in Example 1. Show that

$$u(x,t)=\tfrac{1}{2}f(x+ct)+\tfrac{1}{2}f(x-ct)$$

Hint: $\sin(n\pi x/L)\cos(n\pi t/L)=\tfrac{1}{2}\sin[n\pi(x+ct)/L]+\tfrac{1}{2}\sin[n\pi(x-ct)/L]$

2. In Example 1 suppose that $L=1$ and that $f(x)=\sin 2\pi x$. Find the resulting displacement $u(x,t)$. Sketch the graph of $u(x,t)$ for $t=0$ and $t=1/kc$, where $k=12,8,6,4,3,8/3,12/5,2$.

3. In Example 1 suppose that the string is at its equilibrium position along the x-axis and that the string is struck so as to give it an initial velocity according to

$$\frac{\partial u}{\partial t}(x,0)=g(x)$$

Find the resulting displacement.

4. Show that the solution of the problem

$$\frac{\partial^2 u}{\partial t^2} = c^2 \frac{\partial^2 u}{\partial x^2}$$

$$u(x,0) = f(x) \qquad \frac{\partial u}{\partial t}(x,0) = g(x)$$

is the sum of the solutions to the problems in Example 1 and Exercise 3.

5. Suppose that the plate in Example 2 has the edges $x=0$ and $x=a$ insulated so that there is no heat flow across these edges. Further suppose that the edge $y=0$ is maintained at zero degrees and that the fourth edge $y=b$ is maintained at a temperature distribution $f(x)$. Then the boundary conditions in (7.27) become altered to

$$\frac{\partial T}{\partial x}(0, y) = 0 \qquad \frac{\partial T}{\partial x}(a, y) = 0 \qquad T(x,0) = 0$$

while the boundary condition in (7.28) remains unchanged. Find the resulting temperature distribution in the plate.

6. Consider the metal rod described at the beginning of Section 7.1. Instead of holding both ends at a temperature of zero, suppose that the end $x=0$ is held at a temperature of zero, but the end $x=L$ is insulated so that there is no heat flow across this end. The latter condition leads to the boundary condition

$$\frac{\partial T}{\partial x}(L, t) = 0$$

Find the resulting temperature distribution in the rod.

7. The telegrapher's equation for electrical impulses in a long wire is

$$\frac{\partial^2 u}{\partial t^2} + 2a\frac{\partial u}{\partial t} + bu = c^2\frac{\partial^2 u}{\partial x^2}$$

For the special case $b=a^2$, find the solution of this equation that satisfies the boundary conditions

$$u(0, t) = 0 \qquad u(L, t) = 0$$

and the initial conditions

$$u(x,0) = f(x) \qquad \frac{\partial u}{\partial t}(x,0) = 0$$

(a) Show that the change of variable $u(x, t) = e^{-at}v(x, t)$ transforms the given differential equation into

$$\frac{\partial^2 v}{\partial t^2} = c^2 \frac{\partial^2 y}{\partial x^2}$$

(b) Show that the change of variable in (a) leads to the boundary conditions

$$v(0, t) = 0 \qquad v(L, t) = 0$$

and the initial conditions

$$v(x, 0) = f(x) \qquad \frac{\partial v}{\partial t}(x, 0) = af(x)$$

(c) Show that a solution of

$$\frac{\partial^2 v}{\partial t^2} = c^2 \frac{\partial^2 v}{\partial x^2}$$

satisfying

$$v(0, t) = 0 \qquad v(L, t) = 0 \qquad v(x, 0) = f(x) \qquad \frac{\partial v}{\partial t}(x, 0) = af(x)$$

is the sum of solutions v_1 and v_2 satisfying

$$v(0, t) = 0 \qquad v(L, t) = 0 \qquad v(x, 0) = f(x) \qquad \frac{\partial v}{\partial t}(x, 0) = 0$$

and

$$v(0, t) = 0 \qquad v(L, t) = 0 \qquad v(x, 0) = 0 \qquad \frac{\partial v}{\partial t}(x, 0) = af(x)$$

(d) The solution v_1 was found in Example 1. Use the method of separation of variables to find v_2.

(e) Find a solution u of the telegrapher's equation that satisfies the given boundary and initial conditions.

8. Consider a circular plate of radius a whose center is at the origin of a plane described by polar coordinates. Suppose that the faces of the plate are insulated and that the rim $r = a$ of the plate is maintained at a temperature distribution $f(\theta)$ that does not vary with time. The steady-state temperature

$T(r, \theta)$ at a point (r, θ) in the plate satisfies the equation

$$\frac{\partial^2 T}{\partial r^2} + \frac{1}{r}\frac{\partial T}{\partial r} + \frac{1}{r^2}\frac{\partial^2 T}{\partial \theta^2} = 0 \tag{7.34}$$

and the boundary condition

$$T(a, \theta) = f(\theta) \tag{7.35}$$

for $-\pi < \theta \leqslant \pi$.

(a) Show that a function $T(r, \theta) = R(r)S(\theta)$ is a solution of equation (7.34) whenever there is a constant k such that

$$\frac{d^2 S}{d\theta^2} + k\theta = 0 \tag{7.36}$$

$$r^2 \frac{d^2 R}{dr^2} + r\frac{dR}{dr} - kR = 0 \tag{7.37}$$

(b) Notice that $T(r, \theta)$ must be periodic in θ with period 2π. Use equation (7.36) to argue that k must be the square of a nonnegative integer.

(c) With $k = n^2$, $n = 0, 1, \ldots$, find solutions of equations (7.36) and (7.37). Equation (7.37) is a Cauchy-Euler equation and has solutions of the form $R(r) = r^a$ for appropriate choices of the constant a.

(d) Assuming that the temperature is finite at the center of the disk, show that for each nonnegative integer n, the function

$$r^n(a_n \cos n\theta + b_n \sin n\theta)$$

is a solution of equation (7.34) for any choice of the constants a_n and b_n.

(e) Use Exercise 22 of Section 7.1 to determine the coefficients a_n and b_n so that the function

$$u(r, \theta) = a_0 + \sum_{n=1}^{\infty} r^n(a_n \cos n\theta + b_n \sin n\theta)$$

is a solution of equation (7.34) that satisfies the boundary condition in (7.35).

APPENDIX 1

Complex Numbers

A **complex number** is an expression of the form $a+bi$, where a and b are real numbers and i is a symbol such that $i^2 = -1$. The symbol i may be thought of as representing $\sqrt{-1}$. A complex number $a+bi$ is called **real** if $b=0$ and **imaginary** if $a=0$. The number a is called the **real part** of the complex number $a+bi$ and is denoted by $\mathrm{Re}(a+bi)$. The number b is called the **imaginary part** of the complex number $a+bi$ and is denoted by $\mathrm{Im}(a+bi)$. Hence

$$\mathrm{Re}(a+bi)=a \qquad \mathrm{Im}(a+bi)=b$$

Arithmetic operations are defined for complex numbers as follows:

$$(a+bi)+(c+di)=(a+c)+(b+d)i$$

$$(a+bi)-(c+di)=(a-c)+(b-d)i$$

$$(a+bi)(c+di)=(ac-bd)+(ad+bc)i$$

$$\frac{a+bi}{c+di} = \frac{ac+bd}{c^2+d^2} + \frac{bc-ad}{c^2+d^2}i \quad \text{if } c^2+d^2 \neq 0$$

The rule for multiplication can be derived as follows. Recalling that $i^2 = -1$, we have

$$(a+bi)(c+di)=ac+adi+cbi+bdi^2$$

$$=(ac-bd)+(ad+cb)i$$

The rule for division can be derived in similar fashion. First note that $(c+di)(c-di)=c^2+d^2$. Then, formally,

$$\frac{a+bi}{c+di} = \frac{(a+bi)(c-di)}{(c+di)(c-di)}$$

$$= \frac{(ac+bd)+(bc-ad)i}{c^2+d^2}$$

$$= \frac{ac+bd}{c^2+d^2} + \frac{bc-ad}{c^2+d^2}i$$

▶ **Example 1**

(a) $(2+3i)+(4-7i)=(2+4)+(3-7)i=6-4i$

(b) $(2+3i)(4-7i)=[2(4)-3(-7)]+[3(4)+2(-7)]i=29-2i$

(c) $\dfrac{2+3i}{4-7i} = \dfrac{2(4)+3(-7)}{4^2+(-7)^2} + \dfrac{3(4)-(2)(-7)}{4^2+(-7)^2}i$

$$= -\frac{13}{65} + \frac{26}{65}i$$

$$= -\frac{1}{5} + \frac{2}{5}i$$ ◀

It is possible to represent complex numbers as points in a plane. The vertical axis is called the **imaginary axis** and the horizontal axis is called the **real axis**, as in Figure 1.

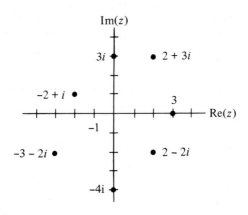

FIGURE 1

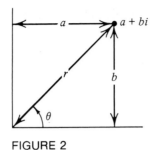

FIGURE 2

Let $z=a+bi$ be any complex number, which for the benefit of Figure 2 has been chosen with $a>0$ and $b>0$. In terms of r and θ we have

$$a=r\cos\theta \qquad b=r\sin\theta$$

so that

$$z=r(\cos\theta+i\sin\theta)$$

This is called the **polar form** of the complex number z. Evidently there are infinitely many choices for the angle θ, while there is only one choice for r, namely

$$r=\sqrt{a^2+b^2}$$

The number r is called the **absolute value** of z and is denoted by $|z|$,

$$|z|=|a+bi|=\sqrt{a^2+b^2}$$

To each complex number $z=a+bi$ there is associated a complex number called the **conjugate** of z, denoted by $\bar{z}$, which is given by

$$\bar{z}=a-bi$$

If z_1 and z_2 are any two complex numbers, then (Exercise 16)

$$\text{(i)} \quad \overline{z_1+z_2}=\bar{z}_1+\bar{z}_2$$

$$\text{(ii)} \quad \overline{z_1z_2}=\bar{z}_1\bar{z}_2$$

Repeated uses of these identities yields

$$\text{(iii)} \quad \overline{z_1+z_2+\cdots+z_n}=\bar{z}_1+\bar{z}_2+\cdots+\bar{z}_n$$

$$\text{(iv)} \quad \overline{z_1 z_2 \cdots z_n} = \overline{z}_1 \overline{z}_2 \cdots \overline{z}_n$$

for any complex numbers $z_1, z_2, \ldots, z_n$. A complex number z is real if and only if $z = \bar{z}$ (Exercise 17).

THEOREM 1 Let $p(x)$ be a polynomial with real coefficients. If z is a complex number such that $p(z) = 0$, then $p(\bar{z}) = 0$.

Proof Let $p(x) = a_n x^n + a_{n-1} x^{n-1} + \cdots + a_1 x + a_0$, with $a_n, a_{n-1}, \ldots, a_0$ real numbers. Then

$$0 = \bar{0} = \overline{p(z)}$$

$$= \overline{a_n z^n + a_{n-1} z^{n-1} + \cdots + a_1 z + a_0}$$

$$= \overline{a_n}\, \bar{z}^n + \overline{a}_{n-1} \bar{z}^{n-1} + \cdots + \overline{a}_1 \bar{z} + \overline{a}_0$$

$$= a_n \bar{z}^n + a_{n-1} \bar{z}^{n-1} + \cdots + a_1 \bar{z} + a_0$$

$$= p(\bar{z}) \quad \blacksquare$$

One of the most useful identities in complex analysis is **Euler's formula**:

$$e^{i\theta} = \cos\theta + i\sin\theta$$

where θ is any real number. This formula may be established as follows:

$$e^{i\theta} = \sum_{n=0}^{\infty} \frac{(i\theta)^n}{n!}$$

$$= \sum_{n=0}^{\infty} i^n \frac{\theta^n}{n!}$$

$$= \sum_{n \text{ even}} i^n \frac{\theta^n}{n!} + \sum_{n \text{ odd}} i^n \frac{\theta^n}{n!}$$

$$= \sum_{m=0}^{\infty} i^{2m} \frac{\theta^{2m}}{(2m)!} + \sum_{m=0}^{\infty} i^{2m+1} \frac{\theta^{2m+1}}{(2m+1)!}$$

$$= \sum_{m=0}^{\infty} \frac{(-1)^m \theta^{2m}}{(2m)!} + i \sum_{m=0}^{\infty} \frac{(-1)^m \theta^{2m+1}}{(2m+1)!}$$

$$= \cos\theta + i\sin\theta$$

Using Euler's formula, we can write a complex number $z=a+bi$ in polar form as

$$z=re^{i\theta}$$

▶ **Example 2**   If $z=1+i$, then $\theta=\pi/4$ and $r=\sqrt{2}$. Hence

$$1+i=\sqrt{2}\,e^{i\pi/4}$$ ◀

Exercises   APPENDIX 1

In Exercises 1–4 solve the given equation for x and y.

1. $2+3i=x+yi$ **3.** $2+3i=(x+y)+(x-y)i$

2. $3-4i=x+yi$ **4.** $2i=(x+yi)^2$

In Exercises 5–14 write the given complex number in the form $a+bi$.

5. $(2+3i)+(6+7i)$ **10.** $\dfrac{1+2i}{2-i}$

6. $(5-4i)-(3-2i)$ **11.** $6\left(\cos\dfrac{\pi}{4}+i\sin\dfrac{\pi}{4}\right)$

7. $(1+i)(3-2i)$ **12.** $3\left(\cos\dfrac{\pi}{3}+i\sin\dfrac{\pi}{3}\right)$

8. $(2-i)(1+2i)$ **13.** $4e^{i\pi/6}$

9. $\dfrac{2-i}{1+2i}$ **14.** $4e^{i\pi/2}$

15. Show that $|z|^2=z\bar{z}$ for any complex number z.

16. Show that $\overline{z_1+z_2}=\bar{z}_1+\bar{z}_2$ and $\overline{z_1z_2}=\bar{z}_1\bar{z}_2$ for any complex numbers z_1 and z_2.

17. Show that a complex number z is real if and only if $z=\bar{z}$.

In Exercises 18–29 write the given complex number in the form $re^{i\theta}$.

18. $1-i$ **22.** -7 **26.** $-5i$

19. $-1-i$ **23.** $2i$ **27.** 6

20. $1+i$ **24.** $\sqrt{3}+i$ **28.** $-4-4\sqrt{3}\,i$

21. $-1+i$ **25.** $1-\sqrt{3}\,i$ **29.** $-1+\sqrt{3}\,i$

30. Show that $\dfrac{d}{d\theta}(e^{i\theta})=ie^{i\theta}$

APPENDIX 2

Determinants

Suppose we wish to solve the system of equations

$$a_{11}x_1 + a_{12}x_2 = b_1$$

$$a_{21}x_1 + a_{22}x_2 = b_2$$

If we solve for x_1 and x_2, we find that

$$(a_{11}a_{22} - a_{12}a_{21})x_1 = b_1a_{22} - b_2a_{12}$$

$$(a_{11}a_{22} - a_{12}a_{21})x_2 = b_2a_{11} - b_1a_{21} \tag{1}$$

To a given matrix

$$\mathbf{M} = \begin{bmatrix} a & b \\ c & d \end{bmatrix}$$

we will associate a number called the determinant of $\mathbf{M}$, denoted by det $\mathbf{M}$,

$$\det \mathbf{M} = ad - bc$$

In terms of the matrices

$$\mathbf{A} = \begin{bmatrix} a_{11} & a_{12} \\ a_{21} & a_{22} \end{bmatrix} \qquad \mathbf{B}_1 = \begin{bmatrix} b_1 & a_{12} \\ b_2 & a_{22} \end{bmatrix} \qquad \mathbf{B}_2 = \begin{bmatrix} a_{11} & b_1 \\ a_{21} & b_2 \end{bmatrix}$$

we are able to write the equations in equation (1) as

$$(\det \mathbf{A})x_1 = \det \mathbf{B}_1$$

$$(\det \mathbf{A})x_2 = \det \mathbf{B}_2$$

so that

$$x_1 = \frac{\det \mathbf{B}_1}{\det \mathbf{A}}, \qquad x_2 = \frac{\det \mathbf{B}_2}{\det \mathbf{A}}$$

whenever $\det \mathbf{A} \neq 0$.

Next we will consider the system of linear equations

$$a_{11}x_1 + a_{12}x_2 + a_{13}x_3 = b_1$$

$$a_{21}x_1 + a_{22}x_2 + a_{23}x_3 = b_2$$

$$a_{31}x_1 + a_{32}x_2 + a_{33}x_3 = b_3$$

If we solve this system for x_1, a lengthy calculation yields

$$Dx_1 = C_1$$

where

$$D = a_{11}a_{22}a_{33} + a_{13}a_{21}a_{32} + a_{12}a_{23}a_{31} - a_{11}a_{23}a_{32} - a_{12}a_{21}a_{33} - a_{13}a_{22}a_{31}$$

and C_1 is obtained from D by substituting b_1, b_2, b_3 for a_{11}, a_{12}, a_{13}, respectively. Further lengthy calculations show that

$$Dx_2 = C_2$$

$$Dx_3 = C_3$$

where C_2 is obtained from D by substituting b_1, b_2, b_3 for a_{12}, a_{22}, a_{32}, respectively, and C_3 is obtained from D by substituting b_1, b_2, b_3 for a_{13}, a_{23}, a_{33}, respectively.

Note that every term in the expression for D has a term of the form $a_{1i}a_{2j}a_{3k}$, where i, j, k is a permutation of $1, 2, 3$. With each term there is a sign $\pm 1 = s(i, j, k)$, called the sign of the permutation i, j, k. If the permutation i, j, k is formed from $1, 2, 3$ by t successive interchanges of pairs of numbers, then $s(i, j, k)=(-1)^t$. Hence

$$s(1,2,3)=(-1)^0= 1 \qquad s(3,1,2)=(-1)^3=-1 \qquad s(2,3,1)=(-1)^2= 1$$

$$s(1,3,2)=(-1)^1=-1 \qquad s(3,2,1)=(-1)^2= 1 \qquad s(2,1,3)=(-1)^1=-1$$

With this notation

$$D= \sum s(i, j, k)a_{1i}a_{2j}a_{3k}$$

where the summation is over all permutations i, j, k of $1, 2, 3$.

The number D is called the determinant of the matrix

$$\mathbf{A} = \begin{bmatrix} a_{11} & a_{12} & a_{13} \\ a_{21} & a_{22} & a_{23} \\ a_{31} & a_{32} & a_{33} \end{bmatrix}$$

We now generalize the concept of determinant to $n \times n$ matrices.

DEFINITION 1   Let the permutation $i_1, i_2, \ldots, i_n$ of $1, 2, \ldots, n$ be formed by t successive interchanges of pairs of numbers. Then $s(i_1, i_2, \ldots, i_n)=(-1)^t$.

▶ **Example 1**   Since

$$3,4,1,2 \rightarrow 3,1,4,2 \rightarrow 3,1,2,4 \rightarrow 1,3,2,4 \rightarrow 1,2,3,4$$

we have $s(3,4,1,2)=(-1)^4=1$.   ◀

DEFINITION 2   The **determinant** of an $n \times n$ matrix $\mathbf{A}=[a_{ij}]$, denoted by $\det \mathbf{A}$, is given by

$$\det \mathbf{A}= \sum s(i_1, i_2, \ldots, i_n)a_{1i_1}, a_{2i_2} \cdots a_{ni_n}$$

where the sum is over all permutations $i_1, i_2, \ldots, i_n$ of $1, 2, \ldots, n$.

Since $1, 2, \ldots, n$ has $n!$ permutations, the calculation of the determinant of an $n \times n$ matrix can be quite a lengthy process. Fortunately, determinants have several properties that ease their calculation. The proofs of these properties follow from the definition of determinant. Since these proofs are tedious and contribute little to the understanding of differential equations, they will be omitted. They can be found in most texts on matrix theory.

THEOREM 1 Let $A = [a_{ij}]$ be an $n \times n$ matrix.

(i) If A is triangular (that is, $a_{ij} = 0$ for all $i < j$ or $a_{ij} = 0$ for all $j < i$), then

$$\det A = a_{11} a_{22} \cdots a_{nn}$$

(ii) If A is obtained from a matrix B by interchanging two rows or columns of B, then $\det B = \det A$.

(iii) If A is obtained from a matrix B by dividing one row or column of B by a constant c, then $\det B = c \det A$.

(iv) If A is obtained from a matrix B by adding a multiple of one row (column) of B to another row (column) of B, then $\det B = \det A$.

Using Theorem 1, we can evaluate a determinant relatively easily. Indeed, Properties (ii), (iii), and (iv) enable us to reduce the matrix A to a triangular matrix, whose determinant can be easily evaluated using Property (i).

▶ **Example 2**

(a) By (ii) we have

$$\det \begin{bmatrix} 1 & 2 \\ 3 & 4 \end{bmatrix} = -\det \begin{bmatrix} 2 & 1 \\ 4 & 3 \end{bmatrix} = \det \begin{bmatrix} 4 & 3 \\ 2 & 1 \end{bmatrix}$$

(b) By (iii) we have

$$\det \begin{bmatrix} 1 & 2 \\ 3 & 4 \end{bmatrix} = 2 \det \begin{bmatrix} 1 & 1 \\ 3 & 2 \end{bmatrix}$$

(c) By (iv), adding -2 times the first row to the second row, we have

$$\det \begin{bmatrix} 1 & 2 \\ 3 & 4 \end{bmatrix} = \det \begin{bmatrix} 1 & 2 \\ 1 & 0 \end{bmatrix} \qquad ◀$$

▶ **Example 3** $cR_i + R_j$ will denote the operation of multiplying the ith row of A by the constant c and adding it to the jth row of A.

$$\det \begin{bmatrix} 3 & 4 & 2 \\ 1 & 2 & 1 \\ 2 & 1 & 4 \end{bmatrix} = -\det \begin{bmatrix} 1 & 2 & 1 \\ 3 & 4 & 2 \\ 2 & 1 & 4 \end{bmatrix} \qquad \begin{array}{l} \text{interchanging the} \\ \text{first two rows} \end{array}$$

$$= -\det \begin{bmatrix} 1 & 2 & 1 \\ 0 & -2 & -1 \\ 2 & 1 & 4 \end{bmatrix} \qquad -3R_1 + R_2$$

$$= -\det\begin{bmatrix} 1 & 2 & 1 \\ 0 & -2 & -1 \\ 0 & -3 & 2 \end{bmatrix} \qquad -2R_1 + R_3$$

$$= -\det\begin{bmatrix} 1 & 2 & 1 \\ 0 & -2 & -1 \\ 0 & 0 & \frac{7}{2} \end{bmatrix} \qquad -\frac{3}{2}R_2 + R_3$$

$$= -(1)(-2)(\tfrac{7}{2}) = 7 \qquad \blacktriangleleft$$

▶ **Example 4**

$$\det\begin{bmatrix} 1 & 2 & 1 \\ 1 & 3 & 2 \\ 2 & 5 & 3 \end{bmatrix} = \det\begin{bmatrix} 1 & 2 & 1 \\ 0 & 1 & 1 \\ 2 & 5 & 3 \end{bmatrix} \qquad -1R_1 + R_2$$

$$= \det\begin{bmatrix} 1 & 2 & 1 \\ 0 & 1 & 1 \\ 0 & 1 & 1 \end{bmatrix} \qquad -2R_1 + R_3$$

$$= \det\begin{bmatrix} 1 & 2 & 1 \\ 0 & 1 & 1 \\ 0 & 0 & 0 \end{bmatrix} \qquad -1R_2 + R_3$$

$$= 0 \qquad \blacktriangleleft$$

There is a commonly used method for computing the determinant of a 3×3 matrix which may be summarized as follows:

1. Write the first two columns of the matrix in their original order to the right of the third column.

$$
\begin{array}{ccc:cc}
a_{11} & a_{12} & a_{13} & a_{11} & a_{12} \\
a_{21} & a_{22} & a_{23} & a_{21} & a_{22} \\
a_{31} & a_{32} & a_{33} & a_{31} & a_{32}
\end{array}
$$

2. In turn, multiply each component of the first row by the other two entries on the diagonal going from left to right. Let S_1 denote the sum of these three products.

$$
\begin{array}{ccccc}
a_{11} & a_{12} & a_{13} & a_{11} & a_{12} \\
a_{21} & a_{22} & a_{23} & a_{21} & a_{22} \\
a_{31} & a_{32} & a_{33} & a_{31} & a_{32}
\end{array}
$$

3. Similarly, multiply each component of the first row by the other two

entries on the diagonal going from right to left. Let S_2 denote the sum of these three products.

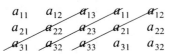

4. The determinant equals $S_1 - S_2$.

A short calculation shows that the above process computes the six terms, along with the appropriate sign, in the definition of the determinant of a 3×3 matrix. Thus the process does compute the determinant of a 3×3 matrix. THIS METHOD WORKS ONLY FOR 3×3 MATRICES. IT DOES NOT WORK FOR $n \times n$ MATRICES WHEN $n > 3$.

Exercises APPENDIX 2

In Exercises 1–12 compute the determinant of the given matrix.

1. $\begin{bmatrix} 1 & 2 \\ 3 & 4 \end{bmatrix}$

2. $\begin{bmatrix} 6 & 5 \\ 1 & 3 \end{bmatrix}$

3. $\begin{bmatrix} a & b \\ c & d \end{bmatrix}$

4. $\begin{bmatrix} 2 & 5 & 1 \\ 2 & 2 & 3 \\ 2 & 5 & 5 \end{bmatrix}$

5. $\begin{bmatrix} 5 & 1 & 2 \\ 10 & 9 & 5 \\ -15 & -3 & -3 \end{bmatrix}$

6. $\begin{bmatrix} 1 & a & a^2 \\ a & a^2 & 1 \\ a^2 & 1 & a \end{bmatrix}$

7. $\begin{bmatrix} 3 & 6 & 7 \\ 3 & 11 & 12 \\ -6 & -7 & -7 \end{bmatrix}$

8. $\begin{bmatrix} 6 & 3 & 7 \\ 11 & 3 & 12 \\ -7 & -6 & -7 \end{bmatrix}$

9. $\begin{bmatrix} 0 & a & 0 \\ a & 0 & b \\ 0 & b & 0 \end{bmatrix}$

10. $\begin{bmatrix} 0 & 1 & 0 & 1 \\ 1 & 0 & 1 & 0 \\ 1 & 1 & 1 & 1 \\ 2 & 1 & 0 & 3 \end{bmatrix}$

11. $\begin{bmatrix} 1 & 2 & 3 & 4 \\ 2 & 2 & 0 & 0 \\ 3 & 0 & 3 & 0 \\ 4 & 0 & 0 & 4 \end{bmatrix}$

12. $\begin{bmatrix} 0 & 5 & 4 & 8 \\ 4 & 21 & 14 & 26 \\ 2 & 8 & 5 & 7 \\ 4 & 6 & 2 & 10 \end{bmatrix}$

13. Let

$$A = \begin{bmatrix} a_{11} & a_{12} \\ a_{21} & a_{22} \end{bmatrix}$$

Show that if $\det A = 0$, then there is a solution other than $x_1 = 0$, $x_2 = 0$ of the system of equations

$$a_{11}x_1 + a_{12}x_2 = 0$$

$$a_{21}x_1 + a_{22}x_2 = 0$$

ANSWERS TO ODD-NUMBERED EXERCISES

1

SECTION 1.1

1. first-order, nonlinear
3. fifth-order, linear
5. third-order, nonlinear.
7. first-order, nonlinear
17. $-1, -4$
19. $-1; 1, 2$

21. $-3, 2$
23. $1-\sqrt{3}, 1+\sqrt{3}$
25. $c_1 = \frac{1}{2}, c_2 = -\frac{1}{2}$
27. $c_1 = 0, c_2 = 0$
29. $x^3/3 + 2$
31. $-\cos x - 1$

SECTION 1.2

1. $1 + ce^{-x}$
3. $x^{-1}(e^x + c)$
5. $(x^2+1)^{-1}\left(\dfrac{x^5}{5} + \dfrac{x^3}{3} + \dfrac{x^2}{2} + \ln x + c\right)$
7. $ce^{1/x} - 1$
9. $e^{-x^3}\left(\dfrac{x^2}{2} + c\right)$
11. $\frac{4}{3} + \frac{11}{3}e^{-3x}$
13. $3x^{-1} - 1 + \ln x$
15. $e^x(\ln x - 1)$
17. $\cot x + \dfrac{\pi^2 - 1}{\sqrt{2}}\csc x$
19. $x \cos x$

21. $\dfrac{x^4}{16} + \dfrac{x^2}{4} + c_1\ln x + c_2$
23. $\dfrac{x^3}{3} - x^2 + 2x + c_1 e^{-x} + c_2$
25. $x + c_1\int e^{\cos x}\,dx + c_2$
27. $xe^x - 2e^x + c_1 x^3 + c_2 x + c_3$
29. $\frac{1}{2}x^2 + c_1 x + c_2 + c_3 e^{-x}$
31. $(1 + ce^x)^{-1}$
33. $(3x^2 + cx^3)^{-1/3}$
35. $(x^{-1}e^x + cx^{-1})^3$
37. $(\pm\tan x + c\sec x)^{-1}$
41. $x + 3e^{x^2/2}\left(1 - 3\int_0^x e^{t^2/2}\,dt\right)^{-1}$

SECTION 1.3

1. 2681 years
3. 1732 years, 1900 years
5. $P_0 e^{kz}$
7. $-.0022 e^{-100t}$
9. $C(t) = \left(C(0) - \dfrac{p}{kV} \right) e^{-kt} + \dfrac{p}{kV}, C(t) \to \dfrac{p}{kV}$ as $t \to \infty$
11. ~~48.33 lb, 1.76 lb/gal~~ 46.73 ibs. , 1.47 ibs/gal
13. $dT/dt = k(T - T_1)$
15. A general solution when $T_0 = 0$ is $T(t) = (-a/b + ce^{3bt})^{-1/3}$. If $b < 0$, then $T(t) \to (-b/a)^{1/3} \neq 0$ as $t \to \infty$, which is impossible.
17. $c_1 e^{Vx/C} + c_2$
19. P_1, P_1, P_1
21. .193 days

SECTION 1.4

1. $\left(\dfrac{3t^2}{2} + c \right)^{1/3}$

3. $\tan \left(\dfrac{t^4}{4} + 5t + c \right)$

5. $\sin x = -\cos t + c$

7. $-\ln|c - e^t|$

9. $\left(\dfrac{3}{2} t^{-2} + c \right)^{-1/3}$

11. $\dfrac{x^2}{2} + \ln|x| = \dfrac{t^2}{2} + t + c$

13. $\sin x = e^t - 1$

15. $2/\cos t$

17. ~~$\ln|e^{-t} + e^{-1}|$~~ $-\ln|e^{-t} - e^{-1} + e^{-e}|$

19. $\ln|te^t - e^t + e^5 + 1|$

21. $\ln|\sec x + \tan x| = 2\cos t - 2$

25. $\left(\dfrac{a}{b}(1 - \alpha)(e^{bt} - 1) + k_0^{1-\alpha} \right)^{1/(1-\alpha)}$

27. $V(0) e^{-a/b}$

SECTION 1.5

1. $t(c - \ln|t|)^{-1}$
3. $ct^2 (1 - ct)^{-1}$
7. $x + \frac{1}{2} \sin(2x + 2t) = t + c$
9. $x + \ln|x + t| = t + c$
11. $h = (sb - rd)/(ad - bc), \quad k = (rc - sa)/(ad - bc)$
13. ~~$(x + 2)(x - 2t + 4) = c$~~ $(2x - t - 1)^2 (x + t + 4) = c$
15. $-\ln|x - 1| + \dfrac{t + 1}{x - 1} = c$

SECTION 1.6

1. $x^3 t^2 + x t^5 + x = c$

3. $t + x(1+t^2)^{1/2} = c$

5. not exact

7. $e^{x+t} + 3x - 2t = c$

9. $x \ln|t| = c$

11. $5x^3 t + 3x t^3 + 2x^2 t^2 + 4xt + x^2 = c$

13. If $\dfrac{dx}{dt} = \dfrac{f(t)}{g(x)}$, then $f(t) - g(x)\dfrac{dx}{dt} = 0$. In the notation of this section, $M(x,t) = f(t)$ and $N(x,t) = -g(x)$. Thus $M_x(x,t) = N_t(x,t) \equiv 0$ and the differential is exact. The converse is not true. For example, see Exercise 1.

SECTION 1.7

1. $x t^{2/3} + t^{2/3} = c$

3. $x^{-1}e^t + t^2 e^t - 2te^t + 2e^t - \int e^t t^{-1} dt = c$

5. If f is only a function of x, then equation (1.65) becomes $f'(x)M(x,t) + f(x)M_x(x,t) = f(x)N_t(x,t)$, so that

$$f'(x) = \frac{N_t(x,t) - M_x(x,t)}{M(x,t)} f(x)$$

7. $t\sin^2 x = c$

9. ~~$x + \ln|xt| = c$~~ $t x e^x = c$

11. In the notation of this chapter, $M(y,x) = p(x)y^{1-n} - q(x)$ and $N(y,x) = y^{-n}$, so that $(M_y - N_x)/N = (1-n)p(x)$. Using equation (1.66) with y and x replacing x and t respectively, we find that $e^{(1-n)\int p(x)\,dx}$ is an integrating factor.

SECTION 1.8

1. t	x	3. t	x
.1	2.2	.1	1.1
.2	2.41	.2	1.21
.3	2.631	.3	1.3320
.4	2.8641	.4	1.4685
.5	3.1105	.5	1.6225
.6	3.3716	.6	1.7980
.7	3.6487	.7	2.0001
.8	3.9436	.8	2.2354
.9	4.2580	.9	2.5130
1.0	4.5937	1.0	2.8454

5. t	x		7. t	x
1.1	.0368		.1	.099
1.2	.0738		.2	.194
1.3	.1112		.3	.282
1.4	.1496		.4	.361
1.5	.1892		.5	.427
1.6	.2305		.6	.480
1.7	.2737		.7	.521
1.8	.3194		.8	.549
1.9	.3678		.9	.565
2.0	.4196		1.0	.570

9. t	x
.1	1.09
.2	1.18
.3	1.26
.4	1.33
.5	1.39
.6	1.45
.7	1.51
.8	1.56
.9	1.61
1.0	1.66

11. $x_{n,h} = \left(1+\dfrac{1}{n}\right)^n$, $\lim\limits_{n\to\infty} x_{n,h}=e=x(1)$

SECTION 1.9

1. t	x		3. t	x
.1	2.2052		.1	1.1054
.2	2.4214		.2	1.2230
.3	2.6499		.3	1.3557
.4	2.8918		.4	1.5071
.5	3.1487		.5	1.6820
.6	3.4221		.6	1.8868
.7	3.7138		.7	2.1305
.8	4.0255		.8	2.4259
.9	4.3596		.9	2.7921
1.0	4.7183		1.0	3.2588

5.

t	x
1.1	.0368
1.2	.0741
1.3	.1120
1.4	.1511
1.5	.1917
1.6	.2342
1.7	.2791
1.8	.3267
1.9	.3776
2.0	.4323

7 and **9.** The answers for Exercises 7 and 9 are the same as those for Exercises 7 and 9 in Section 1.8.

SECTION 1.10

1. yes   **3.** yes   **5.** no

7. all points (a, b) with $a \neq 0$ and $a \neq -b$

9.

11.

13.

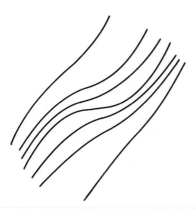

2

SECTION 2.1

3. $-1, -3$

5. $(u+z)'' + a_1(x)(u+z)' + a_0(x)(u+z)$
$= u'' + z'' + a_1(x)(u'+z') + a_0(x)(u+z)$
$= (u'' + a_1(x)u' + a_0(x)u) + (z'' + a_1(x)z' + a_0(x)z)$
$= 0 + f(x) = f(x)$

7. Using Exercise 1, it is easy to verify that $u(x) \equiv 0$ is a solution of the initial-value problem. By Theorem 2.1 this is the only solution.

SECTION 2.2

1. $c_1 e^{2x} + c_2 e^{-2x}$, e^{2x}

3. (a) $c_1 e^{-2x} + c_2 e^{-x}$
  (b) $\frac{1}{2}x - \frac{3}{4}$
  (c) $c_1 e^{-2x} + c_2 e^{-x} + \frac{1}{2}x - \frac{3}{4}$
  (d) $-\frac{9}{4}e^{-2x} + 4e^{-x} + \frac{1}{2}x - \frac{3}{4}$

5. (a) $c_1 x^{-2} + c_2 x^{-1}$
  (b) $\frac{3}{4}x^2$
  (c) $c_1 x^{-2} + c_2 x^{-1} + \frac{3}{4}x^2$
  (d) $\frac{1}{4}x^{-2} + 2x^{-1} + \frac{3}{4}x^2$

7. Show that cy_1 is a solution of the second initial-value problem and use Theorem 2.1 to show that it is unique.

9. Since $(D^2 + a_1 x^{-1}D + a_0)x^r = (r^2 + (a_1 - 1)r + a_0)x^r$, the function $y(x) = x^r$ is a solution of the differential equation if and only if $r^2 + (a_1 - 1)r + a_0 = 0$.

11. $c_1 x + c_2 x^2$

13. $c_1 x + c_2 x^{-1}$

15. $c_1 \cos(\ln x) + c_2 \sin(\ln x)$

17. $c_1 x^{-1/2}\cos\left(\frac{\sqrt{19}}{2}\ln x\right) + c_2 x^{-1/2}\sin\left(\frac{\sqrt{19}}{2}\ln x\right)$

SECTION 2.3

1. $c_1 e^{2x} + c_2 e^{-2x}$

3. $c_1 e^{2x} + c_2 e^{-3x}$

5. $c_1 e^{(-5+\sqrt{13})x/2} + c_2 e^{(-5-\sqrt{13})x/2}$

7. $c_1 e^{2x} + c_2 e^{4x}$

9. $c_1 + c_2 e^{16x}$

11. $c_1 e^x + c_2 x e^x$

13. $c_1 e^{-5x} + c_2 x e^{-5x}$

15. $c_1 e^{(-2+\sqrt{2})x} + c_2 e^{(-2-\sqrt{2})x}$

17. $x e^{-x}$

19. $\frac{5}{2}e^{-x} - \frac{3}{2}e^{-2x}$ $4e^{-x} - 3e^{-2x}$

21. **(b)** $-x^{-1}(c_1 e^{3x} + c_2 e^{-x})^{-1}(3c_1 e^{3x} - c_2 e^{-x})$

23. $c_1 + c_2 e^{\sqrt{x}/C}$

25. $\theta_0 e^{-t\sqrt{g/L}} + \theta_0 \sqrt{\frac{g}{L}} t e^{-t\sqrt{g/L}}$

SECTION 2.4

1. $\begin{cases} 28.00345 - .00345 e^{.00095z} & \text{if } z \leqslant 3.5 \\ \quad 42.3156 e^{-.1176} & \text{if } 3.5 < z \end{cases}$

SECTION 2.5

1. $c_1 \cos 3x + c_2 \sin 3x$

3. $e^{-x/2}\left(c_1 \cos \frac{\sqrt{3}}{2}x + c_2 \sin \frac{\sqrt{3}}{2}x \right)$

5. $e^x(c_1 \cos\sqrt{2}\,x + c_2 \sin\sqrt{2}\,x)$

7. $e^{x/2}\left(c_1 \cos \frac{\sqrt{3}}{2}x + c_2 \sin \frac{\sqrt{3}}{2}x \right)$

9. $e^{-5x}(c_1 \cos 3x + c_2 \sin 3x)$

11. $e^{-x}(c_1 \cos\sqrt{3}\,x + c_2 \sin\sqrt{3}\,x)$

13. $c_n \sin nx$, where $n = 1, 2, 3, \dots$ and c_n is any constant

15. $k\pi h / \sqrt{(E-V)(2m)}$, where $k = 0, 1, 2, \dots$ and $0 < E - V$

SECTION 2.6

1. **(a)** $2e^{-t}\left(\cos\sqrt{5}t + \frac{1}{\sqrt{5}}\sin\sqrt{5}t\right)$ $e^{-2t}\left(2\cos\sqrt{5}t + \frac{4}{\sqrt{5}}\sin\sqrt{5}t\right)$

(b) $\frac{3}{5}e^{-2t}\sin\sqrt{5}\,t$

(c) $e^{-4t}\left(.2\cos\sqrt{5}\,t+\dfrac{6.8}{\sqrt{5}}\sin\sqrt{5}\,t\right)$

3. $e^{-8t}(\cos 6t-\frac{2}{3}\sin 6t)$

SECTION 2.7

1. $c_1\cos x+c_2\sin x+2x+3$

3. $e^{-x}\left(c_1\cos\sqrt{2}\,x+c_2\sin\sqrt{2}\,x\right)+\frac{1}{3}x^4-\frac{8}{9}x^3+\frac{4}{9}x^2+\frac{32}{27}x+\frac{47}{81}$

5. $c_1+c_2e^x+\frac{1}{20}e^{5x}-\frac{1}{3}x^3-4x^2-8x$

7. $c_1e^x+c_2e^{2x}+\frac{1}{6}e^{-x}+\frac{1}{2}x^2+\frac{3}{2}x+\frac{7}{4}$

9. $e^{-3x/2}\left(c_1\cos\dfrac{\sqrt{7}}{2}x+c_2\sin\dfrac{\sqrt{7}}{2}x\right)+(x^2+\frac{11}{16}x+\frac{89}{512})e^{-4x}+\frac{3}{4}$

11. $c_1+c_2e^{-7x}+\frac{1}{14}x^2-\frac{1}{49}x+\frac{1}{8}e^x$

13. $c_1e^{-x}+c_2xe^{-x}+\frac{1}{2}x^2e^{-x}$

15. $c_1e^{-x}+c_2e^{-4x}+\frac{1}{28}e^{3x}+\frac{1}{18}e^{2x}-\frac{1}{4}x+\frac{5}{16}$

17. $c_1e^{-4x}+c_2e^{-x}+\frac{1}{40}e^{4x}+\frac{1}{18}e^{2x}$

19. $c_1+c_2e^{5x}+(-\frac{1}{4}x+\frac{3}{16})e^x+\frac{1}{10}x^2-\frac{24}{25}x$

21. $-3\cos x-\sin x+2x+3$

23. $e^{-x}\left(\frac{115}{81}\cos\sqrt{2}\,x+\frac{131}{81}\sqrt{2}\,\sin\sqrt{2}\,x\right)+\frac{1}{3}x^4-\frac{8}{9}x^3+\frac{4}{9}x^2+\frac{32}{27}x+\frac{47}{81}$

SECTION 2.8

1. $c_1e^x+c_2e^{-x}-\frac{1}{5}\cos 2x$

3. $c_1\cos x+c_2\sin x-\frac{1}{2}x\cos x+x^2-2$

5. $c_1e^{-5x}+c_2e^{2x}-\frac{1}{150}e^{-x}\cos 3x-\frac{7}{150}e^{-x}\sin 3x$

7. $c_1e^{2x}+c_2xe^{2x}+(\underset{25}{6x}+\frac{68}{125})\cos x+(-\underset{25}{8x}-\frac{124}{125})\sin x$

9. $c_1e^{+3x}+c_2e^{+2x}+\frac{1}{52}\cos 2x-\frac{5}{52}\sin 2x+\frac{1}{2}e^{4x}$

11. $c_1e^{x/2}\cos\dfrac{\sqrt{7}}{2}x+c_2e^{x/2}\sin\dfrac{\sqrt{7}}{2}x+(-\frac{9}{106}x+\frac{33}{5618})e^{2x}\cos 3x$
    $+(-\frac{5}{106}x+\frac{177}{2809})e^{2x}\sin 3x$

13. $c_1e^{4x}+c_2e^{-4x}-\frac{1}{17}\cos x-\frac{1}{32}\sin 4x$

15. $c_1e^{-x/2}\cos\dfrac{\sqrt{3}}{2}x)+c_2e^{-x/2}\sin\dfrac{\sqrt{3}}{2}x+x-1-\cos x$

17. $c_1e^{-x}\cos x+c_2e^{-x}\sin x+xe^{-x}\sin x-\frac{3}{2}xe^{-x}\cos x$

19. $c_1\cos x+c_2\sin x+\frac{1}{2}x\sin x+\frac{1}{5}e^{2x}+x^2-2-\frac{1}{3}\sin 2x$

21. $\frac{3}{5}e^x - \frac{2}{5}e^{-x} - \frac{1}{5}\cos 2x$

23. $3\cos x + \frac{3}{2}\sin x - \frac{1}{2}x\cos x + x^2 - 2$

SECTION 2.9

1. (a) $-\dfrac{.1}{3}e^{-t}\cos 3t - .1e^{-t}\sin 3t$, $q(t) \equiv .1$

   (b) $2\sqrt{2}$

3. $e^{-8t} + 8te^{-8t} - 1$

5. $k(t) = c_1\cos c + c_2\sin ct + \bar{k}$ is a general solution. Set $A = c_1^2 + c_2^2$ and choose a so that $\sin a = \dfrac{c_1}{c_1^2 + c_2^2}$ and $\cos a = \dfrac{c_2}{c_1^2 + c_2^2}$. Then $k(t) = A(\sin a\cos ct + \cos a\cos ct) + \bar{k} = A\cos(ct + a) + \bar{k}$.

7. If $w^2 \neq k/m$, then

$$x = c_1\cos\left(\sqrt{\frac{k}{m}}\,t\right) + c_2\sin\left(\sqrt{\frac{k}{m}}\,t\right) + \frac{hw^2}{k - mw^2}\cos wt$$

$$y = c_3\cos\left(\sqrt{\frac{k}{m}}\,t\right) + c_4\sin\left(\sqrt{\frac{k}{m}}\,t\right) + \frac{hw^2}{k - mw^2}\sin wt$$

If $w^2 = k/m$, then

$$x = c_1\cos\left(\sqrt{\frac{k}{m}}\,t\right) + c_2\sin\left(\sqrt{\frac{k}{m}}\,t\right) + \frac{hw^2}{2k}t\sin wt$$

$$y = c_3\cos\left(\sqrt{\frac{k}{m}}\,t\right) + c_4\sin\left(\sqrt{\frac{k}{m}}\,t\right) + \frac{hw^2}{2k}t\cos wt$$

SECTION 2.10

1. $c_1e^{-x} + c_2xe^{-x} + \frac{1}{16}e^{3x}$

3. $c_1e^{-x} + c_2e^{-2x} + 7xe^{-x}$

5. $c_1\cos x + c_2\sin x + e^x$

7. $c_1e^{-3x} + c_2xe^{-3x} + \frac{9}{2}x^2e^{-3x}$

9. $c_1e^{-x/2}\cos\sqrt{3}\,x + c_2e^{-x/2}\sin\sqrt{3}\,x - \frac{8}{73}\cos 3x + \frac{3}{73}\sin 3x$

11. $c_1e^{(-1+\sqrt{6})x} + c_2e^{(-1-\sqrt{6})x} - \frac{1}{20}\cos x - \frac{3}{20}\sin x$

13. $c_1e^{(1+\sqrt{5})x/2} + c_2e^{(1-\sqrt{5})x/2} - \frac{4}{5}\cos x - \frac{2}{5}\sin x$

17. $c_1e^x + c_2e^{2x} + 3xe^{2x} + \frac{1}{3}e^{-x}$

19. $c_1e^x + c_2e^{2x} - 4xe^x - 5xe^{2x}$

21. $c_1\cos x + c_2\sin x + \frac{2}{37}e^{6x} - \frac{1}{3}\sin 2x$

23. $c_1e^{2x} + c_2e^{-2x} + \frac{1}{4}xe^{2x} + \frac{1}{5}\sin x$

25. $c_1e^x + c_2xe^x - \frac{3}{2}\cos x + \frac{1}{2}\sin x$

27. $c_1e^{-2x} + c_2e^{3x} - \frac{1}{50}\cos x + \frac{7}{50}\sin x + \frac{5}{52}\cos 2x + \frac{1}{52}\sin 2x$

29. $c_1e^{(-7+\sqrt{29})x/2} + c_2e^{(-7-\sqrt{29})x/2} + \frac{2}{13}e^x - \frac{4}{197}\cos 2x - \frac{56}{197}\sin 2x$

33. See answer to Exercise 7 of Section 2.9.

SECTION 2.11

1. $c_1x^2 + c_2x^{-2}$

3. $c_1x^{-1} + c_2x^{-1}\ln|x|$

5. $c_1x^3 + c_2x^{-2}$

7. $c_1x^{3/2} + c_2x^{3/2}\ln|x|$

9. $c_1x^2 + c_2x^2\ln|x|$

SECTION 2.12

1. $c_1\cos x + c_2\sin x - (\cos x)\ln|\cos x| + x\sin x$

3. $c_1e^{3x} + c_2xe^{3x} + x^2e^{3x}(\frac{1}{2}\ln|x| - \frac{3}{4})$

5. $c_1x + c_2x^2 + \frac{1}{2}x^3\ln|x| - \frac{3}{4}x^3$

7. It is easy to show that $v_1'(x) = -f(x)\sin x$ and $v_2'(x) = f(x)\cos x$. Thus we may choose v_1 and v_2 to be $v_1(x) = \int_0^x -f(t)\sin t\, dt,\quad v_2(x) = \int_0^x f(t)\cos t\, dt$, so that a particular solution is

$$(\cos x)v_1(x) + (\sin x)v_2(x) = \int_0^x (-\cos x\sin t + \sin x\cos t)f(t)\, dt$$

$$= \int_0^x \sin(x-t)f(t)\, dt$$

9. Since y_1 and y_2 are linearly independent, neither is the zero function. If $W[y_1, y_2](x_0) = 0$, then there is a nonzero vector $\begin{bmatrix} c_1 \\ c_2 \end{bmatrix}$ such that

$$\begin{bmatrix} y_1(x_0) & y_2(x_0) \\ y_1'(x_0) & y_2'(x_0) \end{bmatrix}\begin{bmatrix} c_1 \\ c_2 \end{bmatrix} = \begin{bmatrix} 0 \\ 0 \end{bmatrix}$$

This gives the pair of equations

$$c_1 y_1(x_0) + c_2 y_2(x_0) = 0$$

$$c_1 y_1'(x_0) + c_2 y_2'(x_0) = 0$$

Set $z(x) = c_1 y_1(x) + c_2 y_2(x)$. Then z is a solution of the initial-value problem $[D^2 + a_1(x)D + a_2(x)]y = 0$, $y(x_0) = 0$, $y'(x_0) = 0$. The unique solution of this initial-value problem is $y(x) \equiv 0$. Thus we have arrived at an impossible situation because $z(x) \neq 0$ (since either c_1 or c_2 is nonzero and the functions y_1 and y_2 are linearly independent).

SECTION 2.13

1. $\int_{-\infty}^{\infty} f(t)\delta(t-a)\,dt = \int_{-\infty}^{\infty} f(t+a)\delta(t)\,dt = f(0+a) = f(a)$
3. $g(t)\delta(t) = 0$ for all $t \neq 0$ since $\delta(t) = 0$ for all $t \neq 0$. Using equation (2.104), we have $\int_{-\infty}^{\infty} g(t)\delta(t)\,dt = g(0) = 1$.

SECTION 3.1

1. If $u(x) \equiv 0$, then $u^{(k)}(x) \equiv 0$ for every $k = 1, 2, \ldots$.
5. $0 = (x^3 D^3 + xD - 1)(x^r) = (r-1)^3 x^r$. Therefore r must be 1.

SECTION 3.2

1. (a) $1, 2, -3$
   (b) $c_1 e^x + c_2 e^{2x} + c_3 e^{-3x}$
   (c) $e^x + e^{2x} + e^{-3x}$
3. (a) $1, -1, -5$
   (b) $c_1 x + c_2 x^{-1} + c_3 x^{-5}$
   (c) $2x^{-1} - x^{-5}$
5. (b) $c_1 e^x + c_2 e^{2x} + c_3 e^{3x}$
   (c) $3e^x - 3e^{2x} + e^{3x}$
   (e) $c_1 e^x + c_2 e^{2x} + c_3 e^{3x} - \frac{1}{6}x - \frac{11}{36}$
   (f) $\frac{21}{6} e^x - \frac{117}{36} e^{2x} + \frac{19}{18} e^{3x} - \frac{1}{6}x - \frac{11}{36}$

7. (b) $c_1 + c_2 x + c_3 e^{2x} + c_4 e^{-2x}$
 (c) $1 + x + e^{2x} + e^{-2x}$
 (e) $c_1 + c_2 x + c_3 e^{2x} + c_4 e^{-2x} - x^3$
 (f) $1 - 2x + \frac{3}{4} e^{2x} - \frac{3}{4} e^{-2x} - x^3$

9. $c_1 x + c_2 x^2 + c_3 x^3$

11. $c_1 x + c_2 \cos(\ln x) + c_3 \sin(\ln x)$

SECTION 3.3

1. $W[y_1, y_2](x) = 2x$. No! If y_1 and y_2 are solutions of the same second-order linear differential equation, then the Wronskian $W[y_1, y_2]$ is either identically zero or never zero.

3. $W[y_1, y_2](0) \neq 0$. By Theorem 3.5 the solutions are linearly independent.

SECTION 3.4

1. $c_1 e^{2x} + c_2 e^{-5x} + c_3 e^{-3x}$

3. $c_1 \cos 2x + c_2 \sin 2x + c_3 e^{-5x} + c_4 x e^{-5x}$

5. $c_1 \cos\sqrt{2}\, x + c_2 \sin\sqrt{2}\, x + c_3 e^{\sqrt{2}\, x} + c_4 e^{-\sqrt{2}\, x}$

7. $e^{-3x/2} \left(c_1 \cos \dfrac{\sqrt{7}}{2} + c_2 \sin \dfrac{\sqrt{7}}{2} \right) + e^{x/2}(c_3 x^2 + c_4 x + c_5)\cos \dfrac{\sqrt{3}}{2} x$
 $\quad + e^{x/2}(c_6 x^2 + c_7 x + c_8)\sin \dfrac{\sqrt{3}}{2} x$

9. $c_1 + c_2 e^{-x} + c_3 e^{-2x}$

11. $c_1 e^x + e^{-x/2} \left(c_2 \cos \dfrac{\sqrt{3}}{2} x + c_3 \sin \dfrac{\sqrt{3}}{2} x \right)$

13. $c_1 e^x + c_2 e^{-x} + c_3 x e^{-x} + c_4 e^{2x} + c_5 e^{-3x}$

15. $\left(c_1 x^3 + c_2 x^2 + c_3 x + c_4 \right)\cos\sqrt{5}\, x + (c_5 x^3 + c_6 x^2 + c_7 x + c_8)\sin\sqrt{5}\, x$

17. $c_1 x^5 + c_2 x^4 + c_3 x^3 + c_4 x^2 + c_5 x + c_6 + e^{(-3-\sqrt{5})x}(c_7 x^2 + c_8 x + c_9)$
 $\quad + e^{(-3+\sqrt{5})x}(c_{10} x^2 + c_{11} x + c_{12})$

19. $(c_1 x^2 + c_2 x + c_3)e^x + (c_4 x^2 + c_5 x + c_6)e^{-2x} + (c_7 x^2 + c_8 x + c_9)e^{-3x}$
 $\quad + (c_{10} x^2 + c_{11} x + c_{12})e^{-6x}$

21. $(c_1 x^2 + c_2 x + c_3)e^{-ax} + (c_4 x^2 + c_5 x + c_6)e^{-bx}$

23. $c_1 + c_2 x + c_3 e^{rx} + c_4 e^{-rx}$, where $r = \left[(H+h)/EI \right]^{1/2}$

25. $P^2 < EIK$

SECTION 3.6

1. $c_1e^x + c_2xe^x + c_3e^{-x} + x^2 + 2x + 5$

3. $c_1e^{2x} + c_2xe^{2x} + c_4e^{-x} + \frac{1}{6}x^2e^{2x}$

5. $e^{-x/2}(c_1x + c_2)\cos\dfrac{\sqrt{3}}{2}x + e^{-x/2}(c_3x + c_4)\sin\dfrac{\sqrt{3}}{2}x + \frac{1}{9}e^x + x - 2$

7. $c_1e^{-x} + e^{2x}(c_2\cos x + c_3\sin x) + \frac{3}{25}e^x\cos x - \frac{4}{25}e^x\sin x$

9. $c_1\cos 2x + c_2\sin 2x + c_3e^{2x} - \frac{2}{15}\cos x + \frac{1}{15}\sin x$

11. $c_1e^x + c_2xe^x + c_3x^2e^x + \frac{1}{6}x^3e^x$

13. $c_1 + c_2x + c_3x^2 + c_4e^{3x} + c_5xe^{3x} + \frac{1}{216}x^4 + \frac{1}{81}x^3$

15. $e^{-12t} + 12te^{-12t} + 72t^2e^{-12t} - 1$

17. $e^{Ax/\sqrt{2}}\left(c_1\cos\dfrac{x}{\sqrt{2}} + c_2\sin\dfrac{x}{\sqrt{2}}\right) + e^{-Ax}\left(c_3\cos\dfrac{x}{\sqrt{2}} + c_4\sin\dfrac{x}{\sqrt{2}}\right) + R^2/ET$
   where $A = (12/r^2T^2)^{1/4}$.

19. $c_1\cos 208.6t + c_2\sin 208.6t + c_3\cos 1.9t + c_4\sin 1.9t + 1.35(10)^{-5}\sin t$

SECTION 3.7

1. $c_1e^x + c_2e^{2x} + c_3e^{3x} + \frac{1}{6}e^{4x}$

3. $c_1e^{-x} + c_2e^x + c_3e^{3x} + \frac{5}{4}xe^{3x}$

5. $c_1e^x + c_2e^{-x/2}\cos\dfrac{\sqrt{3}}{2}x + c_3e^{-x/2}\sin\dfrac{\sqrt{3}}{2}x + \frac{1}{7}e^{2x}$

7. $c_1e^{7x} + c_2e^{x/2}\cos\dfrac{\sqrt{7}}{2}x + c_3e^{x/2}\sin\dfrac{\sqrt{7}}{2}x + \frac{1}{58}\cos 3x$

9. $c_1e^x + c_2xe^x + c_3e^{-4x} + \frac{1}{28}x^3 + \frac{6}{17}\cos x + \frac{3}{34}\sin x$

11. $c_1e^{-5x} + c_2e^{(-2+\sqrt{3})x} + c_3e^{(-2-\sqrt{3})x} + \frac{1}{297}e^{4x} - \frac{139}{19721}\cos 2x + \frac{20}{19721}\sin 2x$

13. $c_1e^x + c_2e^{-x} + c_3\cos x + c_4\sin x + \frac{1}{4}xe^x - \frac{1}{4}xe^{-x}$

15. $c_1e^{\sqrt{2}x} + c_2e^{-\sqrt{2}x} + c_3e^{(-1+\sqrt{2})x} + c_4e^{(-1-\sqrt{2})x} + \frac{1}{6}e^x + \frac{5}{42}e^{2x}$
   $\quad - \frac{3}{4}\cos x + \frac{3}{4}\sin x$

SECTION 3.8

1. $c_1 + c_2\cos x + c_3\sin x + \ln|\sec x + \tan x| - x\cos x + (\sin x)\ln|\cos x|$

3. $c_1x + c_2x^{-1} + c_2x^{-2} + \frac{1}{40}x^3\ln x - \frac{38}{1600}x^3$

5. $c_1e^x + c_2xe^x + c_3e^{-x} - \frac{1}{4}e^x\int e^{-x}(2x+1)x^{-1}\,dx$
   $\quad + \frac{1}{2}xe^x\int e^{-x}x^{-1}\,dx + \frac{1}{4}e^{-x}\int e^x x^{-1}\,dx$

7. $c_1e^x + c_2e^{-x} + c_3xe^{-x} + \frac{1}{4}e^x\int e^x f(x)\,dx$
$+ \frac{1}{4}e^{-x}\int(2x-1)e^x f(x)\,dx + \frac{1}{2}xe^{-x}\int e^x f(x)\,dx$

SECTION 3.9

1. The solutions to the true equation have the form $c_1\cos x + c_2\sin x$. In case (a) all solutions tend to zero as $x \to \infty$. The true solutions do not. In case (b) all solutions, except $y(x) \equiv 0$, are unbounded as $x \to \infty$. The true solutions are bounded.

4

SECTION 4.1

1. $2/s^3$

3. $\dfrac{a}{s^2+a^2}$

5. $\dfrac{6}{s} + \dfrac{1}{s^2}$

7. $0 < s$

9. $a < s$

11. $k < s$

13. $e^{-t}\sin 2t$

15. t^3

17. $t\cos t$

19. $e^{-t}\sin t$

21. 1

SECTION 4.2

1. $\dfrac{3}{s} + \dfrac{7}{s^2}$

3. $\dfrac{5s}{s^2+1} + \dfrac{4s}{(s^2+1)^2}$

5. $\dfrac{3}{s^2+9}$

7. $\dfrac{-s}{s^2+2as+2a^2}$

9. $\dfrac{3}{s}(1-e^{-2s})$

11. $\dfrac{1}{s}(5-e^{-3s})$

13. $\dfrac{a}{(s+b)^2+a^2}, \quad \dfrac{2as}{(s^2+a^2)^2}, \quad \dfrac{2a^5-4a^3s^2-6as^4}{(s^2+a^2)^4}$

15. $\dfrac{e^{-s}}{s-1}$

17. $2e^{3t}$

19. $\dfrac{1}{2\sqrt{3}}t\sin\sqrt{3}\,t$

21. $5te^{-3t}$

23. $2\cos\sqrt{3}\,t + \frac{4}{3}\sin\sqrt{3}\,t$

SECTION 4.3

1. $\frac{1}{2}(e^t - e^{-t})$
3. $\frac{5}{4}t - \frac{5}{8}\sin 2t$
5. $-\frac{5}{18}\cos 3t + \frac{1}{18}\sin 3t + \frac{5}{18}e^{3t}$
7. $\begin{cases} 0 & \text{if } 0 \leqslant t \leqslant a \\ 1 & \text{if } a < t \end{cases}$
9. $\cos\sqrt{5}\,t + \sin\sqrt{5}\,t$
11. $e^{-3t}\cos t - 3e^{-3t}\sin t$

SECTION 4.4

1. t

3. $\begin{cases} \dfrac{t^2}{2} & \text{if } 0 \leqslant t \leqslant 3 \\[2mm] 3t - \dfrac{9}{2} & \text{if } 3 \leqslant t \end{cases}$

5. $\begin{cases} 0 & \text{if } t < 2 \\[2mm] \dfrac{t^2}{2} - 2t + 2 & \text{if } 2 \leqslant t \leqslant 3 \\[2mm] t - \dfrac{5}{2} & \text{if } 3 < t \end{cases}$

7. $e^{2t} - e^t$
9. $\frac{1}{4} - \frac{1}{4}\cos 2t + \frac{1}{2}\sin 2t$

11. $\begin{cases} t & \text{if } 0 \leqslant t \leqslant 1 \\ 0 & \text{if } 1 < t \end{cases}$

13. $\begin{cases} \frac{1}{2}(1 - e^{-2t}) & \text{if } 0 \leqslant t \leqslant 3 \\ 0 & \text{if } 3 < t \end{cases}$

SECTION 4.5

1. $-\frac{1}{2}t - \frac{1}{4} + \frac{13}{4}e^{-2t}$
3. $-2e^{-t} - \frac{1}{4}e^{-2t} + \frac{1}{2}t^2 - \frac{3}{2}t + \frac{9}{4}$
5. $-\frac{1}{17}\cos t - \frac{4}{17}\sin t + \frac{1}{17}e^{4t}$
7. $-2e^{-t}\cos t + e^{-t}$
9. $-\frac{1}{8}e^t - \frac{1}{8}e^{-t} + \frac{1}{4}\cos t + \frac{3}{4}\sin t + \frac{1}{4}te^t$

11. **(a)** $e^{-t}-e^{-2t}$, $0<t$ **(b)** $e^{-t}-e^{-2t}+\frac{1}{2}e^{-3t}$, $0<t$

13. $\begin{cases} 1-\cos t & \text{if } 0\leqslant t\leqslant 1 \\ \cos(t-1)-\cos t & \text{if } 1<t \end{cases}$

15. $3-e^{-t}$

17. $\dfrac{P}{6EI}\begin{cases} x^3 & \text{if } 0\leqslant x\leqslant \dfrac{L}{2} \\ x^3-\left(x-\dfrac{L}{2}\right)^3 & \text{if } \dfrac{L}{2}<x \end{cases} -\dfrac{P}{12EI}x^3-\dfrac{P}{16EI}L^2x$

5

SECTION 5.1

1. (a) $\begin{bmatrix} 4 \\ 10 \\ 16 \end{bmatrix}$ **(b)** $\begin{bmatrix} 13 \\ 28 \\ 43 \end{bmatrix}$ **(c)** $\begin{bmatrix} 0 \\ 0 \\ 0 \end{bmatrix}$ **(d)** $\begin{bmatrix} 13 \\ 31 \\ 49 \end{bmatrix}$

7. If $\mathbf{x}=\begin{bmatrix} x \\ y \\ z \end{bmatrix}$, then $\mathbf{A}=\begin{bmatrix} 2 & 3 & -1 \\ 1 & -4 & 5 \\ 0 & 1 & 0 \end{bmatrix}$

9. $\mathbf{x}'=\begin{bmatrix} 0 & 1 \\ -1 & 0 \end{bmatrix}\mathbf{x}+\begin{bmatrix} 0 \\ e^t+t^2 \end{bmatrix}$ where $\mathbf{x}=\begin{bmatrix} y \\ y' \end{bmatrix}$

11. $\mathbf{x}'=\begin{bmatrix} 0 & 1 & 0 & 0 \\ 0 & 0 & 1 & 0 \\ 0 & 0 & 0 & 1 \\ -3 & 2 & -3 & 0 \end{bmatrix}\mathbf{x}+\begin{bmatrix} 0 \\ 0 \\ 0 \\ t \end{bmatrix}$ where $\mathbf{x}=\begin{bmatrix} y \\ y' \\ y'' \\ y''' \end{bmatrix}$

SECTION 5.2

5. $e^{7t}\begin{bmatrix} 4a \\ 5a \end{bmatrix}$ and $e^{2t}\begin{bmatrix} b \\ -b \end{bmatrix}$ are solutions, where a and b are any nonzero numbers.

SECTION 5.3

1.

3. Flow lines are spirals to the origin as indicated in the diagram.

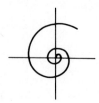

9. (a) $\dfrac{\partial^2 F}{\partial x^2} + \dfrac{\partial^2 F}{\partial y^2} = 2 + (-2) = 0$

(b) The equations for the flow lines are

$$\frac{dx}{dt} = \frac{\partial F}{\partial x} = 2x \qquad \frac{dy}{dt} = \frac{\partial F}{\partial y} = -2y$$

so that $dy/dx = -y/x$. This last equation is separable (see Section 1.4) and has solutions determined by the implicit relationship $xy = c$, where c is an arbitrary constant.

(c)

(d) The flow lines are graphs of the solutions of $\mathbf{X}' = \begin{bmatrix} 2 & 0 \\ 0 & -2 \end{bmatrix} \mathbf{X}$, where $\mathbf{X} = \begin{bmatrix} x \\ y \end{bmatrix}$. If two different particles occupy the same point c at the same time t_0, then there would be two distinct solutions of the given equation satisfying the initial condition $\mathbf{X}(t_0) = c$. This is impossible.

SECTION 5.4

1. $c_1 \mathbf{u}(t) + c_2 \mathbf{v}(t)$

3. $c_1 \mathbf{u}(t) + c_2 \mathbf{v}(t) + c_3 \mathbf{w}(t)$

5. $r = -1$, $c = -d$; $r = 6$, $c = \frac{3}{4}d$; Yes, if neither vector is the zero vector

7. $a = \frac{1}{2}$, $b = -1$, $c = \frac{1}{4}$, $d = 1$; $c_1 \mathbf{u}(t) + c_2 \mathbf{v}(t) + \begin{bmatrix} \frac{1}{2}t - 1 \\ \frac{1}{4}t + 1 \end{bmatrix}$ where $\mathbf{u}(t)$ and $\mathbf{v}(t)$ are as in Exercise 2.

9. Let $U_1(t)$ and $U_2(t)$ be the components of **u** and consider the system of equations

$$c_1 u_{11}(t_0) + c_2 u_{12}(t_0) = U_1(t_0)$$

$$c_1 u_{21}(t_0) + c_2 u_{22}(t_0) = U_2(t_0)$$

By Exercise 8, $u_{11}(t_0)u_{22}(t_0) - u_{21}(t_0)u_{12}(t_0) \neq 0$. Therefore there is a solution c_1, c_2 of this system of equations. Then $c_1 \mathbf{u}_1(t_0) + c_2 \mathbf{u}_2(t_0) = \mathbf{u}(t_0)$. Both $c_1 \mathbf{u}_1 + c_2 \mathbf{u}_2$ and **u** are solutions of the initial-value problem $\mathbf{x}' = A\mathbf{x}$, $\mathbf{x}(t_0) = \mathbf{u}(t_0)$. By Theorem 5.3 the two solutions must be identical.

SECTION 5.5

1. $c_1 e^t \begin{bmatrix} 1 \\ 1 \end{bmatrix} + c_2 e^{-t} \begin{bmatrix} -1 \\ 1 \end{bmatrix}$

3. $c_1 e^{(-3+\sqrt{5})t/2} \begin{bmatrix} 1-\sqrt{5} \\ 2 \end{bmatrix} + c_2 e^{(-3-\sqrt{5})t/2} \begin{bmatrix} -2 \\ 1+\sqrt{5} \end{bmatrix}$

5. $c_1 \begin{bmatrix} 0 \\ 1 \\ -1 \end{bmatrix} + c_2 e^{-2t} \begin{bmatrix} 2 \\ -1 \\ 0 \end{bmatrix} + c_3 e^{-3t} \begin{bmatrix} 1 \\ 0 \\ -1 \end{bmatrix}$

7. $c_1 e^t \begin{bmatrix} 1 \\ -1 \\ 0 \\ 1 \end{bmatrix} + c_2 e^{-t} \begin{bmatrix} 0 \\ 0 \\ 0 \\ 1 \end{bmatrix} + c_3 e^{2t} \begin{bmatrix} 1 \\ 0 \\ 0 \\ 0 \end{bmatrix} + c_4 e^{-2t} \begin{bmatrix} -1 \\ 1 \\ 3 \\ 5 \end{bmatrix}$

9. $e^t \begin{bmatrix} 1 \\ 1 \end{bmatrix} - e^{-t} \begin{bmatrix} -1 \\ 1 \end{bmatrix}$

11. $-\dfrac{1}{2\sqrt{5}} e^{(-3+\sqrt{5})t/2} \begin{bmatrix} 1-\sqrt{5} \\ -2 \end{bmatrix} + \dfrac{1}{2\sqrt{5}} e^{(-3-\sqrt{5})t/2} \begin{bmatrix} 1+\sqrt{5} \\ -2 \end{bmatrix}$

13. $2 \begin{bmatrix} 0 \\ 1 \\ -1 \end{bmatrix} + e^{-2t} \begin{bmatrix} 2 \\ -1 \\ 0 \end{bmatrix} - 2e^{-3t} \begin{bmatrix} 1 \\ 0 \\ -1 \end{bmatrix}$

15. $e^t \begin{bmatrix} 1 \\ -1 \\ 0 \\ 1 \end{bmatrix} - 6e^{-t} \begin{bmatrix} 0 \\ 0 \\ 0 \\ 1 \end{bmatrix} + e^{2t} \begin{bmatrix} 1 \\ 0 \\ 0 \\ 0 \end{bmatrix} + e^{-2t} \begin{bmatrix} -1 \\ 1 \\ 3 \\ 5 \end{bmatrix}$

SECTION 5.6

1. $c_1 \begin{bmatrix} \cos t \\ \sin t \end{bmatrix} + c_2 \begin{bmatrix} -\sin t \\ \cos t \end{bmatrix}$

3. $c_1 \begin{bmatrix} -2\cos\sqrt{5}\,t \\ \cos\sqrt{5}\,t - \sqrt{5}\,\sin\sqrt{5}\,t \end{bmatrix} + c_2 \begin{bmatrix} -2\sin\sqrt{5}\,t \\ \sqrt{5}\,\cos\sqrt{5}\,t + \sin\sqrt{5}\,t \end{bmatrix}$

5. $c_1 e^{3t} \begin{bmatrix} 1 \\ 1 \\ 1 \end{bmatrix} + c_2 e^{-3t/2} \begin{bmatrix} 2\cos\dfrac{\sqrt{3}}{2}t \\ -\cos\dfrac{\sqrt{3}}{2}t + \sqrt{3}\sin\dfrac{\sqrt{3}}{2}t \\ -\cos\dfrac{\sqrt{3}}{2}t - 3\sin\dfrac{\sqrt{3}}{2}t \end{bmatrix}$

$+\, c_3 e^{-3t/2} \begin{bmatrix} 2\sin\dfrac{\sqrt{3}}{2}t \\ -\sin\dfrac{\sqrt{3}}{2}t - \sqrt{3}\cos\dfrac{\sqrt{3}}{2}t \\ -\sin\dfrac{\sqrt{3}}{2}t + \sqrt{3}\cos\dfrac{\sqrt{3}}{2}t \end{bmatrix}$

7. $2\begin{bmatrix} \cos t \\ \sin t \end{bmatrix} + 2\begin{bmatrix} -\sin t \\ \cos t \end{bmatrix}$

9. $-\begin{bmatrix} -2\cos\sqrt{5}\,t \\ \cos\sqrt{5}\,t - \sqrt{5}\sin\sqrt{5}\,t \end{bmatrix} + \dfrac{2}{5}\begin{bmatrix} -2\sin\sqrt{5}\,t \\ 5\cos\sqrt{5}\,t + \sqrt{5}\sin\sqrt{5}\,t \end{bmatrix}$

11. $e^{3t}\begin{bmatrix} 1 \\ 1 \\ 1 \end{bmatrix}$

SECTION 5.7

1. $500\sqrt{3} \approx 866$

3. (a) $c_1 \begin{bmatrix} f_{12} \\ f_{21} \end{bmatrix} + c_2 e^{-(f_{12}+f_{21})t}\begin{bmatrix} 1 \\ -1 \end{bmatrix}$

(b) $c_1 e^{-at}\begin{bmatrix} 1 \\ 1 \end{bmatrix} + c_2 e^{-3at}\begin{bmatrix} 1 \\ -1 \end{bmatrix}$

5. (b) From the hint it follows that the discriminant for the quadratic polynomial in (a) is positive. Hence, its roots are real. Since

$$\sqrt{(c_{01}+c_{12}+c_{21}+c_{31})^2 - 4c_{12}(c_{01}+c_{31})} < \sqrt{(c_{01}+c_{12}+c_{21}+c_{31})^2}$$

$$= c_{01}+c_{12}+c_{21}+c_{31}$$

the roots are negative.

7. $\begin{bmatrix} 0 & 1 & 0 & 0 \\ -\dfrac{k_1+k}{m_1} & \dfrac{r_1}{m_1} & \dfrac{k}{m_1} & 0 \\ 0 & 0 & 0 & 1 \\ \dfrac{k}{m_2} & 0 & -\dfrac{k_2+k}{m_2} & \dfrac{r_2}{m_2} \end{bmatrix}$

9. (a) $\dfrac{dx}{dt} = \begin{bmatrix} 0 & 1 & 0 & 0 \\ -2 & 0 & 1 & 0 \\ 0 & 0 & 0 & 1 \\ 1 & 0 & -2 & 0 \end{bmatrix} x + \begin{bmatrix} 0 \\ E \\ 0 \\ 0 \end{bmatrix}$

(b) $c_1 \begin{bmatrix} \cos t \\ -\sin t \\ \cos t \\ -\sin t \end{bmatrix} + c_2 \begin{bmatrix} \sin t \\ \cos t \\ \sin t \\ \cos t \end{bmatrix} + c_3 \begin{bmatrix} \cos\sqrt{3}\,t \\ -\sqrt{3}\sin\sqrt{3}\,t \\ -\cos\sqrt{3}\,t \\ \sqrt{3}\sin\sqrt{3}\,t \end{bmatrix} + c_4 \begin{bmatrix} \sin\sqrt{3}\,t \\ \sqrt{3}\cos\sqrt{3}\,t \\ -\sin\sqrt{3}\,t \\ -\sqrt{3}\cos\sqrt{3}\,t \end{bmatrix}$

(c) $\begin{bmatrix} -\frac{2}{3}E \\ 0 \\ -\frac{1}{3}E \\ 0 \end{bmatrix}$

(d) $c_1 \begin{bmatrix} \cos t \\ -\sin t \\ \cos t \\ -\sin t \end{bmatrix} + c_2 \begin{bmatrix} \sin t \\ \cos t \\ \sin t \\ \cos t \end{bmatrix} + c_3 \begin{bmatrix} \cos\sqrt{3}\,t \\ -\sqrt{3}\sin\sqrt{3}\,t \\ -\cos\sqrt{3}\,t \\ \sqrt{3}\sin\sqrt{3}\,t \end{bmatrix}$

$+ c_4 \begin{bmatrix} \sin\sqrt{3}\,t \\ \sqrt{3}\cos\sqrt{3}\,t \\ -\sin\sqrt{3}\,t \\ -\sqrt{3}\cos\sqrt{3}\,t \end{bmatrix} + \begin{bmatrix} -\frac{2}{3}E \\ 0 \\ -\frac{1}{3}E \\ 0 \end{bmatrix}$

(e) $\frac{1}{2}E \begin{bmatrix} \cos t \\ -\sin t \\ \cos t \\ -\sin t \end{bmatrix} + \frac{1}{6}E \begin{bmatrix} \cos\sqrt{3}\,t \\ -\sqrt{3}\sin\sqrt{3}\,t \\ -\cos\sqrt{3}\,t \\ \sqrt{3}\sin\sqrt{3}\,t \end{bmatrix} + \begin{bmatrix} -\frac{2}{3}E \\ 0 \\ -\frac{1}{3}E \\ 0 \end{bmatrix}$

SECTION 5.8

1. $x_1(t)=c_1 e^t + c_2 e^{-t}, \quad x_2(t)=c_1 e^t - c_2 e^{-t}$

3. $x_1(t)=2c_1 e^{10t} - 2c_2 e^{-6t}, \quad x_2(t)=c_1 e^{10t} + 3c_2 e^{-6t}$

5. $x_1(t)=c_1 e^t + c_2 e^{5t} + \frac{3}{5}t - \frac{6}{5}, \quad x_2(t)=-c_1 e^t + c_2 e^{5t} - \frac{2}{5}t + \frac{9}{5}$

7. $x_1(t)=c_1 e^t + 2c_2 e^{2t}, \quad x_2(t)=-c_2 e^{2t} + c_3 e^{3t}, \quad x_3(t)=-c_1 e^t - c_3 e^{3t}$

9. $x_1(t)=c_2 \cos t + c_3 \sin t + e^t - t - 1,$
$x_2(t)=c_1 e^{-t} + \frac{1}{2}c_2(\cos t + \sin t) + \frac{1}{2}c_3(-\cos t + \sin t) + \frac{1}{2}e^t - 1,$
$x_3(t)=-c_1 e^{-t} + \frac{1}{2}c_2(\cos t + \sin t) + \frac{1}{2}c_3(-\cos t + \sin t) + \frac{1}{2}e^t - t + 1$

11. $x_1(t)=\frac{5}{2}e^t + \frac{17}{10}e^{5t} + \frac{3}{5}t - \frac{6}{5}, \quad x_2(t)=-\frac{5}{2}e^t + \frac{17}{10}e^{5t} - \frac{2}{5}t + \frac{9}{5}$

13. See the answer to Exercise 1 of Section 7.

15. See the answer to Exercise 3(a) of Section 7.

17. $c_1(t) = a\cos\dfrac{\sqrt{2}}{10}t + b\sin\dfrac{\sqrt{2}}{10}t + 5(10^{-5})$, $c_2(t) =$

$a\left(\cos\dfrac{\sqrt{2}}{10}t + \sqrt{2}\sin\dfrac{\sqrt{2}}{10}t\right) + b\left(-\sqrt{2}\cos\dfrac{\sqrt{2}}{10}t + \sin\dfrac{\sqrt{2}}{10}t\right) + 2.5(10^{-4})$,

where a and b are arbitrary constants.

19. **(c)** Both limits are c/a_2.

SECTION 5.9

1. $x_1 = \frac{1}{2}e^t - \frac{1}{2}e^{-t}$, $\quad x_2 = \frac{1}{2}e^t + \frac{1}{2}e^{-t}$

3. $x_1 = \frac{19}{4}e^{10t} - \frac{7}{4}e^{-6t}$, $\quad x_2 = \frac{19}{8}e^{10t} + \frac{21}{8}e^{-6t}$

5. $x_1 = -\frac{11}{5}e^t - \frac{31}{25}e^{5t} - \frac{3}{5}t + \frac{17}{25}$, $\quad x_2 = \frac{11}{5}e^t - \frac{31}{25}e^{5t} + \frac{2}{5}t - \frac{33}{25}$

7. $x_1 = -2e^t - 2e^{2t}$, $\quad x_2 = -e^{2t}$, $\quad x_3 = -2e^t$

9. $x_1(t) = \sin t + e^t - t - 1$, $\quad x_2(t) = e^{-t} + \frac{1}{2}(-\cos t + \sin t) + \frac{1}{2}e^t - 1$, $\quad x_3(t) = -e^{-t} + \frac{1}{2}(-\cos t + \sin t) + \frac{1}{2}e^t - t + 1$

11. See the answer to Exercise 1 of Section 5.7.

13. See the answer to Exercise 17 of Section 5.8.

15. See the answer to Exercise 5 of Section 5.7.

SECTION 5.10

1.

t	x	y
.1	.895171	1.09484
.2	.781398	1.17874
.3	.659817	1.25086
.4	.531643	1.31048
.5	.398158	1.35701
.6	.260694	1.38998
.7	.120625	1.40906
.8	−.020648	1.41406
.9	−.161716	1.40494
1.0	−.301168	1.38177

3.

t	x	y
.1	.004837	−.004837
.2	.018730	−.018730
.3	.040818	−.040818
.4	.070320	−.070320
.5	.106531	−.106531

t	x	y
.6	.148812	$-.148812$
.7	.196586	$-.196586$
.8	.249329	$-.249329$
.9	.30657	$-.30657$
1.0	.36788	$-.36788$

5. Some typical values are

t	z	z'
.1	1.00588	.117823
.5	1.1518	.627521
1.0	1.67653	1.56456
1.5	2.93142	3.96523
2.0	17.1615	60192.3

7. Some typical values are

t	x	x'
1	.536418	$-.85895$
2	$-.474193$	$-.994315$
3	-1.10887	$-.152747$
4	$-.727503$	.856172
5	.366157	1.15774
6	1.19383	.335454
7	.927933	$-.819526$
8	$-.219396$	-1.30286
9	-1.24523	.54471
10	-1.12831	.746531

SECTION 6.1

1. $\displaystyle\sum_{n=1}^{\infty} \frac{1}{(2n-1)!} x^{2n-1}$

3. $\displaystyle\sum_{n=0}^{\infty} \frac{1}{n!} x^{n}$

5. $\displaystyle\sum_{n=0}^{\infty} (-1)^n x^n$

7. Preceding Theorem 6.1 it is shown that $a_k = \dfrac{f^{(k)}(a)}{k!}$. Exactly the same argument shows that $b_k = \dfrac{g^{(k)}(a)}{k!}$. Since $f(x) = g(x)$ for all x in I, $f^{(k)}(x) = g^{(k)}(x)$ for every x in I. In particular, $f^{(k)}(a) = g^{(k)}(a)$. Hence $a_k = b_k$ for all k and the two series are identical.

9. radius of convergence $= 3$, $-4 < x < 2$

11. radius of convergence $= 1$, $-1 < x < 1$

13. radius of convergence $= \infty$, $-\infty < x < \infty$

15. radius of convergence $= 2$, $-2 \leqslant x < 2$

17. $0, 3, -3$

19. $0, 3, (n + \frac{1}{2})\pi$, where $n = 0, \pm 1, \pm 2, \ldots$

21. analytic for all x

SECTION 6.2

1. $\displaystyle\sum_{n=0}^{\infty} \frac{(-1)^n}{(2n)!} x^{2n}$, $\displaystyle\sum_{n=0}^{\infty} \frac{(-1)^n}{(2n+1)!} x^{2n+1}$

3. $1 + \displaystyle\sum_{n=1}^{\infty} \frac{(-1)^n 1^2 4^2 \cdots (3n-2)^2}{(3n)!} x^{3n}$

$x + \displaystyle\sum_{n=1}^{\infty} \frac{(-1)^n 2^2 5^2 \cdots (3n-1)^2}{(3n+1)!} x^{3n+1}$

5. $1 - \frac{1}{2}x^2 + \frac{1}{6}x^3 + \frac{1}{12}x^4 + \cdots$, $x - \frac{1}{2}x^2 - \frac{1}{6}x^3 + \frac{1}{6}x^4 + \cdots$

7. $1 + 2x^2 - \frac{2}{3}x^3 + \frac{1}{3}x^4 + \cdots$, $x + \frac{1}{3}x^3 - \frac{1}{6}x^4 + \cdots$

9. $1 - \frac{1}{2}x^2 - \frac{1}{3}x^3 - \frac{5}{24}x^4 + \cdots$, $x + \frac{1}{2}x^2 + \frac{1}{6}x^3 + \frac{1}{12}x^4 + \cdots$

11. (a)

$1 + \displaystyle\sum_{n=1}^{\infty} \frac{(p-2n-2)\cdots(p-2)p(p+1)(p+3)\cdots(p+2n-1)}{(2n)!}(-1)^n x^n$

$x + \displaystyle\sum_{n=1}^{\infty} \frac{(p-2n+1)\cdots(p-3)(p-1)(p+2)(p+4)\cdots(p+2n)}{(2n+1)!}(-1)^n x^n$

(b) Suppose that p is an even integer. In the first series we have $a_{n+2} = (n-p)(n+p-1)a_n/(n+2)(n+1)$. When $n = p$, $a_{n+2} = 0$. It follows that $a_n = 0$ for all $n \geqslant p+2$. Therefore the first solution is a polynomial when p

is an even integer. If n is an odd integer, a similar argument shows that the second solution is a polynomial.

13. (a) $1 + \sum\limits_{n=1}^{\infty} \dfrac{k^2(k^2-2^2)(k^2-4^2)\cdots\left(k^2-(2n-2)^2\right)}{(2n)!}x^{2n}$

$\qquad x + \sum\limits_{n=1}^{\infty} \dfrac{(k^2-1)(k^2-3^2)\cdots\left(k^2-(2n-1)^2\right)}{(2n+1)!}x^{2n+1}$

(b) In either series $a_{n+2}=(k^2-n^2)a_n/(n+2)(n+1)$. Thus when $n=k$, $a_{n+2}=0$, so that $a_n=0$ for $n\geqslant k+2$. Hence one of the series is a polynomial.

SECTION 6.3

1. When we insert $y(x)$ into the equation, we find that $\sum_{n=0}^{\infty}(n^2+2n+2)a_n x^n=0$. Hence $(n^2+2n+2)a_n=0$ for every n. Since $n^2+2n+2\neq 0$ for every n, we must have $a_n=0$ for every n. Thus $y(x)\equiv 0$ is the only solution that can be written as a power series.

3. $c_1 x^{-1} \sum\limits_{n=0}^{\infty} \dfrac{(-1)^n}{(1)(3)\cdots(2n-1)n!}x^n + c_2 x^{-1/2}\sum\limits_{n=0}^{\infty}\dfrac{(-1)^n}{(1)(3)\cdots(2n+1)n!}x^n$

5. $x^2 \sum\limits_{n=0}^{\infty}\dfrac{(-1)^n}{2^{2n}n!(n+2)!}x^n$ The method of Frobenius does not yield two linearly independent solutions.

7. $c_1 x^{-1}(e^{-x}+x-1)+c_2(1-x^{-1})$, which can be written in the more compact form $C_1 x^{-1}e^{-x}+C_2(1-x^{-1})$.

9. $c_1 x^{1/2}\left(1+\sum\limits_{n=0}^{\infty}\dfrac{1}{2^{3n}(1)(3)\cdots(2n+1)n!}x^{2n}\right)$

$\qquad + c_2 x^{-1/2}\left(1+\sum\limits_{n=0}^{\infty}\dfrac{1}{2^{3n}(1)(3)\cdots(2n-1)n!}x^{2n}\right)$

11. $c_1 x^{-1}+c_2 x^{-1/2}$

13. $c_1 \sum\limits_{n=0}^{\infty}\dfrac{(-1)^n 2^n}{(1)(4)\cdots(3n+1)n!}x^n + c_2 x^{-1/3}\sum\limits_{n=0}^{\infty}\dfrac{(-1)^n 2^n}{(2)(5)\cdots(3n-1)n!}x^n$

15. $c_1 x^{-1}+c_2 x^{-3}$

17. $\sum\limits_{n=0}^{\infty}\dfrac{(-1)^n}{(n!)^2}x^n$

19. $a_1=0,\ a_2=6550\,a_0,\ a_3=3048\,a_0,\ a_4=1.43(10)^7 a_0$

SECTION 6.4

1. $c_1 x + c_2 x \left(\ln x + \sum\limits_{n=1}^{\infty} \dfrac{(-1)^n}{nn!} x^n \right)$

3. If $y_1(x) = x \sum\limits_{n=0}^{\infty} \dfrac{(-1)^n}{n!} x^n = x e^{-x}$, then a general solution is $c_1 y_1(x) + c_2 (y_1(x) \ln x + x^2 - \frac{3}{4} x^3 + \frac{11}{36} x^4 + \cdots)$

5. If $y_1(x) = \sum\limits_{n=0}^{\infty} \dfrac{(-1)^n}{2^{2n}(n!)^2} x^{2n+1}$, then a general solution is $c_1 y_1(x) + c_2 (y_1(x) - \frac{1}{4} x^3 + \frac{3}{128} x^5 + \cdots)$

SECTION 6.5

1. $\dfrac{d}{dx}(x^p J_p(x)) = \dfrac{d}{dx} \left(\sum\limits_{n=0}^{\infty} \dfrac{(-1)^n}{n!(n+p)!2^{2n+p}} x^{2n+2p} \right)$

$$= \sum\limits_{n=0}^{\infty} \dfrac{(-1)^n}{n!(n+p-1)2^{2n+p-1}} x^{2n+2p-1}$$

$$= x^p J_{p-1}(x)$$

3. From Exercise 1 we obtain $px^{p-1} J_p(x) + x^p \dfrac{d}{dx}(J_p(x)) = x^p J_{p-1}(x)$. Upon multiplying each side by x^{-p} and rearranging the terms, we obtain the desired identity.

5. $J_p(x) = 2^{-p} x^p / p! + $ terms involving x with powers greater than p. Since $\lim_{x \to 0^+} |Y_p(x)| = +\infty$, any solution of Bessel's equation satisfying $y(0) = 0$ must be a constant multiple of $J_p(x)$. A straightforward calculation shows that $\dfrac{d^p}{dx^p}(J_p(x)) = 2^{-p} + $ terms involving x with powers greater than zero. Thus $\dfrac{d^p}{dx^p}(J_p(x))|_{x=0} = 2^{-p}$. It follows that $J_p(x)$ is the solution of the given initial-value problem.

7. $\lim\limits_{x \to 0} x^{-p} J_p(x) =$

$$\lim\limits_{x \to 0} \left\{ \dfrac{1}{2^p p!} + \text{terms involving } x \text{ with powers greater than zero} \right\} = \dfrac{1}{2^p p!}$$

9. $c_1 J_4(2x) + c_2 Y_4(2x)$

11. $c_1 x^2 J_2(x) + c_2 x^2 Y_2(x)$

13. A general solution of the differential equation is $c_1 J_0(2x^2) + c_2 Y_0(2x^2)$. Since $\lim_{x \to 0^+} |Y_0(x)| = +\infty$, we must have $c_2 = 0$. Since $J_0(0) = 1$, we must have $c_1 = 1$. Thus $J_0(2x^2)$ is the unique solution of the given initial-value problem.

15. (b) $c_1 J_0\left(2kx^{1/2}/\sqrt{g}\right)+c_2 Y_0\left(2kx^{1/2}/\sqrt{g}\right)$

   (c) $k=j\sqrt{g}\,L^{-1/2}2$, where j is a positive root of J_0.

17. A general solution of the differential equation is $c_1 x^{-1/2}J_1(2kx^{1/2})+c_2 x^{-1/2}Y_1(2kx^{1/2})$. Since h is finite at $x=0$, $\lim_{x\to0}|x^{-1/2}J_1(2kx^{1/2})|=0$, and $\lim_{x\to0^+}|x^{-1/2}Y_1(2kx^{1/2})|=+\infty$, we must have $c_2=0$. At $t=0$ we have $h(x,0)=\beta x$. In particular, $\beta L=h(L,0)=y(L)=L^{-1/2}c_1 J_1(2kL^{1/2})$ so that $c_1=\beta L^{3/2}(J_1(2kL^{1/2}))^{-1}$.

19. $c_1 x^{1/2}J_1(2kx^{1/2})+c_2 x^{1/2}Y_1(2kx^{1/2})$

SECTION 6.6

1.
$$K_n(x)=\frac{\pi}{2}i^{n+1}\left(J_n(ix)+i\frac{2}{\pi}\left[\left(\gamma+\ln\frac{ix}{2}\right)J_n(ix)\right.\right.$$

$$-\frac{1}{2}\sum_{k=0}^{n-1}\frac{(n-k-1)!}{k!}\left(\frac{ix}{2}\right)^{2k-n}$$

$$\left.\left.+\frac{1}{2}\sum_{k=0}^{\infty}(-1)^{k+1}\frac{\varphi(k)+\varphi(k+n)}{k!(k+n)!}\left(\frac{ix}{2}\right)^{2k+n}\right]\right)$$

$$=\frac{\pi}{2}i^{n+1}\left(J_n(ix)+i\frac{2}{\pi}\left[\left(\gamma+\ln\frac{x}{2}+i\frac{\pi}{2}\right)J_n(ix)\right.\right.$$

$$-\frac{1}{2}\sum_{k=0}^{n-1}\frac{(n-k-1)!}{k!}(-1)^k i^n\left(\frac{x}{2}\right)^{2k-n}$$

$$\left.\left.+\frac{1}{2}\sum_{k=0}^{\infty}(-1)^{k+1}(-1)^k i^n\frac{\varphi(k)+\varphi(k+n)}{k!(k+n)!}\left(\frac{x}{2}\right)^{2k+n}\right]\right)$$

$$=(-1)^{n+1}\left(\gamma+\ln\frac{x}{2}\right)I_n(x)+\frac{1}{2}\sum_{k=0}^{n-1}\frac{(n-k-1)!}{k!(k+n)!}(-1)^k\left(\frac{x}{2}\right)^{2k+n}$$

$$+\frac{1}{2}(-1)^n\sum_{k=0}^{\infty}\frac{\varphi(k)+\varphi(k+n)}{k!(n+k)!}\left(\frac{x}{2}\right)^{2k+n}$$

3. $c_1 x^{1/2}I_0(x^{1/2})+c_2 x^{1/2}K_0(x^{1/2})$

5. $c_1 I_0(\tfrac{2}{3}x^{3/2})+c_2 K_0(\tfrac{2}{3}x^{3/2})$

7. (a) $(3+t)\dfrac{d^2x}{dt^2}-\tfrac{1}{4}x=0$

(b) $(3+t)^{1/2}\left[c_1I_1\big((3+t)^{1/2}\big)+c_2K_1\big((3+t)^{1/2}\big)\right]$

(c) $\dfrac{(3+t)^{1/2}}{2\left[I_1(2)K_1'(2)-I_1'(2)K_1(2)\right]}\left[K_1'(2)I_1\big((3+t)^{1/2}\big)-I_1'(2)K_1\big((3+t)^{1/2}\big)\right]$

9. (a) $(x^2D^4+8xD^3+12D^2-k^4)X=0$

(d) $x^{-1}\left[c_1J_2(2kx^{1/2})+c_2Y_2(2kx^{1/2})+c_3I_2(2kx^{1/2})+c_4K_4(2kx^{1/2})\right]$

7

SECTION 7.1

1. If $k=0$, then $X(x)=ax+b$ for some constants a and b. Since $0=X(0)=b$ and $0=X(L)=a$, we must have $X(0)\equiv0$. If $k<0$, then $X(x)=ce^{-\sqrt{k}\,x}+de^{\sqrt{k}\,x}$ for some constants c and d. We have

$$0=X(0)=c+d$$

$$0=X(L)=ce^{-\sqrt{k}\,L}+de^{\sqrt{k}\,L}$$

A short calculation shows that $c=0$, $d=0$ is the only solution of this system. Hence, $X(x)\equiv0$. If $0<k$, then $X(x)=c_1\cos\sqrt{k}\,x+c_2\sin\sqrt{k}\,x$. We have

$$0=X(0)=c_1$$

$$0=X(1)=c_1\cos\sqrt{k}\,L+c_2\sin\sqrt{k}\,L$$

$$=c_2\sin\sqrt{k}\,L$$

Since we want $X(x)=0$, we must have $c_2\neq0$. Therefore $\sin\sqrt{k}\,L=0$. That is, $\sqrt{k}\,L=0,\ \pm\pi,\ \pm2\pi,\ \ldots$ or equivalently $k=n^2\pi^2/L^2,\ n=0,1,2,\ldots$. Notice that if $k=0$, then $X(x)\equiv0$. Thus, $k=n^2\pi^2/L^2,\ n=1,2,\ldots$.

3. $-\displaystyle\sum_{n=1}^{\infty}\dfrac{2}{n\pi}(-1)^n\sin n\pi x;\quad 0,\ \tfrac{1}{4},\ \tfrac{1}{2},\ \tfrac{3}{4},\ 0$

5. $\displaystyle\sum_{n=1}^{\infty}\dfrac{2n\pi}{n^2\pi^2+1}(1-(-1)^ne)\sin n\pi x;\quad 0,\ e^{1/4},\ e^{1/2},\ e^{3/4},\ 0$

7. $\displaystyle\sum_{n=2}^{\infty} \frac{2n}{\pi(n^2-1)}(1-(-1)^{n-1})\sin n\pi x;$ $0, \sqrt{2}/2, 0, -\sqrt{2}/2, 0$

9. $\displaystyle\sum_{n=1}^{\infty} \frac{2}{n\pi}\left(2(-1)^n - \cos\frac{n\pi}{2}\right)\sin n\pi x;$ $0, \frac{1}{4}, 0, -\frac{1}{4}, 0$

11. $\dfrac{1}{2} + \displaystyle\sum_{n=1}^{\infty} \frac{2[(-1)^n-1]}{n^2\pi^2}\cos n\pi x;$ $\frac{1}{4}, \frac{1}{2}, \frac{3}{4}$

13. $e-1+ \displaystyle\sum_{n=1}^{\infty} \frac{2}{n^2\pi^2+1}(e(-1)^n-1)\cos n\pi x;$ $e^{1/4}, e^{1/2}, e^{3/4}$

15. $\cos \pi x;$ $\sqrt{2}/2, 0, -\sqrt{2}/2$

17. $\displaystyle\sum_{n=1}^{\infty} \left[\frac{2[(-1)^n-1]}{n^2\pi^2} + \frac{2\sin(n\pi/2)}{n\pi}\right]\cos n\pi x;$ $\frac{1}{4}, 0, -\frac{1}{4}$

19.

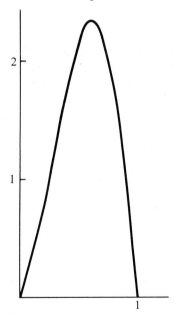

21. (a) and **(b)** From the identity $x=\theta-\sin\theta$ one easily obtains $dx=(1-\cos\theta)\,d\theta$. Using this identity along with $y=1-\cos\theta$, the desired identities are obtained in a straightforward manner.

(c) Use the trigonometric identity $2\cos A \cos B=\cos(A+B)+\cos(A-B)$.

(d) $\dfrac{d}{dx}\left(\displaystyle\int_\theta^\pi \cos(n\theta-x\sin\theta)\,d\theta\right)=-\displaystyle\int_0^\pi \sin(n\theta-x\sin\theta)\sin\theta\,d\theta$

$\dfrac{d^2}{dx^2}\left(\displaystyle\int_0^\pi \cos(n\theta-x\sin\theta)\,d\theta\right)=\dfrac{d}{dx}\left(\displaystyle\int_0^\pi \sin(n\theta-x\sin\theta)\sin\theta\,d\theta\right)$

$=-\displaystyle\int_0^\pi \cos(n\theta-x\sin\theta)\sin^2\theta\,d\theta$

(e) Using (b) we have

$$b_n = \frac{2}{\pi} \int_0^\pi \cos(n\theta - x\sin\theta)(2 - 2\cos\theta - \sin^2\theta)\, d\theta$$

$$= \frac{4}{\pi} \int_0^\pi \cos(n\theta - x\sin\theta)\, d\theta - \frac{4}{\pi} \int_0^\pi \cos(n\theta - x\sin\theta)\cos\theta\, d\theta$$

$$- \frac{2}{\pi} \int_0^\pi \cos(n\theta - n\sin\theta)\sin^2\theta\, d\theta$$

Using (c), (d), and the identity in (7.14) with $x = n$, we have

$$b_n = 4J_n(n) - 4J_n(n) + 2J_n''(n) = 2J_n''(n)$$

(f) $J_n(x)$ is a solution of Bessel's equation of order n. That is, $x^2 J_n''(x) + x J_n'(x) + (x^2 - n^2)J_n(x) = 0$. When $x = n$ we have $n^2 J_n''(n) + n J_n'(n) = 0$ from which the desired result follows directly.

(g) Using (e) and (f), we have $b_n = -2n^{-1}J_n'(n)$ for $n \geq 1$. In (a) we found that $b_0 = \frac{3}{2}$. Therefore

$$y = b_0 + \sum_{n=1}^\infty b_n \cos nx$$

$$= \frac{3}{2} - 2 \sum_{n=1}^\infty n^{-1}J_n'(n)\cos nx$$

SECTION 7.2

1. $u(x,t) = \sum_{n=1}^\infty a_n \sin(n\pi x/L)\cos(n\pi ct/L)$

$$= \sum_{n=1}^\infty a_n \left[\sin\left(n\pi(x+ct)/L\right) + \sin\left(n\pi(x-ct)/L\right) \right]/2$$

$$= \frac{1}{2} \sum_{n=1}^\infty a_n \sin(n\pi(x+ct)/L) + \frac{1}{2} \sum_{n=1}^\infty a_n \sin(n\pi(x-ct)/L)$$

$$= \frac{1}{2}f(x+ct) + \frac{1}{2}f(x-ct)$$

3. $\sum_{n=1}^\infty a_n \sin(n\pi x/L)\sin(n\pi ct/L)$, where $n\pi ca_n/L$ is the nth Fourier sine coefficient of g.

5. $\sum_{n=0}^\infty a_n \cos(n\pi x/a)\sinh(n\pi y/a)$, where $a_n \left[\sinh(n\pi b/a)\right]^{-1}$ is the nth Fourier cosine coefficient of f.

7. (d) $v_1(x, t)$ is the function in (7.25), while

$$v_2(x, t) = \sum_{n=1}^{\infty} a_n \sin(n\pi x/L) \sin(n\pi ct/L)$$

where $n\pi c/aL$ is the nth Fourier sine coefficient of f.
(e) $u(x, t) = e^{-at}[v_1(x, t) + v_2(x, t)]$

Appendix 1

1. $x=2$, $y=3$
3. $x=\frac{5}{2}$, $y=-\frac{1}{2}$
5. $8+10i$
7. $5+i$
9. $-4i/5$
11. $3\sqrt{2} + 3\sqrt{2}\,i$
13. $2\sqrt{3} + 2i$
15. If $z=a+bi$, then $\bar{z}=a-bi$ and we have $z\bar{z}=a^2+b^2=|z|^2$.
17. If $z=a+bi$, then $\bar{z}=a-bi$. If z is real, then $b=0$ and we have $z=\bar{z}$. Conversely, if $z=\bar{z}$, then $bi=-bi$, so b must be zero. Thus z is real whenever $z=\bar{z}$.
19. $\sqrt{2}\,e^{5\pi i/4}$
21. $\sqrt{2}\,e^{3\pi i/4}$
23. $2e^{\pi i/2}$
25. $2e^{-\pi i/3}$
27. $6e^{0i}$
29. $2e^{2\pi i/3}$

Appendix 2

1. -2
3. $ad-bc$
5. 105
7. 30
9. 0
11. 96
13. $\det A = a_{11}a_{22} - a_{12}a_{21}$. If $a_{11}=a_{12}=a_{21}=a_{22}=0$, then $x_1=1$, $x_2=1$ is a solution. If either a_{11} or a_{12} is not zero, then $x_1=a_{12}$, $x_2=-a_{11}$ is a solution. If either a_{21} or a_{22} is not zero, then $x_1=a_{22}$, $x_2=-a_{12}$ is a solution.

INDEX

Numbers in parentheses refer to exercises.